A View of the Offender's World

ABOUT CRIMINALS

Mark Pogrebin Editor

University of Colorado at Denver

SAGE Publications
International Educational and Professional Publisher
Thousand Oaks ■ London ■ New Delhi

For information:

Sage Publications, Inc.
2455 Teller Road
Thousand Oaks, California 91320
E-mail: order@sagepub.com

Sage Publications Ltd.
1 Oliver's Yard
55 City Road
London EC1Y 1SP
United Kingdom

Sage Publications India Pvt. Ltd.
B-42, Panchsheel Enclave
Post Box 4109
New Delhi 110 017 India

Printed in the United States of America

Library of Congress Cataloging-in-Publication data

About criminals : a view of the offender's world / edited by Mark R. Pogrebin.
    p. cm.
Includes bibliographical references and index.
ISBN 0-7619-2816-2 (pbk.)
    1. Criminals—United States. 2. Criminal behavior—United States. 3. Crime—United States.
I. Pogrebin, Mark.
HV6789.A3442 2004
364.3′0973—dc22

2003025481

04   05   06   07   08   10   9   8   7   6   5   4   3   2   1

Acquiring Editor:	Jerry Westby
Editorial Assistant:	Vonessa Vondera
Production Editor:	Sanford Robinson
Copy Editor:	Mary L. Tederstrom
Typesetter:	C&M Digitals (P) Ltd.
Indexer:	Teri Greenberg
Cover Designer:	Michelle Lee Kenny

ABOUT CRIMINALS

CONTENTS

I am very pleased to dedicate this book to the newest member of our family, our granddaughter Elyza Pogrebin Berry. A very beautiful girl who shares two religions, two races, and a variety of loving grandparents. You bring a smile to my face.

PREFACE

The purpose of this book is to present students with recent and important research on criminal behavior. The methodological approach taken is termed naturalistic, one that allows criminals to discuss their offenses and lifestyles from their perspective. Having offenders' voices heard along with the researchers' analyses offers students a real-life view of what, how, and why various criminals behave the way they do. In short, the field studies conducted by the authors for all the articles in this anthology should provide the reader with a realistic portrayal of just what actual offenders say about crime and their participation in it.

In this anthology I have chosen a collection of readings that highlight criminological concepts that are extremely relevant to the study of crime and criminals. The articles explore a wide array of specific types of criminal behavior, covering the majority of topics on criminal types found in almost every standard criminological text. In a very real sense, this book is unique because each type of crime is arranged under a particular subject with different perspectives discussed for similar types of criminal behavior.

The readings feature field study research on property crimes, violent crimes, sex crimes, white-collar crimes, gangs and crime, drugs and crime, gender and crime, and a final section on desistance from crime. Along with the articles, I provide a brief introduction to each category of crime and a description of each article's contents. These short introductions are meant to help students understand the method that each researcher used to conduct the study, as well as to provide an overall description of the study's general contents.

It is my belief that the naturalistic approach through the use of field studies will assist students in gaining a more in-depth understanding of how criminological theories of crime causation actually relate to real-world examples of criminal behavior. I have found from years in the college classroom that the real-world accounts of offenders for their criminality facilitate active learning and tend to motivate student participation in a more meaningful way by connecting abstract theoretical concepts to the real-life experiences of criminals.

About Criminals: A View of the Offender's World can be used in courses as a supplemental anthology in criminology courses and as a separate text for those courses taught from a criminal behavior systems approach. Either way this book is utilized, it should aid students in providing an analytical perspective on the various motivations and explanations for types of criminal behavior. Perhaps being introduced to the different types of criminal offenses and offenders throughout this book may dispel many of the myths or stereotypes we tend to believe about criminals based on images we receive from movies, media outlets, and television and may develop a more realistic understanding of why and how offenders commit particular crimes and how they account for their involvement in such a deviant lifestyle.

ACKNOWLEDGMENTS

First and foremost, I would like to thank the authors who wrote the articles that make up this anthology. Some written in past years, and others more recently, are representative of high caliber research on crime and criminals.

Jerry Westby, criminal justice editor at Sage, deserves my gratitude for his belief in this project and his helpful suggestions throughout the preparation of this book. My thanks to Vonessa Vondera, editorial assistant, for her help with all the important logistics that were necessary to get this book to press. A special thanks to graduate student Patricia Dahl for her computer skills and putting up with me during the final stages of this work, and Jerry Venor, scholar, teacher, and friend, whose suggestions from the onset of this project were extremely helpful. Last but not least, a very special thanks to my wonderful wife, Professor Lyn Taylor, who sat through many a dinner listening to my constant ramblings and ever changing thoughts on ways to improve this work. You are very special.

ABOUT THE EDITOR

Mark R. Pogrebin is a Professor and Director of Criminal Justice in the Graduate School of Public Affairs at the University of Colorado at Denver. He has conducted numerous field studies in the areas of police undercover work, tragic events, Afro-American policewomen, emotion management, women jailers, psychotherapists' deviant behavior with clients, women in prison and on parole, and the strategic uses of humor among police. He has published four books and numerous journal articles and has had more than twenty of his publications reprinted in anthologies.

INTRODUCTION

The vast majority of criminology textbooks break particular types of crimes into groupings or classifications, such as white-collar or property offenses and so on, for purposes of describing and analyzing multiple variations of criminal behavior. Grouping or typifying certain categories of offenses and offenders may serve to highlight types of criminal behavior and allow for systematic study. This anthology attempts to do just that. Its purpose is to provide students with a more in-depth understanding of the types of criminality by utilizing studies that not only break down particular crimes into specific categories but also include the offenders' self-explanations or accounts for their criminal activities and lifestyles.

Some years ago, Clinard and Quinney (1973) attempted to define or distinguish types of criminality because they believed that putting particular crimes into categories allowed for a level of commonality for comparative and systematic analysis. Scarpitti and Nielson (1999, p. 318) note, "It has been shown that criminal typologies are helpful in allowing us to understand the nature of criminal offenses." According to Meier (1989), behavior system typologies can be utilized for organizing data about offenders and are used by researchers to break crimes into smaller, more meaningful categories.

The historical argument for the limitations of approaching the study of crime by looking at separate types of criminal offenses is understood. There is an awareness that most offenders do not generally specialize in a particular type of criminal activity, but rather participate in a variety of offenses. In short, criminals are versatile, not specialized, because the majority of their crimes are not planned, but rather are opportunistic in nature. However, adult criminals are more likely to specialize in a particular type of crime than are juveniles (Kempf, 1987), especially if they are actively participating in ongoing criminal activities.

Perhaps it should be noted that I am cognizant of the debate, pros and cons, strengths and weaknesses, for studying crime typologies. What I am attempting to accomplish with the production of this anthology is to offer criminology students the opportunity to gain insight into crime and criminals by offering them research on actual offenders who, throughout the majority of the pages in this book, actually discuss and provide details of offending behavior and reasons for their participation in crime. That is, their motivations, descriptions of how they operate, their thoughts about victims, and a whole host of "thick" descriptive analysis about their unlawful lifestyles is discussed. The objective is to provide students with the opportunity to draw their own conclusions about crime and criminals based on the interviews and observations of offenders portrayed in the 31 selected studies published in this book.

The ultimate value for the naturalistic use of field studies with criminals is best said by Miethe and McCorkle (2001, p. 17) in their discussion of the importance of letting offenders relate their personal experiences and rationale for their behavior:

One of the most important but often neglected sources of data about crime is offenders' own accounts. Perpetrators of crimes are in the unique position of being able to describe, in their own words, the motivations and causes of crime, the level and nature of crime calculus, and the perceived effectiveness of crime control activities in detecting crime. Narrative accounts by offenders also provide the rich details about the situational dynamics of crime and target selection process.

The articles selected in this book of readings offer a naturalistic method for the study of various types of crimes and offenses. Every article used a fieldwork perspective, which stresses interpretive, ethnographic methods that attempt to provide insightful knowledge at a close range (Daly & Chesney-Lind, 1988) and, further, expose parts of the criminal's social world that often remain hidden by more traditional methodological techniques (Caulfield & Wonders, 1994). The selected studies attempt to have offenders explain how they make sense of the world in which they live (Spradley, 1980).

Any book of readings attempts to offer a comprehensive view of the subject matter's contents. However, it would be prohibitive to select articles that cover every type of crime. So, as author, I purposely based my selection of writings on two major criteria; the first being the perceived interests to students, coupled with my subjective thoughts on the quality and relevance for each study chosen, and the second, the availability of naturalistic, qualitative type articles for particular categories of crime. To my surprise, there remains a genuine void of naturalistic materials for some criminal types, and as a result, there is a bit of overlap between a few categories. On the whole, it is my belief that this publication will bring to life the world of crime and criminals for students who are striving for a connection between theory and reality.

REFERENCES

Caulfield, S., & Wonders, N. (1994). Gender and justice: Feminist contributions to criminology. In G. Barak (Ed.), *Varieties of criminology* (pp. 213–229). Westport, CT: Prager.

Clinard, M., & Quinney, R. (1973). *Criminal behavior systems.* New York: Holt, Rinehart, and Winston.

Daly, K., & Chesney-Lind, M. (1988). Feminism and criminology. *Justice Quarterly, 5,* 497–535.

Kempf, K. (1987). Specialization and the criminal career. *Criminology, 25,* 399–420.

Meier, R. (1989). *Crime and society.* Boston: Allyn & Bacon.

Miethe, T., & McCorkle, R. (2001). *Crime profiles.* Los Angeles: Roxbury Publishing.

Scarpitti, F., & Nielsen, A. (1999). *Crime and criminals.* Los Angeles: Roxbury Publishing.

Spradley, J. (1980). *Participant observation.* New York: Hold, Rinehart, and Winston.

Part I

PROPERTY CRIMES

Property crimes are the most frequent type of offences committed in the United States. As distinguished from violent offenses, property crimes most often do not have a confrontational interaction between offenders and victims. According to the Uniform Crime Reports list of property index crimes, burglary, larceny-theft, motor vehicle theft, and arson are considered the most serious property offenses. However, vandalism, receiving stolen property, shoplifting, and destructive acts against property are included under the definition of property offenses.

From the following selection of readings, it can be seen that some property crimes are committed for monetary gain (burglary), whereas others cause destruction of property (graffiti) but fulfill other nonmonetary purposes for offenders. All in all, I have attempted to provide an introspective view of various property crimes. Although far from being comprehensive of every type of crime that comes under this offense category, the readings, I hope, will offer some insight into those law violators who commit these crimes as well as the rationale for their offenses.

The first chapter, by Richard Wright, describes the process of committing residential burglary. Wright describes the balance that maximizes rewards against the degree of risk that offenders must constantly be cognizant of during the burglary. In his analysis of burglars, Wright discusses the tactics they use to search the residential homes once they have broken in without being discovered. He found that burglars have a script they follow once in the residence, based on past experiences, that guides them in determining where the most valuable goods are in the home.

How do people who commit monetary-induced property offenses turn their stolen goods into money? Here, Cromwell, Olson, and Avary report on amateur fences that receive and sell stolen property, which was provided to them by property offenders. The authors vividly portray a description of the decision strategies involved in selling stolen items to people in the community who are willing to purchase them at bargain prices. The authors go on to analyze the perspectives of the property offender and the fences that purchase stolen goods. The fact that Cromwell and his colleagues found property thieves and fences to be ordinary people, novices, who buy and sell stolen goods on a part-time basis, offers a new perspective on the type of person who is involved in this illegal activity.

Jack Katz explores and analyzes the seductions of crime by using shoplifters as his example. He explores what he terms "sneaky thrills," which result from taking something (shoplifting) that belongs to someone else without their knowing it. This act can be intrinsically gratifying for the shoplifter and often results in providing a euphoric high comparable to taking psychoactive drugs, which according to Katz, can psychologically surpass the material rewards of the illegal act.

The final selection offers a discussion of a property crime that has no material gain for its offenders. Jeff Ferrell provides an in-depth study of urban graffiti writers whose purpose is to resist the perceived control they experience in city environments and simultaneously undermine the authorities who attempt to control them. Although it is illegal to deface property with graffiti, these youthful writers do not perceive themselves as committing an unlawful offense, but rather see themselves as fighting against political repression and social control agents.

1

Searching a Dwelling: Deterrence and the Undeterred Residential Burglar

Richard Wright

In an attempt to learn about the strategies and tactics that burglars use to search residences they enter for purposes of stealing items of value, Wright and his colleague Scott Decker interviewed 105 burglars in St. Louis, Missouri. They found that most of those burglars who break into people's homes want to search the residence as quickly as possible in order to avoid being discovered. They follow a mental script developed from past experience to assist them in finding the maximum amount of money and items in the short time they have to commit the burglary. The script they follow allows them to search rooms almost automatically without having to make time-consuming decisions once in the dwelling. By following such a cognitive strategic plan, they are able to find items of value quickly and avoid detection.

In most jurisdictions, a residential burglary has been completed in the eyes of the law the moment an offender enters a dwelling without permission, intending to commit a crime therein. But seen through the eyes of the burglars themselves, a break-in is far from complete at this point. Indeed, the offense has just begun. They must still transform their illicit intentions into action—which, in practice, almost invariably involves searching for goods and stealing them—and escape from the scene

without getting caught, injured, or killed. But in doing this offenders are on the horns of a dilemma. On the one hand, the more time they spend searching a residence, the better chance they stand of realizing a large financial reward. On the other hand, the longer they remain inside a target, the greater risk they run of being discovered. Having entered a dwelling, then, offenders must strike a deceptively complex, subjective balance that maximizes reward within the limits of acceptable risk. How is such a balance actually struck? Criminologists interested in the decision-making of residential burglars have devoted almost no attention to this process, despite the fact that such offenders obviously continue to make decisions throughout the commission of their break-ins. An examination of this matter is crucial to the development of a fuller understanding of the decision-making calculus of property offenders.

In an attempt to learn more about the tactics used by offenders to search dwellings, a colleague, Scott Decker, and I located and interviewed 105 currently active residential burglars in St. Louis, Missouri. The residential burglars were recruited through the efforts of a field-based informant—an ex-offender with a solid reputation for integrity and trustworthiness in the criminal underworld. Working through chains of street referrals, the field recruiter contacted active residential burglars, convinced them to take part in the project, and sat in on interviews that lasted two hours or more. In the pages that follow, I report just a small portion of what the offenders said during those interviews, focusing on how they search dwellings once they have broken into them.

Once inside a target, the first concern of most offenders is to reassure themselves that no one is at home. They do this in a variety of ways. Some run through the dwelling and take a quick glance into every room. Others remain still and silent, listening for any sound of movement. Still others call out something along the lines of "Is anybody home?" More than anything, such actions probably represent an attempt by offenders to put worries about being attacked by an occupant behind them so that they can devote their attention to searching for cash and goods.

> When you first get inside, you go through all the rooms to make sure no one's home. Once [you see] there's no one home, that's when you start gettin' busy, doin' your job.

Thus having reassured themselves, offenders often experience a sudden realization that everything inside the residence is theirs for the taking. One female offender said this realization made her feel as if she was in Disneyland— calling to mind a magical world in which fantasy had become reality. Another offender

likened the feeling he got inside an unoccupied dwelling to being in a fashionable shopping mall, "except you don't have to pay for stuff; just take it." Such comments suggest that, during this phase of the burglary process, offenders perceive themselves to be operating in a world that is qualitatively different from the one they inhabit day-to-day. Jack Katz refers to this world as "an enchanted land," the phenomenological creation of a mind bent on crime.[1]

Inside dwellings, shielded from public view, many offenders also experience a marked reduction in anxiety. They already have broken in, there is no turning back, and it makes little sense to agonize over the potentially negative consequences of their actions. Recognizing this, offenders have a tendency to settle down and turn their attention to searching the residence. This is not to suggest that the burglars stop worrying about the risks altogether. Most of them continue to be somewhat fearful, with the length of time they are willing to spend inside targets providing a rough indication of the gravity of their concern.

THE BRIEF SEARCH

The outside world does not stand still while burglars are searching dwellings and, as noted above, their vulnerability to discovery increases the longer they remain inside them. Occupants may return unexpectedly. Neighbors may become suspicious and call the police. Patrolling officers may spot a broken window and stop to investigate. Burglars are well aware of these risks, and the vast majority of them try to limit the time they spend in places.

> When you first get in [to a dwelling], do whatever you gon do; do it quick and get on out of there!
>
> When you doin' [a burglary], you work fast. You go straight to what you want to get and then you come out of there.
>
> I know three minutes don't seem like a long time, but in a house that's long! You just go straight and do what you got to do; [three minutes is] a long time.

The ability of the burglars to locate goods without undue delay is facilitated by strict adherence to a cognitive script that guides their actions almost automatically as they move through dwellings; this script allows them to flow through the search process without periodically having to stop to calculate their next step. Virtually all of the burglars reported having a tried-and-true method of searching residences which, they believe, produces the maximum yield in cash and goods per unit of time invested. The search pattern varies somewhat from offender to offender, largely as a function of the time an individual is prepared to remain inside a given target. With few exceptions, however, the burglars agree that, upon entering a dwelling, one should make a beeline for the master bedroom; this is where cash, jewelry, and guns are most likely to be found. These items are highly prized because they are light, easy to conceal, and represent excellent pound-for-pound value.

> I'm hittin' that bedroom first. I'd rather hit that bedroom than the living room cause it's more valuables in the bedroom than it is in the living room—jewelry, guns and money. So everything is in the back [of the dwelling] somewhere; most likely in the bedroom.
>
> The first thing you always do when you get to a house is you always go to the bedroom. That's your first move . . . [b]ecause that's where the majority of people keep they stuff like jewelry or cash. You know it's gon be a jewelry box in the bedroom; you know you ain't gon find it in the living room. Guns, you ain't gon find that too much in the living room.

The burglars believed that searching the master bedroom first also enhances their personal safety, especially when this room is located on the second floor. They feel particularly vulnerable upstairs because their only escape route—the stairway—can easily be cut off should an occupant return unexpectedly. As a result, most of them want to begin their search on the second floor and get downstairs as quickly as possible.

The burglars reported searching four main places within the master bedroom. The first stop for most of them is the dresser, where they quickly go through each drawer—often dumping the contents onto the floor or bed—looking primarily for cash and jewelry.

> [Y]ou got to look around, you got to ransack a little. You got to realize too that you don't have very much time to ransack neither . . . You always start in the bedrooms, in the drawers; that's where they keep the money and the jewelry.

They typically turn next to the bedside table, hoping to find a handgun and, perhaps, some cash and jewelry. From there, many of them search the bed itself because some people, especially those who are elderly or poor, continue to keep their savings under the mattress. Lastly, the burglars usually will rummage through the bedroom closet, looking mainly for cash and pistols hidden in a shoebox or similar container.

If the search of the main bedroom has been moderately successful, some of the burglars will not bother to look through the remaining rooms, preferring simply to make good their escape. They know full well that a more exhaustive search of the premises could net them a larger financial return, but are unwilling to assume the increased risk entailed in spending longer inside the target. As one burglar who typically searches only the master bedroom explained: "You miss a lot, but it's all gravy if you get away." A majority of the burglars, however, conduct at least a cursory search of the rest of the dwelling before departing. A number of them said they usually have a quick look around the kitchen. Surprisingly, most of these burglars are not searching for silverware or kitchen appliances so much as for cash and jewelry; they claimed that such items sometimes can be found hidden in a cookie jar or in the refrigerator/freezer.

> [The valuables are located in] either the freezer, the icebox, in they bedroom . . . in the dresser drawer or under the mattress . . . People put money in plastic bags behind the meat in they freezer.

Some of the burglars also search the bathroom, concentrating on the medicine cabinet where they hope to find not only psychoactive prescription drugs (e.g. valium), but perhaps hidden money and valuables.

> Like a lot of times I've [gone through] the medicine cabinets, quite a few people leave money in

the medicine cabinet. I don't know if you've ever heard of that, but I . . . found about forty dollars in there once and the second time I found about twenty-five dollars.

Few burglars, however, bother to rummage about in bedrooms occupied by young children because, in their view, such rooms are unlikely to contain anything worth stealing.

I ain't gon even worry about the kids' bedrooms. Mom and Dad have all the jewelry and stuff in they room. I know the little kids ain't got nothing in they room.

Little kids' rooms, I don't usually go in there because they don't usually have much. I don't have any kids so ain't no sense me goin' in there unless one of my little nieces or nephews might want something and I might keep an eye out for them.

Most of the burglars usually search the living room just prior to leaving the target because the items kept there tend to be heavy or bulky (e.g., television sets, videocassette recorders, stereo units); hence they best are left to the last minute. Indeed, some offenders do not bother to search the living room because carrying out the cumbersome goods located therein is likely to draw the attention of neighbors. As one offender explained: "I don't carry no T.V.s . . . because that's an easy bust." And another added: "You never take nothing big. You look for [something small]; something won't nobody see you bringin' out."

By employing such strategies to search dwellings, the burglars usually can locate enough cash and other valuables to meet their immediate needs in a matter of minutes. As one said: "You'd be surprised how fast a man can go through your house." Occasionally, however, the predetermined search strategies used by offenders fail to yield the expected results. This may cause some of them to depart from their normal modus operandi, remaining in the residence for longer than they feel is safe in order to find something of value.

[I] go straight to the main spots where I think the main stuff is in. Never mess around with, well, at least try not to, stay out the petty spots.

Bathrooms, you know, what you goin' in there for? I may go in there if the house ain't no good, you know, [isn't] what I thought it was; then I start gettin' desperate and stuff and lookin' everywhere. Gotta get somethin'. Let me see if they got a gold toothbrush, you know, just anything.

In breaking into the dwelling, these offenders already have assumed considerable risk and they are determined to locate something worth stealing, if only to justify having taken that risk in the first place. Add to this the fact that most of them are under pressure to obtain money quickly, and their decision to carry on searching despite the increasing risk of being discovered seems more sensible still; to abandon the offense would require quickly finding and breaking into an alternative target, with all of the attendant hazards.

On other occasions, offenders are tempted to linger in residences when they discover something that convinces them that an especially desirable item must be hidden elsewhere in the building. Put differently, they are enticed into accepting a higher level of risk, believing that the extra time devoted to searching is justified by the potential reward.

Then when I'm goin' through the dresser drawer or somethin', I might find shell boxes; they got a gun! Definitely! And I'm gon find it.

It is at this point that we can begin to glimpse the danger of allowing oneself to be seduced by the possibility of realizing a large financial gain. As noted above, the burglars are operating in a world that is qualitatively different from the one they inhabit day-to-day; a world in which they can take whatever catches their fancy. It is easy to see how they could get carried away by the project at hand, trying to take everything and disregarding the risks altogether. Generally speaking, the burglars are aware of this threat and try their best to avoid becoming so wrapped up in the offense as to throw caution to the wind. They do this by focusing on the items they originally intended to steal, resolving not to get greedy.

[M]ost of the time you want to get in and get out as quick as possible. See, that's how a lot of people get caught; they get greedy. You go in and get what you first made up your mind to get. When you take the time to ramble for other things, and look through this and look through that, you taking a chance.

I put my mind on one thing. That's what I'm a get. I ain't gon be ransackin' all through there . . . See, I don't get greedy once I go in and do a burglary.

From the perspective of these burglars, it makes little sense to steal more than necessary to meet their immediate needs because doing so involves additional risk. Besides, most of them could not transport more than they already had. As one put it: "I don't want to be in [the dwelling] that long. I ain't gon be able to carry all that stuff anyway."

Not surprisingly, some offenders did report getting carried away during particular offenses, being seduced by the allure of the available goods such that they forgot all about the risks of lingering in the target.

The stuff that was in there, it just had this attraction to your eyes. It made you feel like, 'God, I need that!,' you know. So that's what, so we just kept on . . . everything just attracted our eyes.

I was downstairs just lookin' around cause I was real choicy; I was real choicy this last job I had. [My partner] came downstairs and said, 'Man, you better get what you gon get and come on!' 'Man, what time is it?' I said, 'I got about fifteen minutes.' He said, 'Man, you better hurry up!' You know, he was rushin' me then cause usually I be rushin' them. 'Alright then, I'm a take this V.C.R., this T.V., let's go.'

Cases such as these, though, are the exception rather than the rule; almost all of the burglars usually adhere to a well-rehearsed cognitive script in searching targets. Admittedly, not all of them enter dwellings intending to steal specific items, preferring instead to allow themselves to be seduced by whatever catches their eye. But even these offenders typically stick to a set pattern in moving through residences.

When you go in [to a target], the first thing you do is go straight to the back. As you go to the back, you already lookin'. While you lookin', you pick certain things out that you gon take with you that you know you can get. You can pick them out just by lookin' as you walkin'. Then, when you turn around to come out of there, you already know what you gon get . . . It depends on what I spot and if I think it's of value; not no particular things.

Many residential burglars, of course, commit their offenses acting in concert with others. As Neal Shover has observed, co-offending appeals to many burglars because "it facilitates management of the diverse practical demands of stealing."[2] Foremost among these demands is avoiding detection. Working with others allows offenders to locate and transport goods more quickly, thereby reducing the risk of being caught in the act.

[I have searched dwellings more extensively], but that was only when I do it with friends. Cause I have more time; while I'm downstairs, he's upstairs. It's all timing. Fast as you do it and then get out of there.

I guess [I work with others] because it would take so much time for me to have to look for everything all by myself. If it's two or three of us it will be that much quicker.

A lot of times the places that I normally pick, it's quite a few items there . . . But then I use someone else so we can get the job done and move on out; get away as soon as possible instead of making a lot of trips to get everything [out of the dwelling].

[I commit burglaries with a partner] just in case they have lots of stuff in there I want. We can hurry up an' get it out.

A number of the burglars also reported that working with one or more accomplices is safer than operating alone because it permits them to post a look-out.

It's always good to go in a house with at least three dudes. You know, two of you'll get the stuff together. The other one look out the window; he be the watch.

[I]t's almost always a little safer to have someone else with you . . . Because if you got someone outside, they can always give a little signal and let you know when someone's coming or whatever. If you're alone, you can't hear these things.

These offenders believe that using a look-out not only is objectively safer, it also carries the subjective benefit of eliminating concerns about being caught by surprise and thus enhances their ability to concentrate on searching for goods to steal. As one of them noted: "By yourself, you never know who behind you."

Several offenders pointed out that accomplices represent ready assistance should they encounter unanticipated resistance.

> Sometime you want somebody with you . . . Sometimes I like to do [residential burglaries] with someone because I like to have protection in case something do happen.
>
> [I work with someone else] because we spent five and a half years together and we can handle ourselves real well. I can trust him; if I get in a tough situation, he would kick ass for me. If a guy with a big baseball bat is going to come kill me, he'll come to the rescue type of thing.

A few offenders chose to work with others in the belief that, should the police arrive unexpectedly, this increases the odds of at least one of them getting away.

> [Working with others decreases the risk] because if it's more than one [offender] and the police do come, then somebody is bound to get away. I guess if you get caught, you know, [the police] gonna catch one of yas, to put it that way, if you're gonna get caught, they're gonna catch one of ya, one of yas always gotta chance of getting away.

The logic underlying this belief is that two or more offenders can split up, running in different directions so that officers have to decide who to chase and who to allow to escape. The delusional aspect of this position is self-evident; why should the police elect not to pursue these offenders in favor of catching their accomplices? Here we are confronting a force that transcends rationality. The burglars acknowledge the risk of apprehension but believe that, because luck is on their side, they personally will avoid such a fate. That said, it remains true that by working with others, offenders may well reduce their individual risk of being caught during a police chase. One subject referred specifically to this fact: "I work with some partners because it's a better chance of gettin' away. They might get caught and I might get away." Along the same lines, another burglar said that he liked working with others because witnesses have a tendency to confuse the features of multiple offenders, making it difficult for them to provide the police with a good description of individual suspects.

Quite a few of the burglars who commit their break-ins with others reported doing so just in case they do get caught. Most of them simply want the reassurance of knowing that, should they be arrested, their co-offenders will be there to share the guilt and shame.[3]

> [I work with others] cause I feel if I get caught, I want them to get caught with me. I mean, I don't want to get caught by myself . . . They gon get caught with me . . . I feel I won't be so guilty if I have somebody with me. Then I won't feel so guilty, I'll feel kind of safe.

Two experienced female offenders, however, do not want to share the blame with accomplices, but rather hope to shift it entirely onto their associates. Both of these offenders work exclusively with men, believing that the police will show leniency toward a woman who claims that she was coerced into an offense by her male partner.

> I think down in my mind, when I first started doing burglary I saw a show and in it the woman claimed mental incompetent; that she'd been brainwashed. And I guess I feel like if we ever got caught that I could blame it on him. It's a pretty shit attitude, but . . . I don't know, kind of feel like I'm smarter than [the male burglars] are.

There may be a grain of truth in this belief.[4] Be that as it may, the important point for our purposes is that these women are convinced this is the case, and thus are able to mentally discount the threat of arrest and punishment. This allows them to get on with the business of searching dwellings, unimpeded by concerns about getting caught.

In short, the burglars who choose to work with others in committing their residential break-ins do so not only for practical reasons, but for psychological ones as well; the company of co-offenders dampens their fear of apprehension and bolsters their confidence.

> Like I told you, I know it sounds strange, but I be scared when I do [a residential burglary]. Then if I have somebody with me and they say, 'Ah, you can do it,' they boost me up and I go on and do it.
> [I work with someone else] when I don't really know about that place, you know, I'm kind of nervous about it. So I feel like I wouldn't be as nervous by me havin' somebody with me on a place that I don't know too much about.

While co-offending has a number of potential advantages, working with other criminals inevitably entails certain risks. In the best of circumstances, such individuals are of dubious reliability, and the pressures inherent in offending can undermine their trustworthiness still further. Several of the burglars said they are becoming increasingly reluctant to work with others, having been let down in the past by co-offenders who failed to carry their weight during offenses.

> See, you can't depend on no one else. That's why I'm goin' to court now . . . I had this so-called buddy of mine supposed to be watchin' this house for me [while I was inside]. And I told him to stand across the street so when I come out I can look across the street and see him. When I came out, he was gone and I had merchandise up under my arm. So I said, 'Let me get on out of here.' I don't know what happened, he might have just left me. So I was gettin' nervous and I just went on and left. And there the police was! Walked right into they arms!

Even those who continue to work with others typically believe that, should something go wrong, their crime partners might well let them down. Many expressed doubts about the ability of accomplices to withstand police interrogation without naming them as co-participants.

> You never know, it just ain't no sure thing. I just say, 'Do unto others what's done unto you.' So I'm not banking on [my partner's ability to remain silent]; if he get caught, then maybe that's damn near my ass is probably caught too. Cause I know if the police say, 'Was somebody with you?' he'll

probably say, 'Yeah, yeah, oh yes.' Police talk to you, you know; [partners] start spillin' they guts. They scared and then all they thinkin' about is themselves. So to be truthful with you, yeah, I never bank on [silence].

Most of these offenders appeared to accept the potential for duplicity among their colleagues with equanimity. As they see it, the police put pressure on arrestees to inform on their partners, and it is naïve to expect them to remain silent. One put it this way: "[My co-offenders] would probably tell on me, but, to be honest, I'd probably tell on them too."

Perhaps the aspect of co-offending that the burglars find most irksome involves what they perceive as a tendency for accomplices to "cuff" (that is, neglect to tell them about) some of the loot found during the search of a target. In their eyes, it is bad enough having to split the proceeds of their crimes, without having to deal with the possibility that their co-workers will try to cheat them. Some offenders attempt to reduce the risk of being cheated by working only with those they know well. For a few, this strategy seems to pay off.

> Well, like they say, 'There's no honor among thieves.' That's what they say, [but] I believe that they really are wrong about that because this here is really loyal. We done did one house, man, where this guy had [a pistol he found during the search] in his pocket already and we didn't know. But he came out and told us when we was splittin' everything up. He took it out and he put it with the rest of the shit to be split up. I wouldn't a did it; I would've kept that.

But most of them continue to believe that even close acquaintances might try to deceive them by cuffing booty.

> That's why I work with the same people, you know what I mean? He don't know what I got from downstairs. I might have found a ten thousand dollar diamond ring. Of course, you got to trust these damn fools, you know? It's easy to do man; it's easy for somebody to rip you off.

One offender reported that he and his usual burglary partner have an agreement whereby they always search the master bedroom together. The logic behind their agreement is that the most easily concealed valuables tend to be found in this room; going through it together represents a means of "keeping each other honest."

Despite the risks of co-offending, the fact remains that a majority of the burglars continue to work with others. As much as anything, their decision to do so undoubtedly reflects the powerful influence of routine. The burglars are used to co-offending. Many of them work with regular partners, and have developed cognitive scripts that incorporate roles for their accomplices. These roles are well understood by their co-offenders and therefore break-ins can be carried out efficiently, with a minimum of confusion or conflict.

> I work with him because when you get a [regular] partner, it's like two pieces of a machine; two gears clicking together and that's the way me and him work.
>
> Everybody has they routine. I check upstairs and then they stay downstairs and get the VCRs and everything. That's the routine we been doing for years.

When the pressures that give rise to burglaries intensify, then, it makes little sense for offenders to deviate from their typical modus operandi and set off alone. These pressures often arise in the context of partying—a group activity—where a shared desire to obtain fast cash for more alcohol or drugs precipitates a decision among those present to commit an offense.[5] As one burglar explained: "We be gettin' high together anyway, I might as well go [with] them. I come back by myself, they gon get high with me anyway."

THE LEISURELY SEARCH

On any given residential burglary, the safest course of action for offenders is to search the target quickly and then leave without delay. Adopting this approach, however, means that they seldom will come away from offenses having stolen more than is necessary to meet their immediate financial needs; there is unlikely to

be anything left over to help them deal with their next monetary crisis. The price of reducing the risks in the short term, therefore, may well be a foreshortening of the time between break-ins and a consequent increase in the frequency with which those risks must be taken. Some burglars, albeit a small minority, are unwilling to pay this price, preferring instead to remain in targets long enough to make certain they have found everything of value.

> [I take] the whole day going through the whole house, sitting down and eating and things of that nature . . . On a burglary, you get all that you can. Some people will just go get certain items, [but] I can just take everything cause everything has a value.

These burglars claim to understand the schedules kept by occupants of their targets and to have a clear idea about how long residents will be out. Thus they can proceed unimpeded by concerns about being discovered in the act of searching places.

> When I do a burglary, I don't go in there and come back out. I go in there and stay! I go in there and stay for a couple of hours. I know these people won't be back home until about five in the evening if they leave at seven in the morning. I be done ransacked the house by then.

Even offenders who do not routinely linger in targets occasionally succumb to the temptation to stay longer when they know that the occupants will be away for some time. Their reasons for doing so, however, often seem to transcend the desire for greater financial rewards. Indeed, many devote this extra time almost wholly to relaxation and entertainment.

> I usually go straight to the bedroom and then I walk around to the living room. I have set at people's house and cooked me some food, watched T.V. and played the stereo . . . I knew they wouldn't be there. But I usually go straight to the bedroom.

The offenders recognize that these offenses are special, being largely free of the temporal constraints that circumscribe most of their break-ins. They respond by taking full advantage of the situation and making themselves at home. In effect, they are acting out the widely held adolescent fantasy of having the run of a place without the obligation to answer to anyone. Some of them, of course, are adolescents. But even among those who are older, few have any experience of being in full control of their living space; the majority still stay with their mothers or have no fixed address. It is easy to appreciate why they enjoy having someplace to themselves. The irony is that they seldom do anything very outrageous. Like teenagers left alone for the weekend by their parents, most simply help themselves to whatever alcohol and food is available and take pleasure in not having to clean up afterward. One, for instance, reported: "Sometimes I cook me some breakfast, but I never wash the dishes."

A number of the burglars said that they sometimes urinate or defecate inside their targets. They attributed their need to do so to the emotional pressures involved in offending. Contrary to popular media accounts, however, they generally do not use the carpet for this purpose; most of them reported using the toilet, sometimes not flushing it afterwards because the resulting noise can drown out the warning sounds of approaching danger. A couple of the burglars did admit to sometimes relieving themselves in rooms other than the bathroom, but they explained this action in terms of safety—they do not want to get cornered, literally with their pants down, in a small space with just one exit—rather than attributing it to any special contempt for the residents.[6] At the same time, these burglars seemed untroubled about the distress this might cause their victims; their sole concern is for their own well-being.

SUMMARY

The vast majority of residential burglars want to search dwellings as quickly as possible in the belief that the longer they remain inside, the more chance they stand of being discovered. They do this by adhering to a cognitive script developed through trial and error to assist them in locating the maximum amount of cash and goods per unit of time invested. Using this script, burglars can proceed almost automatically, without having to make complicated decisions at each stage of the search process. Although the script varies from one offender to the next, it usually calls for them to search the master bedroom first; this is where money, jewelry, and guns are most likely to be found. The living room typically is searched last because the items kept there tend to be difficult to carry and hence are best left until the last minute. Many burglars work with others to expedite the search process; co-offenders can explore one part of the residence while they look through another. By employing a consistent, well-rehearsed modus operandi in searching dwellings, burglars often can locate enough valuables to meet their immediate needs in a matter of minutes. Having successfully done so, most leave without delay.

By the time offenders have entered a target with the intention to steal, a burglary has been committed; the offense can no longer be deterred or prevented. It is at this point that criminologists and crime prevention experts show a marked tendency to lose interest, ceding the field to victimologists and police investigators. This is unfortunate because the activities of offenders during break-ins also may have implications both for decision-making theory, especially in regard to deterrence, and for crime prevention policy. To be sure, the burglars have not been deterred by the threat of sanctions, but that threat nevertheless seems to have a pronounced effect on their actions as they search targets. Most are unwilling to remain inside for long, foregoing the possibility of greater rewards in favor of reducing the risk of being discovered. In fact, for actual offenders this may be where deterrence operates most effectively. And while it is too late to prevent the burglaries, there is still an opportunity to limit the loss of cash and goods if we can understand the cognitive scripts used by burglars to search dwellings well enough to be able to disrupt those scripts.

NOTES

1. Jack Katz. 1988. *Seductions of Crime: Moral and Sensual Attractions in Doing Evil.* New York: Basic Books.

2. Neal Shover. 1991. "Burglary." In Michael Tonry *Crime and Justice: An Annual Review of Research*. Chicago: University of Chicago Press.

3. Ibid.

4. Rita Simon. 1975. *Women and Crime*. Lexington, MA: D.C. Heath.

5. Richard Wright and Scott Decker. 1994. *Burglars on the Job: Streetlife and Residential Break-Ins*. Boston: Northeastern University Press.

6. Dermot Walsh. 1980. "Why Do Burglars Crap on the Carpet?" *New Society* 54:10–1.

2

WHO BUYS STOLEN PROPERTY? A NEW LOOK AT CRIMINAL RECEIVING

PAUL F. CROMWELL, JAMES N. OLSON, AND D'AUNN W. AVARY

Cromwell, Olson, and Avary conducted a field study of 30 active burglars and shoplifters along with an analysis of 190 arrested burglars' and shoplifters' arrest reports written by police. They further gained knowledge about those persons who received stolen goods by interviewing professional and nonprofessional fences who bought illegally attained property. They found that property criminals often sell their illegally attained items to amateur receivers. Although novice receivers do not purchase stolen property to the degree that professional fences do, they provide a large market for illegal goods based on the number of them involved in this business.

The role of the fence in initiating and sustaining property crime has been recognized for centuries. It is the professional fence, however, who has attracted the attention of researchers, policy makers, and law enforcers (Blakey & Goldsmith, 1976; Klockars, 1974; Senate Select Committee, 1973; Steffensmeier, 1986; Walsh, 1977). There is little reliable and valid information regarding the extent of the fencing activities among nonprofessional receivers of stolen property or the degree to which these amateur fences contribute to the initiation and continuing support of property crime. Some earlier studies concluded that thieves are "unable to deal directly with the consuming public and must therefore operate through middlemen who have the financial resources to purchase stolen goods and the contacts to help in their redistribution" (Blakey & Goldsmith, 1976, p. 1515). Indeed, this is true in large-scale theft where a thief must dispose of a truckload of television sets or a collection of fine jewelry. Most property crime, however, involves smaller quantities of stolen goods, of lesser value. Television sets, car stereos, most jewelry, handguns, VCRs, microwave ovens—the items that constitute the loot of the average burglar or shoplifter—may be redistributed

---

EDITOR'S NOTE: From Cromwell, P., Olson, J., & Avary, D. W., "Who buys stolen property?" in *Journal of Crime and Justice, 16,* pp. 75–95. Copyright © 1992. Reprinted with permission from Anderson Publishing Co.

without the assistance of a professional fence. The thief may sell many of these items directly to the ultimate consumer, to individuals who know or suspect that the items they buy are stolen property. Some items may be traded for drugs; [others may be] resold to part-time receivers—those whose primary business activity is something other than buying and selling stolen property. Other stolen merchandise may be sold in pawn shops, flea markets, and garage sales to consumers who do not know or suspect that it was stolen.

The nonprofessional receiver has not been overlooked completely. Hall (1952) included part-time receivers in his typology of fences; he identified the "lay receiver," who buys for personal consumption, and the "occasional receiver," who purchases for resale, but only infrequently. Stuart Henry (1978), who studied property crimes committed by ordinary people in legitimate jobs, concluded that receiving stolen property is not exclusively the province of professional criminals but is an "everyday feature of ordinary people's lives." He states,

> The artificial distinction between "honest" and "dishonest" masks the fact that the hidden economy is the on-the-side, illegal activity of "honest" people who have legitimate jobs and who would never admit to being dishonest. (p. 12)

Steffensmeier (1986) concluded that there is no way of knowing what proportion of stolen goods passes through professional fences. He writes,

> Not all stolen goods are fenced: Some thieves keep the goods or sell them directly to a customer who plans to use them. In other cases thieves may "peddle" their stolen wares on street corners, in bars, or at flea markets. (p. 18)

McIntosh (1976) described the practice of selling directly to the consumer as "self-fencing." She concluded that many thieves sell or pawn their goods to lay receivers, friends, neighbors, acquaintances, police officers, shopkeepers, and other thieves. . . .

FINDINGS

Interviews With Burglars and Professional Fences

Professional burglars must have reliable outlets for their stolen merchandise. Most sell their goods to one or more professional fences; dealing directly with the consumer is too irregular and an uncertain way of doing business. Other burglars, however, have limited access to the professional fence. Novice burglars, juveniles, and drug addicts often find it hard to establish regular business relationships with fences. Novices and juvenile burglars do not often steal "quality" merchandise and have not been "tested" regarding their trustworthiness. Drug addicts have a similar handicap in marketing their goods. They are considered unreliable and untrustworthy because of their drug habits. While several addict/burglars reported that they occasionally sold their stolen merchandise to a professional fence, most had to seek less rewarding and more risky alternative channels for their goods. Many resorted regularly to direct sales to the consuming public. One young burglar, who regularly sold his stolen goods directly to consumers, said:

> I hear about somebody who want a TV or a VCR. I ask 'em how much they want to pay, and then I go get them one. If I already got some stuff, I ask around if anybody want to buy it.

A heroin addict/burglar reported:

> I sell my stuff to [a local fence] when I can. Sometimes he buys stuff from me. Most of the time he don't. He don't trust addicts.

Another informant told the interviewer that he had regular customers [amateur receivers] for his merchandise. He described his "self-fencing" in the following manner:

> There is this lady who buys big dresses—like bigger than size 16. She pays good too. Another lady will buy jewelry and stuff if I have it. I know about 10 people who buy meat. Whatever is on the price tag, they give me half price. There is even a policeman—he used to be a policeman—who buy guns if I get one.

Most of the burglars interviewed would have preferred to sell their goods to a fence. They believed the fence to be a more reliable and less risky market. However, of the 30 informants in the study, only 7 (23%) reported that they could absolutely depend on the professional fences in the community to take their goods. Others reported only occasional business dealings with professional fences. The following statement was typical:

> If I get guns. Not junk—like Saturday-nite specials—stuff like Smith and Wessons—I can sell to the fences. I sold some big diamond rings and a Rolex to [local fence] last year. They don't buy TVs and VCRs though.

Professional fences reported a strong aversion to doing business with drug addicts and juveniles. However, one fence, while expressing his contempt for drug addicts, bought stolen items from several obvious drug users while we were observing. He explained that, "So many thieves these days are addicts that you got to do a little business with them or you go broke." Another fence, more adamant in his refusal to do business with drug addicts, posted a sign over his cash register. It stated, "NO ADDICTS."

The fences reported that amateurs, drug users, "kids," and other "flakes," could not be trusted. They "snitch" and turn in their buyers when arrested. Professional thieves do not so readily "give up" their meal tickets—their market for stolen goods. Several reported that they would give up their co-defendants before their fence. One expressed his attitude, thusly:

> Shit! Thieves are easy to find. I can get somebody to help me do a crime anywhere. Fences—they harder to replace. You turn in [local fence] and nobody gonna do no business with you after that.

Interviews With Nonprofessional Receivers

We also interviewed 19 nonprofessional receivers; persons who had bought stolen items for personal consumption and/or for resale, but who did not depend upon fencing for all or most of their livelihood. Some had bought stolen property only once or twice. Others were regular consumers of stolen goods. One of these fence/consumers, a college professor who had been buying clothing for himself and his family for 20 years, described his activities as follows:

> I go to [department store] and pick out what I want, and tell [thief]. He brings it around in a few days. I pay one-third of the price tag.

A college student said:

> I know this guy. He's a pothead. He gets speakers, and CD players, and all kind of stuff like that. I've bought stuff from him a lot.

A homemaker in a low-income neighborhood explained how she became a customer of a shoplifter:

> My friend said she bought meat from this drug addict. She said she could get me some meat at half price. First, I bought some steaks for half price. Now he comes by my house every payday and we get all our meat from him. Other stuff too, sometimes.

More than one-half (n = 11) of the nonprofessional receivers we interviewed own or are employed by legitimate businesses, and occasionally buy stolen property at their place of business, primarily for resale. But, unlike the professional fence, they do not rely on buying and selling stolen property as their principal means of livelihood. To them, fencing is a part-time enterprise, secondary to, but usually associated with their primary business activity.

One of these part-time receivers justified his regular purchases from a thief, saying:

> It's not like you have anything to do with the guy stealing the stuff. He has already stole it. If I didn't buy it someone else would. I just take advantage of a good deal when I can. It's good business.

Another explained, saying:

> I don't even know for sure the stuff is hot. All I know is I can buy brand new tires for 20 bucks a piece. The last ones I got from him were Michelins and I sold 'em for $100 each. I'm going to turn that down?

A dry-cleaner whose sideline was buying and selling stolen mens suits from shoplifters reported:

> I can put them in a bag and run them in with my regular cleaning. I don't make a big profit. Some weeks I don't buy anything. Mostly I sell to some friends and family. It helps out when business is slow.

Analysis of Arrest Reports

We analyzed 190 statements (confessions) given to police by arrested burglars, in which the burglars told police where and how they had disposed of the stolen property. The analysis revealed that only 21% of the stolen property was sold to professional fences. More than half (56.4%) was sold to nonprofessional receivers, including drug dealers. Only 12.1% was sold to pawn shops. The remaining 10.5% was reported to have been kept by the thief for personal consumption, thrown away, given away, or recovered by police before redistribution. The following statements are representative:

> We took the microwave to [address deleted] and we sold it to an elderly Mexican lady for $30.
>
> I traded this stuff [cartons of cigarettes] to a man named Mario on the south side for heroin.
>
> The place where we took the guns was a house on [street name deleted] in [town]. A man named [name deleted] lives there. He is in a motorcycle gang named the Outlaws. They buy guns.
>
> Then I went back to the 7–11 and sold the VCR to [a customer in the parking lot].
>
> I sold the disc player and the VCR to [name deleted] at [name deleted] Liquor Store.

DISCUSSION

Why Buy Stolen Property?

Buying stolen property involves many different motives. It represents a means of livelihood for some people—individuals who earn all or a significant proportion of their income from fencing. For others, as Steffensmeier suggests, "The fencing they do more or less helps keep them afloat, get over the hump in their legitimate business, or gives them a little extra pocket money" (1978, p. 118). For many others, buying (and occasionally selling) stolen property is a means of economic adaptation. Henry (1978) and Smith (1987) refer to this activity as constituting an "informal economy" or "hidden economy." Participation means more than stretching the dollar. It may, as Gaughan and Ferman (1987) suggest, be a means of economic survival. They write:

> A number of case studies have shown that low income communities rely on informal economic resources. The importance of hustling in the black ghetto, the persistence of tight kinship networks in working-class urban communities, and the increasing visibility of street peddlers and entertainers testify to this." (p. 23)

Several informants told us that they could not survive [economically] without "hustling." One reported:

> I buy my baby's clothes from this booster [shoplifter]. He sells lots of stuff—even deodorant, aspirin, and medicine. I get cigarettes from another thief. My mother gets all her meat from a booster.

A housewife/informant reported that she occasionally bought and then resold various items from a burglar of her acquaintance. She stated that the income from this source allowed her to stay at home with her children, rather than having to seek outside employment.

While for some, buying and/or selling stolen property was purely economic activity, we were unable to differentiate subjective motivations from economic motivations in many transactions. Psychosocial dynamics were often inextricably bound with economic motives. Some informants reported that they simply could not resist a bargain. Both thieves and receivers reported that "getting a good deal" was an important motivation for the buyer of stolen goods. Some reported buying items for which they had no immediate use, because the "price was right," or that they liked "beating the system." For still others, the occasional purchase of stolen property provided excitement in an otherwise pedestrian existence. One lawyer, for instance, asserted that he bought stolen property not just because he wanted certain items or for the money he might make reselling the items, but also for the "insider feeling" he got through these associations (see Shover, 1971, p. 153). Another "amateur" receiver reported that she was a member of a group of office workers who bought clothing from several shoplifters who made the "rounds" in various offices and businesses in the area. The women frequently placed orders for items of clothing and paid a pre-arranged price for the items when the shoplifter returned with the goods. Our informant revealed that the items they purchased were often bought as gifts for friends and relatives. She described a sort of "party atmosphere" around the office on the day the shoplifter was due to arrive with the goods they had ordered days before. As described by our informant, the transaction appeared to be more social than economic in nature.

Neutralizations

One theme which appeared to characterize nonprofessional receivers was the tendency to neutralize or rationalize their involvement in the purchase of stolen property. Almost all of those interviewed disassociated themselves from the theft, and by extension, the victim(s). Many rationalized that "It was already stolen. If I didn't buy it, someone would." They appeared to view purchasing stolen property as victimless crime, if crime at all. Many neutralized their purchases as "Simply getting a bargain," or rationalized that the victim was an insurance company or a big business that "expects to lose a certain amount of merchandise," and makes up for the loss by increasing prices. Many reported that they did not know for sure that the items were stolen. One informant justified her purchase of a new VCR for $50, stating:

> O.K., it's maybe too good a deal to be completely honest. I asked him if it was stolen and he said, "No," and I took his word for it. That's all I can do. I don't want to know either.

Burglars and other thieves appeared to intuitively understand the psychology of selling stolen property directly to the consumer. Most reported that the items must not be explicitly represented as stolen, yet the buyer must believe them to be illegally obtained, and therefore a "good deal." One burglar/informant occasionally purchased cheap costume jewelry from a discount store and sold it as genuine on street corners to bypassers. While he did not specifically represent the jewelry as stolen, he implied that it was. He reported that he usually made a good profit from this scam. He concluded:

> People are basically dishonest. They just don't like to admit it to themselves.

Shover's (1971) informant described the same phenomenon stating:

> It's the excitement of buying a piece of stolen goods. If you told them . . . that it was legitimate they wouldn't buy it. (p. 153)

Several burglars reported that they devised elaborate stories about the source of their stolen items. They explained that buyers like a good story, even if they don't really believe it. The cover story serves to relieve the buyer's anxiety over buying stolen property. An articulate burglar, a college graduate who turned to burglary after becoming addicted to heroin, analyzed the citizen-receivers he did business with, saying:

> People need to feel good about themselves. Most folks can't accept that they are as crooked as us [burglars]. You gotta help 'em out a little. Give 'em a story about the stuff. They know you're lyin'. Doesn't matter. They need it to keep you and them separated in their minds. You're the thief and they're the good guys.

A Typology of Receivers

It is impossible to characterize those who buy stolen property as a homogeneous group. Rather, they are a diverse group ranging from professional criminals with ties to organized crime (Klockars, 1974; Steffensmeier, 1986) to respected citizens such as schoolteachers, businesspersons and office workers who buy stolen goods for personal consumption (Henry, 1978). They may be differentiated by (1) the frequency with which they purchase stolen property,

(2) the scale or volume of purchases of stolen property, (3) the purpose of purchase (for personal consumption or for resale), and, (4) the level of commitment to purchasing stolen property. Upon the basis of these criteria, we distinguished six levels of receivers or fences:

1. Professional fences

2. Part-time fences

3. Associational fences

4. Neighborhood hustlers

5. Drug-dealers who barter drugs for stolen property

6. Amateurs

Professional Fence. The professional fence is one whose principal enterprise is the purchase and redistribution of stolen property (Blakey & Goldsmith, 1976; Chappell & Walsh, 1974; Klockars, 1974; Steffensmeier, 1986). Professional receivers may transact for any stolen property for which there is a resale potential, or may specialize in stolen property that they can commingle with their legitimate stock or legitimate business (jewelry, dry cleaning, appliance sales or service). The professional receiver generally makes purchases directly from the thief and almost exclusively for resale. These receivers operate proactively; they establish a reliable and persistent flow of merchandise, buying regularly and on a large scale. As a result of this commitment, the professional receiver acquires "a reputation among law breakers, law enforcers and others in the criminal community" (Klockars, 1974, p. 172). Although professional fences frequently operate "legitimate" businesses as fronts for their fencing activities, fencing is their primary occupation.

Part-time Fence. The part-time fence functions in a somewhat nebulous domain between the true professional fence and other categories of receivers. The part-time fence is differentiated from the "professional" fence by frequency of purchase, volume of business, and degree of commitment to the fencing enterprise. Usually they do not buy as regularly as do professional fences, nor do they buy in volume. Further, part-time fences do not depend on fencing as their principal means of livelihood.

There appear to be two general sub-types of part-time fences: (a) *The Passive Receiver*—who purchases stolen goods either for personal use, or for resale to an undifferentiated secondary consumer; and, (b) *The Proactive Receiver*—who buys for resale only, and who may take an active role in the theft by placing orders for specific merchandise and providing offenders with information about potentially lucrative targets for burglary in the same way as professional fences.

Passive buyers are known (by thieves) to be buyers of stolen goods. They are "passive" because they do not actively solicit thieves as suppliers, nor do they contract to buy certain items from thieves. Their commitment as receivers is only at the level of being occasionally available to buy certain items of stolen property when offered by a thief. We identified several passive buyers during the study, including a truck stop operator who bought stolen tires and tools, the manager of a video rental business who bought stolen VCRs and videotapes, and a jewelry store proprietor who bought gold jewelry and silverware. Like all part-time receivers, the passive receiver does not buy in volume, and does not depend on fencing as a principal means of livelihood. He/she may integrate the stolen items into their regular stock, or may personally use the stolen goods—as with the case of an automobile mechanic who buys stolen tools. Like all part-time fences, they are not considered reliable outlets by thieves. They buy only when they have funds available, and/or when they need the particular item(s) offered by the thief. They usually do not have in mind a specific customer to whom they might resell the merchandise.

Proactive receivers mimic professional fences in many respects; however, they do not rely on buying and selling stolen goods for the major part of their livelihood. Their fencing activity is an on-the-side activity—part-time crime. Like most part-time receivers, they generally identify with the dominant values of society and do not consider themselves to be criminals. And, except to a small cadre of thieves (possibly only one or two persons) with whom they work, they are not known as fences in either the criminal subculture or in the community-at-large.

The proactive buyer may contract with a thief for certain items for which he/she has a market. They may actually "take orders" from customers for certain items and arrange to purchase those items from a thief or thieves. Several burglars' and shoplifters' informants reported that they occasionally stole certain items "on order" from both professional fences and from part-time receivers.

The part-time receiver might even have greater access than the professional fence to strategic information regarding potential victims. Because they are otherwise legitimate citizens and businesspersons, they may be trusted by their colleagues and friends with information regarding their possessions, schedules, and security precautions.

We identified three "proactive" part-time fences. They were each otherwise legitimate businesspersons. One owned a jewelry store. The majority of his income was from the legitimate profits of the store. However, he occasionally contracted to buy certain specific items of jewelry from a professional thief. In many cases the jeweler had originally sold the items to the victim—only to steal them back. Another "proactive" part-time receiver, a gunsmith, gained extensive knowledge about a customer's gun collections through his profession, and used that information to provide inside information to a thief about security arrangements and particular items to be stolen. In another case, a pawn broker provided a thief with descriptions of a customer's jewelry and the details of the customer's vacation plans, which the customer had revealed to him during a conversation.

Associational Fence. Associational fences are persons whose legitimate occupations place them in close association and interaction with thieves, as in the case of police officers, criminal defense attorneys, or bail bond agents. Associational receivers may operate from a different economic motivation than other receivers in that they stand to lose financially by refusing to participate in the redistribution of stolen property. This is particularly true for bail bond agents and criminal defense attorneys, who may provide legitimate professional services to property offenders who cannot pay for these services with anything but stolen property (or the

proceeds from their illegal activities). Thieves constitute a significant market for the services that these receivers provide legitimately. To refuse such trade would eliminate these "customers" and would severely curtail earnings. For some it is but a small step from accepting stolen property in return for professional services, to placing orders for items to be stolen.

During the study, a bail bond agent showed the interviewer a matched pair of stainless steel .357 magnum revolvers that had been stolen for him "on order" by a client for whom he had posted bond. He freely acknowledged accepting stolen property occasionally in exchange for his services, justifying his actions by saying, "When they are in a bind and don't have any money, I try to help out." Another associational receiver, a criminal defense lawyer, was completely open about his occasional purchases of stolen property from clients, enthusiastically describing items he had received in exchange for his legal services. He described how he had agreed to represent a burglar in a criminal case, telling him that he wanted a gold Rolex in exchange for his services. He proudly displayed the $12,000 watch to the interviewer and stated, "This is a special order" (Cromwell et al., 1991, p. 79).

Neighborhood Hustler. The neighborhood hustler buys and sells stolen property as one of many "hustles," small-time crime and confidence games, which provide a [usually] marginal living outside the conventional economic system. The neighborhood hustler may be a small-time fence, or he/she may be a "middleman" who brings thieves and customers together, earning a percentage of the sale for service rendered. The neighborhood hustler may also be a burglar who, on occasion, tries a hand at marketing stolen items for others. By definition, he/she is a small-time operator (Blakey & Goldsmith, 1976). Most do not have a place of business, as such. Instead, they work out of the trunk of a car, or from their home. One such entrepreneur described himself, thusly:

> I'm a hustler. I can get you what you want. You got something you want to sell. Tell me. I know where to go and who to see. Ain't nothing happens over here in the Flats [the area of town where he lived] that I don't know about.

While he was, according to others who knew him, grossly exaggerating his abilities, he was an almost stereotypical neighborhood hustler. Few experienced thieves would trust him to buy directly or to sell their merchandise for them. Several informants described him as a "snitch" whose hustles included "giving up" the thieves with whom he did business. He was therefore limited to buying from juveniles, drug addicts and novice thieves who could not market their goods in a more reliable manner.

For some, hustling involves both buying stolen goods for personal consumption, and dealing in stolen merchandise. One such neighborhood hustler among our informants bought cigarettes, food, and clothing for personal use, and bought other items for resale. At the time of our interview, she had recently bought 20 rolls of roofing tarpaper and 10 gallons of house paint from a thief and resold the items to a building contractor for a $75 profit. Her "hustles" also included some low level street drug dealing which occasionally involved bartering drugs for stolen goods, which she then resold.

Drug-Dealer Fence. Although not every drug dealer will barter for drugs, our interviews with thieves and fences suggest that many street level dealers consider fencing and drug sales to be logically compatible enterprises. There are two apparent economic motivations for their willingness to barter: (1) bartering increases their drug sales, opening their market to those with stolen property but without cash, and (2) they can increase their profits by marketing the stolen property at a price well above that given in trade to the addict/thief. One such fence, in discussing the advantages of the arrangement, said:

> I know how to talk to a junkie. I can get it [stolen property] for nothing. I'll give them $30 or $40 [in drugs] for a VCR and turn around and turn it around [resell] for $100.

Another concluded that he sold more drugs because of his willingness to barter. He said:

> Lot of people come to me 'cause I'll take trades. Won't take no junk or TVs or shit like that. If they got guns, jewelry—then we can do business.

Several of the burglar/drug users we interviewed regularly bartered stolen property for drugs. Rather than searching for an outlet for the goods, they went directly to the drug dealer and obtained drugs in exchange. Although they reported that they did not receive the best possible price for their merchandise from the drug dealer, the speed and efficiency of the operation made the arrangement attractive. One subject said, "This is like one-stop shopping."

Amateur Receivers. With the exception of the professional thieves in the sample (n = 5) all had sold their goods directly to consumers on one or more occasions. The least experienced, the juvenile, and drug-using thieves were most likely to sell directly to the consumer on a regular basis.

We distinguished two general categories of amateur receivers: a. Strangers approached in public places [and] b. Persons with whom the thief has developed a relationship and who buys more or less regularly, for personal consumption.

Approaching strangers in public places is risky behavior with a relatively low success rate. It is looked upon with contempt by almost all thieves, and practiced only by those with no other available outlets for their merchandise. Juveniles and drug addicts are most likely to use this technique for disposing of their merchandise. While many thieves in our study expressed their disdain for selling stolen items in this manner, the analysis of confessions given to police by burglars and other thieves revealed that much of the "booty" of these thieves was sold in this manner. Furthermore, many of the nonprofessional receivers we interviewed reported that they had previously bought stolen items from thieves who approached them in a public place. This suggests that the practice may be more widespread than was indicated in the interviews with burglars.

A second category of amateur receivers includes those who have developed relationships with one or more thieves and buy stolen property with some regularity. Most buy primarily for personal consumption. Others occasionally resell the stolen property they buy. Several reported reselling their "hot" merchandise at garage sales or through flea markets.

Two of the "amateurs" resold the merchandise to friends and co-workers. One amateur fence, a public schoolteacher, began her criminal career when she was approached by a student who offered her a "really good deal" on a microwave oven. She stated that she originally bought the oven to help the student, whose family was suffering financial problems. Afterward, the student began to offer her "bargains" regularly, and she became a frequent customer. Eventually she began to offer her colleagues the opportunity to "get in on a good deal," and even posted the following note in the teachers' lounge:

NEED A TV, VCR, MICROWAVE, ETC. ????? SEE ME BEFORE YOU BUY. ½ OFF RETAIL.

Usually, the teacher did not profit financially in this exchange. Instead she garnered the goodwill and appreciation of those to whom she afforded merchandise at well below wholesale prices. Although she admitted to the interviewer that she "probably knew, deep down inside" that the items were stolen, she had never previously admitted it to herself. In explaining her motivation for purchasing goods in such an unconventional manner, she ironically described them as "a real steal."

Some individuals begin as amateur fences and become more deeply involved as a result of the irresistible gains and the virtual absence of sanctions entailed in purchasing stolen property. The overwhelming increase in profits and the thrill of "beating the system" (or at least making a good deal) tempt them into increasing their participation in the distribution of stolen property.

One such amateur, turned part-time fence, a social worker, began her fencing activities when her husband purchased a household appliance from a thief he met in the course of his business as a plumber. At first their purchases were for their own consumption; later they bought Christmas presents for family members. Eventually they established a thriving family business buying stolen property from thieves and selling it in garage sales and flea markets, as well as to amateur receivers cultivated by the husband through his business colleagues and

customers. The informant told the interviewer that she and her husband had put their son through college with the proceeds (Cromwell et al., 1991, p. 77).

Summary

Burglars, shoplifters, and other thieves may sell a substantial proportion of the property they steal to amateur and other nonprofessional receivers. Indeed, only 27% of the stolen property from 190 arrest reports was sold to professional fences. Fences themselves reported that they preferred not to do business with juveniles, inexperienced thieves, or drug addicts—groups which comprise a substantial percentage of the thieves in a community. Although these nonprofessional receivers do not purchase with the frequency, volume, or commitment of the professional fence, our study suggests that they represent a large market for stolen goods, compensating for lack of volume with their sheer numbers.

Conclusions

The extent to which fences contribute to the incidence of property theft has been a central issue of debate. In the late 18th century, English magistrate Patrick Colquhoun (1797) called for vigorous action against fences in London, stating, "Nothing can be more just than the old observation, 'that if there were no Receivers there would be no Thieves'" (p. 298).

Colquhoun's observation continues to have currency. Many believe that if fences could be put out of business, property crime rates would be dramatically reduced (Blakey & Goldsmith, 1976; Walsh, 1977). The fence is portrayed as not only providing a market for stolen goods, but also serving as an instigator and initiator of property theft.

This perspective has been criticized by other observers who argue that the "conception of the fence's role in property theft is bigger and more important than it ought to be, and that the involvement of other participants in an illegal trade is overlooked" (Steffensmeier, 1986, p. 285). Stuart Henry (1977) argues that viewing the fence as the prime mover of property theft "rests on pretending that real crimes are committed by "real" criminals, not by ordinary people, and certainly not by oneself (cited in Steffensmeier, 1986, p. 286).

The extent to which "ordinary" people participate in the hidden economy (buying and selling stolen goods) is yet undetermined; however, our findings suggest that this "part-time crime" is ubiquitous (see also Henry, 1978). Unlike the professional fence, these individuals do not perceive themselves as criminal, or as part of the impetus for property crime. Yet, they provide a ready market for stolen property, particularly for the young, inexperienced, and drug-addicted thieves who lack "connections" with professional fences.

These relatively unstudied channels of redistribution of stolen property may have important implications for crime control. Research that identifies these channels and determines the extent to which stolen property is purchased directly by the consuming public or by "otherwise honest" businesspeople and citizens will assist in the development of strategies to inhibit and disrupt the distribution process.

REFERENCES

Blakey, R. and M. Goldsmith (1976). Criminal Redistribution of Stolen Property: The Need for Law Reform. *Michigan Law Review* 74: 1511–1613.

Brewer, J. and A. Hunter (1989). *Multimethod Research: A Synthesis of Styles*. Newbury Park, CA: Sage.

Cameron, M. (1964). *The Booster and the Snitch*. New York: Macmillan.

Chappell, D. and M. Walsh (1974). Receiving Stolen Property: The Need for Systematic Inquiry into the Fencing Process. *Criminology*, 11:484–497.

Colquhoun, P. (1797). *A Treatise on the Police of the Metropolis*. London, ENG.

Cromwell, P. F., J. N. Olson and D. W. Avary (1991). *Breaking and Entering: An Ethnographic Analysis of Burglary*. Newbury Park, CA: Sage.

Ferman, L. A., S. Henry and M. Hoyman (1987). Issues and Prospects for the Study of Informal Economics: Concepts, Research Strategies and Policy. *Annals of the American Academy of Political and Social Sciences* 493:154–172.

Gaughan, J. P. and L. A. Ferman (1987). Toward an Understanding of the Informal Economy. *Annals of the American Academy of Political and Social Science* 493:15–25.

Hall, J. (1952). *Theft, Law and Society* (2nd ed.). New York: Bobbs-Merrill.

Henry, S. (1978). *The Hidden Economy*. London, ENG: Martin Robertson.

Klockars, C. B. (1974). *The Professional Fence*. New York: Free Press.

Livingston, J. (1992). *Crime and Criminology*. Englewood Cliffs, NJ: Prentice-Hall.

McAlister, A., P. Puska, K. Koskela, U. Pallonen and N. Maccoby (1980). Mass Communication and Community Organization for Public Health Education. *American Psychologist* 35:375–379.

McIntosh, M. (1976). Thieves and Fences: Markets and Power in Professional Crime. *British Journal of Criminology* 16:257–266.

O'Keefe, G. J. (1986). The McGruff National Media Campaign: Its Public Impact and Future Implications. In D. Rosenbaum (ed.), *Community Crime Prevention: Does it Work?* Newbury Park, CA: Sage.

Rosenbaum, D. P. (1986). *Community Crime Prevention: Does it Work?* Newbury Park, CA: Sage.

Senate Select Committee on Small Business (1973). *Hearings on Criminal Redistribution (Fencing) Systems before the Senate Select Committee on Small Business, 93rd Congress. t Session.* Washington, DC: U.S. Gover t Printing Office.

Smith, J. D. (1987). Measuring the Informal Economy. *Annals of the American Academy of Political and Social Sciences* 493:83–99.

Steffensmeier, D. (1986). *The Fence: In the Shadow of Two Worlds*. Totowa, NJ: Rowman and Littlefield.

Walsh, M. (1977). *The Fence: A New Look at the World of Property Theft*. Westport, CT: Greenwood.

3

SNEAKY THRILLS

JACK KATZ

Katz, in his exploration of shoplifting, contends that, without the risks of detection, there would be few reasons to steal. The shoplifting act itself is fraught with danger, and the thrill comes by getting away with the stolen goods without detection. Most of the items shoplifters steal are not of any real monetary value, comparatively speaking. However, as Katz points out, the real reward is getting away with it; sort of stealing for stealing's sake. What Katz hypothesizes is an emotional perspective of illegal activity, one that emphasizes motivations of power and challenges that are presented as obstacles for the shoplifter to overcome in the commission of the crime. This, for Katz, is the real seduction for illegal activity.

Various property crimes share an appeal to young people, independent of material gain or esteem from peers. Vandalism defaces property without satisfying a desire for acquisition. During burglaries, young people sometimes break in and exit successfully but do not try to take anything. Youthful shoplifting, especially by older youths, often is a solitary activity retained as a private memory.[1] "Joyriding" captures a form of auto theft in which getting away with something in celebratory style is more important than keeping anything or getting anywhere in particular.

In upper-middle-class settings, material needs are often clearly insufficient to account for the fleeting fascination with theft, as the account by one of my students illustrates:

[82] I grew up in a neighborhood where at 13 everyone went to Israel, at 16 everyone got a car and after high school graduation we were all sent off to Europe for the summer. . . . I was 14 and my neighbor was 16. He had just gotten a red Firebird for his birthday and we went driving around. We just happened to drive past the local pizza place and we saw the delivery boy getting into his car. . . . We could see the pizza boxes in his back seat. When the pizza boy pulled into a high rise apartment complex, we were right behind him. All of a sudden, my neighbor said, "You know, it would be so easy to take a pizza!". . . . I looked at him, he looked at me, and without saying a word I was out of the door. . . . got a pizza and ran back. . . . (As I remember, neither of us was hungry, but the pizza was the best we'd ever eaten.)

EDITOR'S NOTE: From *Seductions of Crime* by Jack Katz. Copyright © 1988 by Jack Katz. Reprinted by permission of Basic Books, a member of Perseus Books, L.L.C.

It is not the taste for pizza that leads to the crime; the crime makes the pizza tasty.

Qualitative accounts of *initial* experiences in property crime by the poorest ghetto youths also show an exciting attraction that cannot be explained by material necessity. John Allen, whose career as a stickup man living in a Washington, D.C., ghetto . . . recalled his first crime as stealing comic books from a junkyard truck: "we destroyed things and took a lot of junk—flashlights, telephones." These things only occasionally would be put to use; if they were retained at all, they would be kept more as souvenirs, items that had acquired value from the theft, than as items needed before and used after the theft.[2]

What are these wealthy and poor young property criminals trying to do? A common thread running through vandalism, joyriding, and shoplifting is that all are sneaky crimes that frequently thrill their practitioners. Thus I take as a phenomenon to be explained the commission of a nonviolent property crime as a sneaky thrill.

In addition to materials collected by others, my analysis is based on 122 self-reports of university students in my criminology courses.[3] Over one-half were instances of shoplifting, mostly female; about one-quarter described vandalism, almost all male; and the rest reported drug sales, nonmercenary housebreaking, and employee theft. In selecting quotations, I have emphasized reports of female shoplifters, largely because they were the most numerous and sensitively written.

The sneaky thrill is created when a person (1) tacitly generates the experience of being seduced to deviance, (2) reconquers her emotions in a concentration dedicated to the production of normal appearances, (3) and then appreciates the reverberating significance of her accomplishment in a euphoric thrill. After examining the process of constructing the phenomenon, I suggest that we rethink the relationships of age and social class to devious property crime.

FLIRTING WITH THE PROJECT

In the students' accounts there is a recurrent theme of items stolen and then quickly abandoned or soon forgotten. More generally, even when retained and used later, the booty somehow seems especially valuable while it is in the store, in the neighbor's house; or in the parent's pocketbook. To describe the changing nature of the object in the person's experience, we should say that once it is removed from the protected environment, the object quickly loses much of its charm.

During the initial stage of constructing a sneaky thrill, it is more accurate to say that the objective is to be taken or struck by an object than to take or strike out at it. In most of the accounts of shoplifting, the shoplifters enter with the idea of stealing but usually do not have a particular object in mind.[4] Indeed, shoplifters often make legitimate purchases during the same shopping excursions in which they steal. The entering mood is similar to that which often guides juveniles into the short journeys or sprees that result in pranks and vandalism.[5] Vandals and pranksters often play with conventional appearances; for example, when driving down local streets, they may issue friendly greetings one moment and collectively drop their pants ("moon") to shock the citizenry the next. The event begins with a markedly deviant air, the excitement of which is due partly to the understanding that the occurrence of theft or vandalism will be left to inspirational circumstance, creative perception, and innovative technique. Approaching a protected property with disingenuous designs, the person must be drawn to a particular object to steal or vandalize, in effect, inviting particular objects to seduce him or her. The would-be offender is not hysterical; he or she will not be governed by an overriding impulse that arises without any anticipation. But the experience is not simply utilitarian and practical; it is eminently magical.

Magical Environments

In several of the students' recountings of their thefts, the imputation of sensual power to the object is accomplished anthropomorphically. By endowing a thing with human sensibilities, one's reason can be overpowered by it. To the *Alice in Wonderland* quoted below, a necklace first enticed—"I found the one that outshone the rest and begged me to take it"—and then appeared to speak.

[15] There we were, in the most lucrative department Mervyn's had to offer two curious (but very mature) adolescent girls: the cosmetic and jewelry department. . . . We didn't enter the store planning to steal anything. In fact, I believe we had "given it up" a few weeks earlier; but once my eyes caught sight of the beautiful white and blue necklaces alongside the counter, a spark inside me was once again ignited. . . . Those exquisite puka necklaces were calling out to me, "Take me! Wear me! I can be yours!" All I needed to do was take them to make it a reality.

Another young shoplifter endowed her booty, also a necklace, with the sense of hearing. Against all reason, it took her; then, with a touch of fear, she tossed it aside in an attempt to exorcise the black magic and reduce it to a lifeless thing.

[56] I remember walking into the store and going directly to the jewelry stand. . . . This is very odd in itself, being that I am what I would consider a clothes person with little or no concern for accessories. . . . Once at home about 40–45 minutes after leaving the store, I looked at the necklace. I said "You could have gotten me in a lot of trouble" and I threw it in my jewelry box. I can't remember the first time I wore the necklace but I know it was a very long time before I put it on.

The pilferer's experience of seduction often takes off from an individualizing imputation. Customers typically enter stores, not to buy a thing they envisioned in its particularities but with generic needs in mind. A purchased item may not be grasped phenomenally as an individualized thing until it is grasped physically. Often, the particular ontology that a possession comes to exhibit—the charm of a favorite hat or an umbrella regarded as a treasure—will not exist while the item sits in a store with other like items; the item will come to have charm only after it has been incorporated into the purchaser's life—only when the brim is shaped to a characteristic angle or the umbrella becomes weathered. But the would-be thief manages to bring the particular charm of an object into existence before she possesses it. Seduction is experienced as an influence emanating from a particular necklace, compact, or chapstick, even though the particular object one is drawn to may not be distinguishable from numerous others near it.

In some accounts, the experience of seduction suggests a romantic encounter. Objects sometimes have the capacity to trigger "love at first sight."[6] Seduction is an elaborate process that begins with enticement and turns into compulsion. As a woman in her mid-thirties recalled:

A gold-plated compact that I had seen on a countertop kept playing on my mind. Heaven knows I didn't need it, and at $40 it was obviously overpriced. Still, there was something about the design that intrigued me. I went back to the counter and picked up the compact again. At that moment, I felt an overwhelming urge.[7]

Participant accounts often suggest the image of lovers catching each other's eyes across a crowded room and entering an illicit conspiracy. The student next quoted initially imagines herself in control and the object as passive—she is moving to put it in her possession; but at the end of her imagining, the object has the power to bring her pocket to life.

[67] I can see what I want to steal in plain sight, with no one in the aisle of my target. It would be so easy for me to get to the chapstick without attracting attention and simply place it in my pocket. . . . I'm not quite sure why I must have it, but I must. . . .

Some of the details that would make the deviant project hard or easy really are not up to the would-be shoplifter. In part the facility of the project is a matter of environmental arrangements for which she has no responsibility. While she is appreciating the object as a possible object of theft, she considers it at a particular angle. She will approach it from this side, with her back to that part of the scene, taking hold of it at just that part of its surface. In her experience that "It would be so easy," she is mobilizing herself to concentrate on the tangible details of the object. Thus the would-be shoplifter's sense of the facility of the project is constituted not as a feature of her "intent" or mental plan, but as a result of the position of the object in the store and the posture the object takes toward her.

To specify further how the would-be shoplifter endows the inanimate world with a real power to move her, we might consider why the initial stage of magical provocation is part of the project of sneaky thrills and not of other, equally fascinating, forms of deviance. Not all projects in deviance begin with the seductive sense, "It would be so easy." Indeed, some projects in deviance that are especially attractive to young people begin with an appreciation of the difficulties in becoming seduced to them. . . .

As the budding shoplifting project brings the object of deviance to life, the person and the object enter a conspiratorial relationship. "It would be so easy" contains a touch of surprise in the sudden awareness that no one else would notice. The tension of attraction/hesitation in moving toward the object is experienced within a broader awareness of how others are interpreting one's desires. For all *they* know, one's purposes are moral and the scene will remain mundane. The person's situational involvement in sneaky property crimes begins with a *sensual concentration on the boundary between the self as known from within and as seen from without.* . . .

THE REEMERGENCE OF PRACTICED REASON

Independent of the would-be shoplifter's construction of a sense that she might get away with it are any number of contingencies that can terminate the process. For example, the sudden attentions of a clerk may trigger an intimidating awareness of the necessity to produce "normal appearances."[8]

At some point on the way toward all sneaky thrills, the person realizes that she must work to maintain a conventional, calm appearance up to and through the point of exit. The timing of this stage, relative to others in the process, is not constant. The tasks of constructing normal appearances may be confronted only after the act is complete; thus, during the last steps of an escape, vandals may self-consciously slacken their pace from a run to normal walking, and joyriders may slow down only when they finally abandon the stolen car.

In shoplifting, the person occasionally becomes fascinated with particular objects to steal only after appreciating an especially valuable resource for putting on normal appearances. In the following recollection of one of my students, the resource was a parent:

[19] I can clearly remember when we coaxed my mom into taking us shopping with the excuse that our summer trip was coming up & we just wanted to see what the stores had so we could plan on getting it later. We walked over to the section that we were interested in, making sure that we made ourselves seem "legitimate" by keeping my mom close & by showing her items that appealed to us. We thought "they won't suspect us, two girls in school uniforms with their mom, no way." As we carried on like this, playing this little game, "Oh, look how pretty, gee, I'll have to tell Dad about all these pretty things." Eventually a necklace became irresistible.

Whichever comes first, the pull of the person toward the object to be stolen or the person's concentration on devices for deception, to enact the theft the person must bracket her appearance to set it off from her experience of her appearance, as this student's account shows:

[19] My shoplifting experiences go back to high school days when it was kind of an adventurous thing to do. My best friend & I couldn't walk into a store without getting that familiar grin on our faces. . . . Without uttering a word, we'd check out the place. . . . The whole process pretty much went about as if we were really "shopping" except in our minds the whole scene was different because of our paranoia & our knowledge of our real intentions.

Sensing a difference between what appears to be going on and what is "really" going on, the person focuses intently on normal interactional tasks. Everyday matters that have always been easily handled now rise to the level of explicit consciousness and seem subtle and complex. The thief asks herself, "How long does a normal customer spend at a particular counter?" "Do innocent customers look around to see if others are watching them?" "When customers leave a store, do they usually have their heads up or down?" The recognition that all these questions cannot possibly be answered correctly further stimulates self-consciousness. As one student expressed it,

[19] Now, somehow no matter what the reality is, whether the salesperson is looking at you or not, the minute you walk in the store you feel as if it's written all over your face "Hi, I'm your daily shoplifter."

Unless the person achieves this second stage of appreciating the work involved—if she proceeds to shoplift with a relaxed sense of ease—she may get away in the end but not with the peculiar celebration of the sneaky thrill. Novice shoplifters, however, find it easy to accomplish the sense that they are faced with a prodigious amount of work. "Avoiding suspicion" is a challenge that seems to haunt the minute details of behavior with an endless series of questions— How fast should one walk? Do customers usually take items from one department to another without paying? and so on.

To construct normal appearances, the person must attempt a sociological analysis of the local interactional order. She employs folk theories to explain the contingencies of clerk-customer interactions and to guide the various practical tasks of the theft. On how to obscure the moment of illicit taking:

[44] The jewelry counter at Nordstroms was the scene of the crime. . . . I proceeded to make myself look busy as I tried on several pairs of earrings. My philosophy was that the more busy you look the less conspicuous.

On where to hide the item:

[15] Karen and I were inside the elevator now. As she was telling me to quickly put the necklace into my purse or bag, I did a strange thing. I knelt down, pulled up my pants leg, and slipped the necklace into my sock! I remember insisting that my sock was the safest and smartest place to hide my treasure. I knew if I put it in an obvious place and was stopped, I'd be in serious trouble. Besides, packages belonging to young girls are usually subject to suspicion.

Some who shoplift clothes think it will fool the clerks if they take so many items into a dressing room that an observer could not easily keep count, as this student recalled:

[5] We went into a clothing shop, selected about six garments apiece (to confuse the sales people), entered separate dressing rooms and stuffed one blouse each into our bags.

Others, like the following student, think it sufficiently strategic to take two identical garments in, cover one with the other, and emerge with only one visible:

[46] We'd always take two of the same item & stuff one inside the other to make it seem like we only had one.

Many hit on the magician's sleight of hand, focusing the clerk's attentions on an item that subsequently will be returned to hide their possession of another.

[56] [While being watched by a clerk] I was now holding the green necklace out in the open to give the impression that I was trying to decide whether to buy it or not. Finally, after about 2 minutes I put the green necklace back but I balled the brown necklace up in my right hand and placed my jacket over that hand.

In its dramatic structure, the experience of sneak theft has multiple emotional peaks as the thief is exposed to a series of challenges to maintaining a normal appearance. The length of the series varies with the individual and the type of theft, but, typically, there are several tests of the transparency of the thief's publicly visible self, as one student indicated:

[122] I can recall a sneak theft at Penney's Dept. store very well. I was about 12 years old. . . . I found an eyeshadow kit. I could feel my heart pounding as I glanced around to make sure that others weren't watching. I quickly slipped the eyeshadow in my purse and sighed heavily with relief when I realized that no one had seen. I nervously stepped out of the aisle and once again was relieved when I saw that there was no one around the corner waiting to catch me. I caught my

friend's eye; she gave me a knowing glance and we walked to the next section in self-satisfaction for having succeeded so far.

When the person devises the deviant project in advance, even entering the store normally may be an accomplishment. Having entered without arousing suspicion, the would-be shoplifter may relax slightly. Then tension mounts as she seizes the item. Dressing rooms provide an escape from the risk of detection, but only momentarily, as in this student's account:

[19] So, here we were, looking at things, walking around & each time getting closer to the dressing room. Finally we entered it & for once I remember feeling relieved for the first time since I'd walked into the store because I was away at last from those "piercing eyes" & I had the merchandise with me. At this point we broke into laughter. . . . We stuffed the items in our purses making sure that they had no security gadgets on them & then we thought to ourselves "well we're half way there." Then it hit me, how I was safe in the dressing room, no one could prove anything. I was still a "legitimate" shopper.

Then a salesperson may come up and, with an unsuspecting remark, raise the question of transparency to new heights:

[19] I remember coming out of the dressing room & the sales lady looking at me & asking me if I had found anything (probably concerned with only making her commission). I thought I would die.

Finally there is the drama of leaving the store:

[19] Walking out the door was always a big, big step. We knew that that's when people get busted as they step out & we just hoped & prayed that no one would run up to us & grab us or scream "hey you"! The whole time as we approached the exit I remember looking at it as a dark tunnel & just wanting to run down it & disappear as I hung on to my "beloved purse."

Once they have hidden the booty and so long as they are in the store, the would-be-shoplifters must constantly decide to sustain their deviance. Thus, the multiple boundaries of exposure offer multiple proofs not only of their ability to get away with it but of their will toward immorality:

[5] We went into the restroom before we left and I remember telling Lori, "We can drop all this stuff in here and leave, or we can take it with us." Lori wanted to take everything, and as we neared the exit, I began to get very nervous.

Many of these shoplifters understand that clerks or store detectives may be watching them undercover, in preparation for arresting them at the exit door. They also believe that criminal culpability is only established when they leave with the stolen goods. As they understand it, they are not irretrievably committed to be thieves until they are on the other side of the exit; up to that point, they may replace the goods and instantly revert from a deviant to a morally unexceptional status. Were they to believe that they were criminally culpable as soon as they secreted the item, they would continue to face the interactional and emotional challenges of accomplishing deception. But because they think they are not committed legally until they are physically out of the store, they experience each practical challenge in covering up their deviance as an occasion to reaffirm their spiritual fortitude for being deviant. One student described the phenomenon this way:

[56] I guess I had been there so long that I started to look suspicious. I was holding a bright lime green necklace in my left hand and a brown Indian type necklace in my right. A lady, she must have been the store manager, was watching me. She was about 20 ft. away from me and on my left. I could feel her looking at me but I didn't look directly at her. . . . I remember actually visualizing myself putting back both the necklaces and walking out the store with pride and proving this bitch she was wrong and that I was smarter than her, but I didn't. . . . I started out the store very slowly. I even smiled at the lady as I passed by the cash register. It was then that she started toward me and my mind said okay T. what are you going to do now. There was a table full of sweaters on sale near me and I could have easily drop the

necklace on the table and continued out the door. I knew I could and I considered it but I wouldn't do it. I remember just holding the necklace tighter in my right hand. As she was coming toward me I even thought of dropping the necklace and running out the door but I continued in a slow pace even though the thought of them calling my mother if I was caught and what she would do to me was terribly frightening.

In addition to focusing on the practical components of producing a normal appearance, the would-be shoplifter struggles not to betray the difficulty of the project. This is the second layer of work—the work of appearing not to work at practicing normal appearances. The first layer of work is experienced as the emergence of a novel, analytical attention to behavioral detail; the second, as a struggle to remain in rational control, as the following statement by a student illustrates:

[19] You desperately try to cover it up by trying to remember how you've acted before but still you feel as if all eyes are on you! I think, that's the purpose of settling in one area & feeling everything & everyone out. It's an attempt to feel comfortable so that you don't appear obvious. Like maybe if I'm real cool & subtle about it & try on a few things but don't seem impressed w/ anything, I can just stroll out of here & no one will notice. . . .

BEING THRILLED

Usually after the scene of risk is successfully exited, the third stage of the sneaky thrill is realized. This is the euphoria of being thrilled. In one form or another, there is a "Wow, I got away with it!" or an "It was so easy!" A necklace shoplifter stated:

[56] Once outside the door I thought Wow! I pulled it off, I faced danger and I pulled it off. I was smiling so much and I felt at that moment like there was nothing I couldn't do.

After stealing candy with friends, another student recalled:

[87] Once we were out the door we knew we had been successful! We would run up the street . . . all be laughing and shouting, each one trying to tell just how he pulled it off and the details that would make each of us look like the bravest one.

The pizza thief noted:

[82] The feeling I got from taking the pizza, the thrill of getting something for nothing, knowing I got away with something I never thought I could, was wonderful. . . . I'm 21 now and my neighbor is 23. Every time we see each other, I remember and relive a little bit of that thrill. . . .

In a literal sense, the successful thieves were being thrilled: they shuddered or shook in elation, often to the rhythms of laughter. For many, whether successful or not, the experience of youthful shoplifting was profoundly moving, so moving that they could vividly recall minute details of the event years later. . . .

NOTES

1. In her study of some 4,600 shoplifters arrested by detectives in a Chicago department store, Mary Cameron found that all the female shoplifters who were arrested at age 10 were shoplifting in groups, but 85 percent of the 19-year-olds who were arrested had been alone. She concluded that shoplifting becomes progressively more a solitary activity with age. See Mary Cameron, *The Booster and the Snitch* (New York: Free Press, 1964), pp. 102–3.

2. John Allen, *Assault with a Deadly Weapon*, ed. Dianne Hall Kelly and Philip Heymann (New York: McGraw-Hill, 1978), p. 13.

3. Bracketed numbers link quotes to cases in a data set, "Autobiographical Accounts of Property Offenses by Youths, UCLA, 1983–1984," no. 8950, Inter-university Consortium for Political Social Research, Ann Arbor, Michigan. The accounts were volunteered by students in three offerings of my criminology class during 1983 and 1984. Because of the age range of the students and because no time limit was placed on the recollection, the events recounted cover acts by 7-year-olds and middle-aged office workers, with most falling within early adolescence.

4. Cameron (*Booster and Snitch*, p. 82) found that males and females shoplift in the same way as they shop. Males go in for particular items and commonly come out with just one new possession, which makes them harder to catch. Females more often "go shopping" when they shoplift, moving through the store to find items that will be especially appealing.

5. For an example of creative middle-class vandalism, see William J. Chambliss, "The Saints and the Roughnecks," *Society* 11 (November-December 1981): 24–31.

6. "Love-at-first-sight" consumer experiences may be a relatively recent phenomenon, or at least enhanced by the modern social structure of the department store. A student of shoplifting in England reports that "the post-Second World War invitation to shoppers to 'walk around without obligation' makes it very much easier for the shoplifter. Before this time it was never assumed that people entering a shop did not know what they wanted, or that they would leave without making a purchase, or would expect to be allowed to wander without supervision and intervention." D. P. Walsh, *Shoplifting* (London: Macmillan, 1978), p. 22. The supervision of a helpful attendant clerk not only makes sleight-of-hand more difficult, it interferes with the privacy that is so helpful in getting good romances started. Merchandisers' complaints about being burned by shoplifting usually neglect to mention their encouragement that consumers play with possessive passions.

7. L. B. Taylor, "Shoplifting: When Honest Ladies Steal," *Ladies Home Journal*, January 1982, pp. 88–89.

8. On the raising of normal appearances to a problematic level in the process of becoming deviant, see Matza, *Becoming Deviant*.

4

URBAN GRAFFITI: CRIME, CONTROL, AND RESISTANCE

JEFF FERRELL

Drawing on 4 years of fieldwork inside the Denver, Colorado, graffiti underground and on research in other American and European cities, Ferrell explores the various ways in which graffiti writers attempt to resist the controls of the legal and political authorities. Ferrell, after careful examination of hip hop graffiti, concludes that when youthful writers resist authority, their graffiti becomes confrontational in nature and they counterattack, which transforms pressure from official authorities to that of illegal pleasure through their writings.

Over the past two decades, a new form of youthful graffiti—graffiti "writing," as its young practitioners call it—has spread from its origins in New York City to cities throughout the United States, Europe, and other world regions. This article examines this emerging form of graffiti and explores the moments of resistance embedded in it. Specifically, it investigates the lived dynamics of graffiti writing and the lives of youthful graffiti writers in the context of legal and political power, social control, and writers' resistance to them. This examination aims not at reducing the complex processes of graffiti writing, social control, and resistance to a neat grid of cause and effect, but instead at tracing the many moments in which they intersect and interweave. It also aims to reveal the various ways in

which youthful activities like graffiti writing not only shape resistance to existing arrangements but construct alternative arrangements as well.

The methodological framework for this examination of contemporary graffiti writing incorporates both intensive field research inside a particular urban graffiti subculture and comparative field and document research in various other urban settings. Certainly, the foundation for this study is the 4 years (1990–1993) that I conducted ongoing field research and participant observation inside the Denver, Colorado graffiti underground. This research process began, as might be expected, with a trial period during which contacts with the underground were made and expanded, and I was subjected to a series of informal tests, primarily as to my willingness to place myself in the same

EDITOR'S NOTE: From Ferrell, J., "Urban graffiti: Crime, control, and resistance," in *Youth and Society, 27,* pp. 73–92. Copyright © 1995. Reprinted with permission from Sage Publications, Inc.

situations of risk as those encountered by the writers. This preliminary research blossomed into active participant observation inside the underground, involving not only participation in various informal gatherings, parties, and paint-buying trips, but also innumerable graffiti-writing forays in Denver's railyards and alleys (see Ferrell, 1993a). The research culminated, so to speak, in my arrest and trial on charges of "graffiti vandalism."

To develop a comparative perspective on this intensive field research, interviews were also conducted with legal agents, political officials, and others in Denver; and sites of graffiti activity were visited in cities throughout the United States and Europe. Although these visits did not, of course, produce the intensity of information generated in the Denver case, they did provide opportunities for extensive observation, and in some cases, interviews with local writers and those that oppose them. This comparative information was in turn supplemented by newspaper searches and other forms of document research in various U.S. cities.

FORMS OF GRAFFITI AND FORMS OF RESISTANCE

In a remarkable variety of world settings, kids (and others) employ particular forms of graffiti as a means of resisting particular constellations of legal, political, and religious authority. Through an array of painted images, for example, young artists quite thoroughly transformed the political meaning of the Berlin Wall by the time of its destruction (Waldenburg, 1990); and in the former Soviet Union, the graffiti of urban youth cultures emerged as a channel of resistance essential to the undermining of Soviet authority (Bushnell, 1990). In London, feminists, animal rights activists, and others aggressively alter offensive billboards (Posener, 1982); in Northern Ireland, young Catholics paint wall murals that memorialize (and encourage) resistance to British rule, and Protestants and the British military counter-attack through the same medium (Rolston, 1991). Similarly, Nicaraguan youth groups have for years painted street images of Sandino as a form of political resistance and dialogue; post-Sandinista officials

now respond with "mural death squads" (Kunzle, 1993; Sheesley & Bragg, 1991). Toronto street artists develop works that attack colonialism and urge political resistance (Kummel, 1991); and, denied access to radio or newspaper, young Palestinian militants in the occupied lands employ wall painting as their primary form of communication and resistance to Israeli authority (Hedges, 1994; see Ferrell, 1993b).

A particular form of graffiti writing has, during the past 20 years, also emerged out of the economic, political, and ethnic inequalities endemic to the United States. "Hip hop" graffiti—the focus of this study—grew out of the Black neighborhood cultures of New York City in the early and mid-1970s as part of a larger, homegrown, alternative youth culture that included new forms of music (rap, sampling, scratching) and dancing (Brewer & Miller, 1990; Castleman, 1982; Chalfant & Prigoff, 1987; Cooper & Chalfant, 1984; Ferrell, 1993a; Hager, 1984; Lachmann, 1988; Miller, 1994; Stewart, 1987). This highly stylized form of nongang graffiti writing—which includes the "tagging" of subcultural nicknames on city walls and the creation of large illegal murals ("piecing") by "crews" of writers—has today fanned out into large and small cities across the United States and to Europe, Mexico, Central America, and elsewhere (Brett, 1991; Chalfant & Prigoff, 1987; Riding, 1992; Rodriguez, 1994; Rotella, 1994). Its remarkable growth also increasingly incorporates kids from outside the ethnic and economic frameworks of its originators. In Denver, for example, youths from the suburbs and from small towns regularly seek out the urban hip hop graffiti underground; and in Boston, a substantial portion of the city's hip hop graffiti is in fact now produced by crews made up of young Anglo males and based in the suburbs (Jacobs, 1993, p. 1). In southern California, the participation of young people of all sorts in graffiti writing is such that the Los Angeles County Sheriff's Department lists some 800 known graffiti crews; the Los Angeles Rapid Transit District alone spends $13 million a year on clean-up, and the California Department of Transportation budgets up to $5 million for 1994; and authorities now find hip hop (and gang) graffiti inside

Los Angeles City Hall, in abandoned World War II bunkers, and even in the San Gabriel Mountains (Haldane, 1993; Hudson, 1993; MacDuff & Valenzuela, 1993; Maxwell & Porter, 1993; Sahagun, 1992; Tobar, 1993). The members of a national anticrime organization thus recently named graffiti their biggest concern (Ching, 1991, p. A1).

What, though, is the larger cultural and political context in which this wildly popular style of graffiti writing proliferates? And precisely what forms of authority does this graffiti writing resist?

URBAN AUTHORITY, SOCIAL CONTROL, AND THE WRITING OF RESISTANCE

Contemporary graffiti writing occurs in an urban environment increasingly defined by the segregation and control of social space. As Schiller (1989), Soja (1989), M. Davis (1990, 1992a, 1992b), Sorkin (1992), S. Davis (1992), Guterson (1993) and others have shown, major U.S. cities today are systematically fractured by ethnic, class, and consumer segregation— segregation built into skyscrapers and skyways, freeways and transit routes, walled residential enclaves and secured shopping malls, private streets and parks. The caretakers of these physically segregated cities control (or destroy) public space and public communities through privatization and physical insulation, and they employ extensive public and private police power and sophisticated control technologies to enforce their spatial restrictions. Young people who wish to work or wander in these environments face, in addition to these spatial controls, an increasingly aggressive criminalization of their activities by local and state authorities. In recent years, city after city has enacted strict curfews and a multitude of ordinances against loud music, car cruising, and other youthful pleasures (Ferrell, 1993a; LeDue, 1992; Reuter, 1994b). In negotiating the contemporary city, kids are largely walled in and boxed out.

The writing of hip hop graffiti disrupts this orderly latticework of authority, reclaims public space for at least some of those systematically excluded from it, and thus resists the confinement of kids and others within structures of social and spatial control. Hip hop graffiti writers work almost exclusively at night, and in so doing use the cover of darkness to evade curfew restrictions and urban surveillance. In that they gain subcultural status from tagging over as large an area as possible, they also wander widely throughout the city; mobility—and trespass—are essential. Because further status derives from the difficulty of a tag's placement, writers also regularly jump razor wire fences, climb freeway standards or skyscrapers ("tagging the heavens"), and otherwise violate the city's spatial sorting. And time and again, writers talk and tag in such a way as to make clear their resistance to urban control. In Los Angeles, 13-year-old tagger Creator (CRE8) reports that "most of the time I get up (tag) on stop signs and city-owned stuff" (Quintanilla, 1993, p. E6). In Denver, legendary graffiti "king" Rasta 68 likewise announces that, "Personally, I want to hit on city stuff, like bridges, rather than some other person's property. They build the boringest crap around, so why not beautify it?" (Will, 1994, January 2, p. 13). And in Boston, local writer Relm emphasizes in a newspaper interview that he doesn't bomb (tag) individuals, cars, or houses, but only large businesses, public buildings, and other urban symbols of the system he opposes (Jacobs, 1993, p. 28).

If, as alluded to earlier, authority and resistance dance together, the next moment in this tango of urban control and graffiti writing is not difficult to anticipate: The same legal structures, policing powers, and technological safeguards that regulate the city at large are in turn brought down on graffiti writers, and with a vengeance. The array of control technologies and techniques aligned against graffiti writing is itself impressive. Today, legal authorities and corporate sponsors in Los Angeles, San Bernardino, CA, New York, Denver, Las Vegas, Fort Worth, and other cities create police and citizen surveillance teams armed with two-way radios, home video cameras, remote control infrared video cameras, and night-vision goggles; send out antigraffiti helicopter patrols; secure freeway signs and bridges with razor wire and commercial buildings with special graffiti-resistant coatings; and arrange toll-free telephone hotlines for watchful residents and motorists with cellular phones (Bennet, 1992; Ching, 1991; Colvin, 1993b; Fried, 1992; Rainey, 1993; "2 teens," 1991; Valenzuela, 1993; "Writing on the Wall," 1993). They also use U.S. Marines in antigraffiti operations, deploy undercover transit and police officers in the guise of high school students and journalists, stake out popular graffiti-writing areas, and set up sophisticated sting operations to apprehend graffiti writers and stop those who sell spray paint to them ("Albuquerque Police," 1992; Baker, 1991; Carr, 1993; Henderson, 1994; "Lure of Fame," 1994; Molloy & Labahn, 1993; National Graffiti Information Network, 1990; "Sting," 1991; "Teaching Teen," 1994).

These sorts of physical control are backed by growing militancy among antigraffiti activists and by increasingly severe legal sanctions. New York's new police commissioner targets graffiti and other "quality of life" crimes; Los Angeles's mayor Richard Riordan campaigns aggressively against graffiti and now recommends boot camps as punishment for writers; another Los Angeles mayoral candidate suggests "chop[ping] a few fingers off" (Simon, 1993, July 9, p. B3); and Denver's mayor deflects a recall campaign with a vitriolic antigraffiti campaign of his own (Ferrell, 1993a; "These Guys," 1994). A California assemblyman introduces a bill requiring that kids convicted of writing graffiti be publicly paddled; and in St. Louis, an alderman proposes public caning (Bailey, 1994; Gillam, 1994; Henderson, 1994). Other antigraffiti campaigners in Los Angeles and Denver cheer suggestions of lopping off hands, and speak of "hanging, shooting, and castrating" (Colvin, 1993a, p. B4) and publicly spray-painting writers' genitals (Kreck, 1993; Martin, 1992).

In this climate, southern California authorities arrest the parents and grandparents of alleged writers on charges of contributing to the delinquency of minors and sue or otherwise bill other parents for tens of thousands of dollars in damages (Goldman, 1993; Lozano, 1994; MacDuff & Valenzuela, 1993; Valenzuela, 1993). In Los Angeles, writers themselves now face multiple $1,000 civil fines in addition to criminal penalties of $50,000 and 1 year in jail (Simon, 1993, July 9). Business owners in cities around the country confront statutes that

regulate or ban the sale of spray paint and markers to minors and others and that force businesses to clean graffiti from their buildings ("Building Owners," 1994; Fong, 1992; Hanley, 1992; Hynes, 1993; Smith, 1994; Tobar, 1993). And in Denver, Los Angeles, and other cities, aggressively entrepreneurial vigilantes, high school "bounty hunters," and others now receive thousands of dollars in cash awards for turning in writers (Ferrell, 1993a; Reuter, 1994a; Schwada & Sahagun, 1992).

Graffiti writers, of course, counterpunch with new forms of resistance and increased militancy of their own. In the early years of hip hop graffiti, legendary New York City writer Lady Pink said, "Graffiti means 'I'm here.' . . . They want to snub us, but they can't" (Mizrahi, 1981, p. 20), and contemporary writers facing the full force of urban authority echo this sentiment. An 18-year-old Los Angeles tagger arrested six times says, "They want to wipe us out. But graffiti will never die" (Colvin, 1993a, p. B4); and a Compton tagger tells city officials, "You can lock me up, but you're not going to arrest all of us. How are you guys going to make us stop? You don't know how" (Tobar, 1993, p. B3). To prove their point, writers decorate, and desecrate, the very control structures in which they are caught. Kids involved in a city work program at Los Angeles City Hall reach for "the heavens" by tagging the top floor of the city hall tower (Sahagun, 1992). In response to the Denver mayor's antigraffiti campaign, Voodoo paints a "Recall" piece and poem along the bike path where the mayor jogs. A Boston writer on trial for graffiti affixes tagged stickers—an increasingly popular form of pre-fabricated tagging—throughout the courthouse and, remarkably, on the back of the prosecutor's legal pad (Jacobs, 1993). And Chaka—southern California's most notorious and prolific tagger—is arrested for tagging a courthouse elevator while visiting the probation officer supervising his previous conviction for tagging (MacDuff & Valenzuela, 1993; Martin, 1992).

To avoid later detection, writers in Las Vegas, Denver, and other cities also increasingly wear latex gloves when they tag or piece and take other practical measures to avoid apprehension. But for writers, the most remarkable and insidious form of resistance to increased repression is not a practical measure but a pleasurable response. This is the adrenalin rush. Writers consistently report to me and to others that their experience of tagging and piecing is defined by the incandescent excitement, the adrenalin rush, that results from creating their art in a dangerous and illegal environment—and that heightened legal and police pressure therefore heightens this adrenalin rush as well. In Los Angeles, Creator says, "I bomb because I like the chase, the getting up [tagging] without getting caught.... Catch me if you can" (Quintanilla, 1993, p. E1); and in San Bernardino, an ex-tagger adds, "I miss the rush. It's a rush because you're taking a chance of getting caught. You do it to see if you can get away with it. It's like an addiction—you can't stop" (MacDuff & Valenzuela, 1993, p. A11). Well-known Denver writers like Z13, Rasta 68, Eoosh, and Voodoo also speak regularly of "that rush" one gets from graffiti, its links to illegality, and the ways in which increased police pressure means, for them, increased excitement; as Voodoo says, with regard to piecing, "Right before you hit the wall, you get that rush. And right when you hit the wall, you know that you're breaking the law, and that gives that extra adrenalin flow" (Ferrell, 1993a, p. 82). A Denver street artist thus concludes, "Doing graffiti is a real adrenalin rush. That provides a lot of the pull and draw to the taggers. The city doesn't understand that the more they publicize the crackdown, the more active the taggers will become" (Ferrell, 1993a, p. 148). A Las Vegas "hip hop shop" owner summarizes the situation succinctly: "The harder the city comes down on them, the more fun it is for them" ("Writing on the Wall," 1993, p. 4C).

As the adrenalin rush shows, graffiti writers resist the pressure brought against them not only by fighting it, but by using it for their own purposes and by transforming political pressure into personal and collective pleasure. Here again we see the dance of authority and resistance and the strange steps that it follows—in this case, the authorities' role in amplifying the meaning and intensity of the very activity they wish to suppress. In this ongoing interplay, we also begin to see the magnitude of the battle between graffiti writers and urban authorities. This battle is certainly, as headline writers are

wont to put it, a "war of the walls"; in doing graffiti, writers challenge the "aesthetics of authority" (Ferrell, 1993a, pp. 178–186) that govern the city, invent new visual conventions, and give lie by their tags and pieces to the vision of a city under firm political control. But this war of the walls is, more profoundly, a war of the worlds. For graffiti writing not only confronts and resists an urban environment of fractured communities and segregated spaces; it actively constructs alternatives to these arrangements as well.

Resistance, Identity, and Alternative Arrangements (Graffito Ergo Sum)

The writing of graffiti is an inherently collective activity. Although writers tag and piece against the controls of the city, they also tag and piece for one another, and in so doing build alternative structures of meaning and status. Tagging goes on as a collective conversation among writers, a process of symbolic interaction by which writers challenge, cajole, and surprise one another. Like his counterparts in cities throughout the United States, Los Angeles writer Rival emphasizes that he tags for the respect of "other taggers. Who cares about adults?" (Glionna, 1993, p. B4). Writers also piece primarily for one another. Writers' pieces are executed and evaluated within elaborate subcultural conventions of color, proportion, and design; and although writers may hope that their pieces will be seen by the public, they can be certain that they will be seen and judged by other writers. In this sense, tagging and piecing create an alternative system of public communication for kids who otherwise have little access to avenues of urban information. And in this sense, like their Palestinian counterparts across the Atlantic, U.S. graffiti writers paint a complex system of subterranean signs directly onto the walls of cities that otherwise would render them invisible.

In tagging and piecing for one another, writers also construct alternative systems of status and identity. Both for those kids increasingly shut out of traditional channels of achievement and for those who, through ethnicity or education, retain some modicum of choice, graffiti writing provides a powerful alternative process for shaping personal identity and gaining social status. Black, Latino, and Anglo boys in the southern California graffiti crew TIKs, for example, have quit high school chess teams and spurned advanced placement classes to devote as much time as possible to graffiti. The result is not only status among other writers, but invitations to parties and relationships with girls who also write; as one TIK says, "without graffiti, what do I got?" (Glionna, 1993, p. B4). A young female tagger from East L.A. likewise points out, "You know how rich people have their names on their houses or something? Well, tagging is like that. People see your name.... It makes people feel good" (Diaz, 1992, p. B5). The power of these alternative systems of status and identity can be seen in the intensity with which writers do graffiti. Rasta 68 claims that "I eat, sleep, and breathe graffiti" (Will, 1994, p. 12); Chaka not only tags the courthouse, but maps locations and tags for 7 hours each night; writers jump razor wire and climb billboards to earn status by "tagging the heavens"; and, in southern California, businesses are tagged, repainted, and tagged again four times in a day (MacDuff & Valenzuela, 1993; Quintanilla, 1993).

As graffiti writing shapes youthful identities, it also builds alternative communities. The crews to which writers belong not only tag and piece together, but form deep social bonds as their members share time and resources, construct collective artistic orientations, and defend one another from enemies real and imagined. In Los Angeles, Creator notes that, "It's like a family to belong to a crew. They watch your back, you watch theirs. You kick it everyday with them.... You get friendship, love, supplies, everything" (Quintanilla, 1993, p. E1). Similarly, the 80 or so kids who belong to the FBI crew in southern California emphasize the "sense of family the crew has brought to taggers' lives" (Nazario & Murphy, 1993, p. B1) and mourn the deaths of seven crew members in a car crash; as one tagger says, "It was family, love, tagging, everything" (Nazario & Murphy, 1993, p. B4). In Denver, crews like Syndicate hold regular "art sessions" to work on collective designs, share the "piecebooks" in which they draw their designs, and often pool their talents to work on large, elaborate pieces. As Rasta 68

says, Syndicate is "ten people with ten brains and twenty eyes to watch out for opposing authority or enemy and to get down with the brain waves thrown down on the wall" (Ferrell, 1993a, p. 36).

Significantly, the alternative communities that writers create often violate the city's everyday ethnic segregation by incorporating kids of various ethnic backgrounds; as seen previously, southern California's TIK crew is multiethnic, and Denver crews are often made up of both Anglo and Latino kids. These crews also provide an important, street-level alternative to gangs and gang membership. Writer after writer in Denver, Los Angeles, and elsewhere reports that graffiti writing and crew membership led him or her away from gang identity and activity. The members of Denver's largely Latino NC (No Claims) crew emphasize that hip hop culture generally, and hip hop graffiti writing specifically, exist for them as lived alternatives to participation in Latino street gangs. And as the members of FBI say, "A lot of people want to gang-bang, but we focus on just being together as one, trying to keep out of trouble.... We aren't hoodlums—these guys were like brothers. We all care for each other. Many of us don't get any support from our parents" (Nazario & Murphy, 1993, pp. B1, B4; see Donnan & Alexander, 1992; Hubler, 1993; Martin, 1992).

These small communities of writers also contribute to the larger communities of which they are a part. In Denver, writers have painted pieces commenting on local politics, war, and AIDS, and have been commissioned to paint drug awareness and "stay in school" murals. And in New York City, drug dealers and others pay writers to paint large "Rest In Peaces"— murals that commemorate those who have died on the streets (Marriott, 1993; Sanchez, 1993). Clearly, graffiti writers and crews serve as the folk artists of urban communities; day-to-day chroniclers of urban life and death, they represent the worlds they help create. As Lady Pink says, in recalling the early years of hip hop graffiti, "We were like sixties radicals, rebelling against the system. I was dodging bullets in the service of folk art, bringing art to the people" (Siegel, 1993, p. 68).

As the "Rest in Peaces" begin to show, graffiti also contributes to alternative economic arrangements and underground economies. Hip hop graffiti shops in Denver, Los Angeles, Las Vegas, and elsewhere now sell magazines, videos, spray tips, markers—and lines of clothing designed and produced by writers (Sipchen, 1993; Will, 1994; "Writing on the Wall," 1993). In New York, Los Angeles, and Denver, writers pass out business cards to those who admire their pieces, execute commissioned murals for home and shop owners, and even parlay exposure in antigraffiti mural painting programs into commissioned art work (Horovitz, 1992; Marriott, 1993; Pool, 1992). Increasingly, graffiti writing provides for top writers some hope of economic survival and economic self-determination in an environment that alternates unemployment with minimum wage work. It also creates for writers avenues of artistic development and entrepreneurship outside the restricted circles of gallery art (Ferrell, 1993a).

As they piece and tag, then, graffiti writers not only alter the look of the city and resist its structures of authority, but at the same time create elaborate urban alternatives. Engaging in what anarcho-syndicalists of the early 20th century called "direct action," and punks of the later 20th century dubbed "D.I.Y." (do it yourself), graffiti writers invent out of their own activities alternative systems of aesthetics, representation, identity, and meaning. In a world of dead-end jobs and declining career opportunities, they construct new channels for achieving status and earning money. In cities partitioned by ethnicity and social class, they assemble new lines of transurban communication and build new communities that bridge ethnic and class divisions. As they wander the city, they invent new forms of social organization inside the all-too-orderly rubble of the old.

YOUTH AND RESISTANCE

A careful examination of hip hop graffiti writing begins to reveal the many ways in which young graffiti writers resist the structures of authority under which they are placed. By the very nature of their activities and associations, youthful graffiti writers violate the sorts of spatial controls that constipate the contemporary

city and confine kids and others to prearranged patterns of social isolation. When these violations precipitate further controls, graffiti writers counterattack, not only with directly confrontational styles of writing but with a shared "adrenalin rush" that transforms legal pressure into illicit pleasure. And, as graffiti writers participate in this dance of urban control and resistance, they at the same time construct elegantly alternative arrangements that shape both individual identities and communities of support and meaning.

The various forms of resistance embedded in youthful graffiti writing in turn remind us of the sort of approach scholars might productively take toward larger issues of youth and resistance. Neither dreamy romanticism nor theoretical rigidity will suffice; both distance us from the subjects of our study, leave us dependent on secondhand stereotypes, and ultimately demean kids' actions and identities. Carefully situating our research in young people's daily lives, on the other hand, broadens our scope to include the many and varied manifestations of authority and resistance entangled there and pushes us to pay attention to the particular meanings of authority and resistance in the everyday, collective experience of youth. In employing this methodology of attentiveness, we are likely to find in kids' lives forms of resistance far more remarkable than those that romanticism imagines or rigidity imposes—forms of resistance that both confront structures of authority and begin to build alternatives in and around them. And like graffiti writing, these various moments of youthful resistance—too often dismissed as mindlessly destructive—in fact merit our attention not only for undermining contemporary social arrangements but for imagining new ones as well. The words of the Russian anarchist Michael Bakunin echo in the everyday lives of young people, and off the graffiti-covered walls of the contemporary city: "The passion for destruction is a creative passion, too" (Lehning, 1974, p. 58).

REFERENCES

Abbott, K. (1994, February 8). Big neighbor is watching. *Rocky Mountain News*, pp. 3D, 5D.

Adler, P. A., & Adler, P. (1993). The coming of age of crack cocaine. *Contemporary Sociology, 22*, 848–851.

Albuquerque police impersonate journalist. (1992). *The News Media and the Law, 16*, 14.

Atlanta, C., & Alexander, G. (1989). Wild style: Graffiti painting. In A. McRobbie (Ed.), *Zoot suits and second-hand dresses* (pp. 156–168). London: Macmillan.

Bailey, E. (1994, June 23). Paddling bill puts Conroy in hot seat of national debate. *Los Angeles Times*, pp. A1, A12.

Baker, A. (1991, October 27). Anti-graffiti youth crew repaints bridge. *Fort Worth Star-Telegram*, p. 30.

Bennet, J. (1992, September 27). A new arsenal of weapons to tag graffiti artists. *New York Times*, p. E2.

Brett, P. (1991). Flourishing graffiti art leads to credit at Parisian "worker's university." *Chronicle of Higher Education, 37*, A34.

Brewer, D. (1992). Hip hop graffiti writers' evaluations of strategies to control illegal graffiti. *Human Organization, 51*, 188–196.

Brewer, D., & Miller, M. (1990). Bombing and burning: The social organization and values of hip hop graffiti writers and implications for policy. *Deviant Behavior, 11*, 345–369.

Building owners may face fines for leaving graffiti. (1994, May 30). *New York Times*, p. 20.

Bushnell, J. (1990). *Moscow graffiti: Language and subculture*. Boston: Unwin Hyman.

Carr, C. (1993). Operation "Gung Ho." *Marines, 22*, 7–9.

Castleman, C. (1982). *Getting up: Subway graffiti in New York*. Cambridge: MIT Press.

Chalfant, H., & Prigoff, J. (1987). *Spraycan art*. London: Thames and Hudson.

Ching, S. (1991, August 30). S.B. Valley cities find graffiti to be growing, ugly problem. *San Bernardino Sun*, p. A1.

Colvin, R. (1993a, September 1). Taggers debate their critics, who see no art in graffiti. *Los Angeles Times*, pp. B1, B4.

Colvin, R. (1993b, April 13). Teaming up on taggers. *Los Angeles Times*, p. B1.

Cooper, M., & Chalfant, H. (1984). *Subway art*. London: Thames and Hudson.

Davis, M. (1990). *City of quartz*. London: Verso.

Davis, M. (1992a). Fortress Los Angeles: The militarization of urban space. In M. Sorkin (Ed.), *Variations on a theme park* (pp. 154–180). New York: Hill and Wang.

Davis, M. (1992b, June 1). In L.A., burning all illusions. *The Nation, 254*, 743–746.

Davis, S. (1992, August 31/September 7). Streets too dead for dreamin'. *The Nation, 255*, 220–221.

de Certeau, M. (1984). *The practice of everyday life*. Berkeley: University of California Press.

Diaz, C. (1992, December 14). "I did it out of boredom." *Los Angeles Times*, p. B5.

Donnan, S., & Alexander, D. (1992, December 14). "Spray cans don't kill." *Los Angeles Times*, p. B5.

Ferrell, J. (1993a). *Crimes of style: Urban graffiti and the politics of criminality*. New York: Garland.

Ferrell, J. (1993b). The world politics of wall painting. *Social Justice, 20*, 188–202.

Fiske, J. (1991). An Interview with John Fiske. *Border/Lines, 20/21*, 4–7.

Fong, T. (1992, January 30). Northglenn fights gang graffiti. *Rocky Mountain News*, p. 14.

Fried, J. (1992, April 6). Watch out, scrawlers, you're on graffiti camera. *New York Times*, p. B3.

Gang unit may get Gulf War equipment. (1992, April 7). *Rocky Mountain News*, p. 15.

Gillam, J. (1994, May 24). Assembly bill would make graffiti a paddling offense. *Los Angeles Times*, p. A12.

Glionna, J. (1993, March 10). Leaving their mark. *Los Angeles Times*, pp. B1, B4.

Goldman, A. (1993, November 13). Elderly couple arrested in tagging case. *Los Angeles Times*, p. B8.

Gomez, M. (1993). The writing on our walls: Finding solutions through distinguishing graffiti art from graffiti vandalism. *University Michigan Journal of Law Reform, 26*, 633

Graffiti on church. (1993, May 19). *Rocky Mountain News*, p. 14A.

Greenberg, D. (1993). *Crime and capitalism*. Philadelphia: Temple.

Guerrero, G. (1993, July 19). Beyond the drug war. *The Nation, 257*, pp. 113–115.

Guterson, D. (1993, March/April). Home, safe home. *Utne Reader*, pp. 62–67.

Hager, S. (1984). *Hip hop*. New York: St. Martin's.

Haldane, D. (1993, November 14). Bunkers hold hieroglyphics of modern youth. *Los Angeles Times*, pp. A3, A26.

Hanley, R. (1992, June 11). Jersey city escalates graffiti war. *New York Times*, pp. B1, B8.

Hedges, C. (1994, January 24). To read all about it, Palestinians scan the walls. *New York Times,* p. 12B.

Henderson, A. (1994). Graffiti. *Governing, 7,* 40–44.

Horovitz, B. (1992, July 9). Graffiti central. *Los Angeles Times,* pp. D1, D3.

Hubler, S. (1993, November 18). Tag lines. *Los Angeles Times,* pp. B1, B4.

Hudson, B. (1993, April 24). Scrawl of the wild. *Los Angeles Times,* pp. B1, B2.

Hutchison, R. (1993). Blazon nouveau: Gang graffiti in the barrios of Los Angeles and Chicago. In S. Cummings & D. J. Monti (Eds.), *Gangs* (pp. 137–171). Albany: State University of New York Press.

Hynes, M. (1993, March 31). County graffiti ordinance takes effect. *Las Vegas Review-Journal,* p. 1B.

Jacobs, S. (1993, December 5). Suburban kids "bomb" city graffiti scene. *Boston Sunday Globe,* pp. 1, 28.

Kreck, D. (1993, July 26). Don't spray it, say it when it comes to cleaning up graffiti. *Denver Post,* p. B1.

Kummel, P. (1991). Beyond performance and permanence. *Border/Lines, 22,* 10–12.

Kunzle, D. (1993, April). The mural death squads of Nicaragua. *Z Magazine,* pp. 62–66.

Lachmann, R. (1988). Graffiti as career and ideology. *American Journal of Sociology, 94,* 229–250.

LeDue, D. (1992, October 11). Community curfews attempt to rein in the time of young life. *Philadelphia Inquirer,* p. B7.

Lehning, A. (Ed.). (1974). *Michael Bakunin: Selected writings.* New York: Grove.

Lozano, C. (1994, January 13). Parents get tagged with $38,000 bill. *Los Angeles Times,* p. B8.

Lure of fame traps graffiti suspects. (1994, June 11). *New York Times,* p. 8.

Lyng, S. (1990). Edgework: A social psychological analysis of voluntary risk taking. *American Journal of Sociology, 95,* 851–886.

Macdonald, B. (1992). Citti politti: Cultural politics in Los Angeles. In G. Riposa & C. Dersch (Eds.), *City of angels* (pp. 15–30). Dubuque, IA: Kendall/Hunt.

MacDuff, C., & Valenzuela, C. (1993, February 28). Price of tagging. *San Bernardino Sun,* pp. A1, A11.

Marriott, M. (1993, October 3). Too legit to quit. *New York Times,* p. B8.

Martin, H. (1992, April 14). A clean slate. *Los Angeles Times,* p. B3.

Maxwell, J., & Porter, D. (1993). *Report on taggers.* Los Angeles: Los Angeles County Sheriff's Department.

Miller, I. (1994). Piecing: The dynamics of style. *Calligraphy Review, 11,* 20–33.

Mizrahi, M. (1981, October 21). Up from the subway. *In These Times,* pp. 19–20.

Molloy, J., & Labahn, T. (1993). "Operation GETUP" targets taggers to curb gang-related graffiti. *Police Chief, 60,* 120–123.

National Graffiti Information Network. (1990). *National Graffiti Information Network December update.*

Nazario, S., & Murphy, D. (1993, July 4). A "family" lost. *Los Angeles Times,* pp. B1, B4.

Pool, B. (1992, August 13). Youths wielding spray-paint cans learn difference between vandalism and art. *Los Angeles Times,* pp. B1, B8.

Posener, J. (1982). *Spray it loud.* London: Pandora.

Quintanilla, M. (1993, July 14). War of the walls. *Los Angeles Times,* pp. E1, E6.

Rainey, J. (1993, June 2). Surveillance teams to help fight graffiti. *Los Angeles Times,* pp. A1, A11.

Respect. (1993, June/July). *The Seed,* pp. 34–37.

Reuter. (1994a, January 6). High school cuts crimes with bounties. *Rocky Mountain News,* p. 36A.

Reuter. (1994b, January 19). Miami OKs teen curfew to cut crime. *Rocky Mountain News,* p. 31A.

Riccardi, N. (1994, June 22). 7 arrested in $100,000 freeway tagging spree. *Los Angeles Times,* p. B1, B8.

Riding, A. (1992, February 6). Parisians on graffiti: Is it vandalism or art? *New York Times,* p. A6.

Rodriguez, L. (1994, May 8). Los Angeles' gang culture arrives in El Salvador, courtesy of the INS. *Los Angeles Times,* p. M2.

Rolston, B. (1991). *Politics and painting: Murals and conflict in Northern Ireland.* London: Associated University Presses.

Rotella, S. (1994, March 20). Border lines. *Los Angeles Times,* pp. A3, A26.

Sahagun, L. (1992, April 2). Rash of city hall graffiti. *Los Angeles Times,* pp. B1, B4.

Sanchez, R. (1993, September). Drug dealers are new city arts patrons. *Prize Press,* p. 5.

Sanko, J. (1994, August 31). Romer unveils "kid's crusade." *Rocky Mountain News,* p. 8A.

Schiller, H. (1989). *Culture, Inc.: The corporate takeover of public expression.* New York: Oxford University Press.

Schwada, J., & Sahagun, L. (1992, August 11). Graffiti reward program nearly out of money. *Los Angeles Times*, pp. B1, B4.

Scott, J. (1990). *Domination and the arts of resistance*. New Haven, CT: Yale University Press.

Sheesley, J., & Bragg, W. (1991). *Sandino in the streets*. Bloomington: Indiana University Press.

Siegel, F. (1993, March/April). Lady Pink: Graffiti with feminist intent. *Ms. Magazine*, pp. 66–68.

Simon, R. (1993, July 9). Riordan OKs $1,000 penalty for tagging. *Los Angeles Times*, pp. B1, B3.

Sipchen, B. (1993, March 4). Scrutinizing work of taggers, gangsters, street artists. *Los Angeles Times*, pp. E2, E3.

Smith, J. (1994, July 27). City approves anti-graffiti regulations. *Fort Worth Star-Telegram*, pp. 21, 28.

Soja, E. (1989). *Postmodern geographies*. London: Verso.

Sorkin, M. (Ed.). (1992). *Variations on a theme park: The new American city and the end of public space*. New York: Hill and Wang.

Stewart, S. (1987). Ceci tuera cela: Graffiti as crime and art. In J. Fekete (Ed.), *Life after postmodernism* (pp. 161–180). New York: St. Martin's.

Sting ensnares graffiti stars. (1991, November 22). *Denver Post*, p. 8B.

Teaching teen "write" from wrong. (1994, June 26). *Rocky Mountain News*, p. 22A.

These guys do windows. (1994, January 17). *Newsweek*, p. 48.

Tobar, H. (1993, March 3). County OKs new graffiti crackdown. *Los Angeles Times*, p. B3.

2 teens spotted by helicopter held in vandalism of wall. (1991, April 30). *San Bernardino Sun*, p. B3.

Valenzuela, C. (1993, January 10). "We're tired of graffiti." *San Bernardino Sun*, pp. A1, A4.

Waldenburg, H. (1990). *The Berlin Wall*. New York: Abbeville.

Weber, B. (1994, April 28). City council candidates talk crime. *Rocky Mountain News*, p. 16A.

Will, E. (1994, January 2). Painting the town: The battle over graffiti in Denver. *Denver Post*, pp. 10–13.

Wilson, W. (1992, December 28). Emergence of outsider art throughout L.A. *Los Angeles Times*, pp. F6, F7.

Writing on the wall. (1993, February 21). *Las Vegas Sun*, pp. 1C, 4C.

Part II

VIOLENT CRIMES

Violent crimes include any illegal behavior that threatens to or actually does harm another through physical means. Such criminal acts include murder, robbery, rape, and assault. Violent acts are viewed as the most serious of crimes by our society and are punished more severely than any other type of criminal offense by our justice system.

Violent crime is more feared by the public than other types of crime and receives more mass media publicity than most other types of crime. It is usually an urban problem, but does occur less frequently in nonurban areas, as one of the articles in the section illustrates. Crimes involving violence are most often committed by urban males, 15 to 24 years of age, who come from neighborhoods that can best be characterized as lower socioeconomic in nature and are located in the inner city with predominantly minority populations. Excluding robbery, most acts of violence can be described as crimes of passion that are motivated by people who are most often known to each other. Emotions play a big part in the occurrence of violent behavior, and offenders usually have a history of previous violence in their background.

Ray and Simons offer an exception to violent crime as mainly an urban phenomenon. In their study of homicide in small towns, they describe how such murders occurred and the circumstances that led up to them. The authors conclude that the majority of their homicide respondents cited a vocabulary of motives used by murderers to explain their violent crime. The majority of their homicide study subjects pointed to reasons involving excuses that appealed to conventional moral perceptions for the need to commit such a heinous act. Excuses included impaired judgment due to alcohol and narcotic use, stress, accidental crimes, and so on. For those whose motives justified their actions, self-defense was the appeal they used and by doing so attempted to show a commitment to conventional morality rather than to the norms of a violent subculture.

In his study of rape, Hale explores the rationale of men who have committed this violent act. The author explores the different ways that rapists think about carrying out the crime and their intentions once they have selected a victim to sexually assault. Men who perpetrate this violent crime do think about the rewards as well as the possibility of being caught and punished for their assaulting behavior. However, concerns about being apprehended usually do not outweigh their desire to inflict physical harm to the victim. Hale concludes that violence stands out as a leading factor for rape. Most men who rape women do it more as an act of revenge or to punish the victim than for any other reason.

Jacobs and Wright focus on the motives of active armed robbers by attempting to understand the decision-making process that they utilize in their real-life environments. The authors claim that their major purpose was to comprehend how armed offenders go from not thinking about committing a robbery with a weapon to the determination to actually commit the crime. By interviewing a number of active offenders, the authors conclude that although a perceived need for instant money is what motivates robbers in their decision to commit this type of violent crime, the actual decision is formed by their past and present participation in street culture, which serves to link motivation for criminal activity to past experiences and future expectations.

Carjacking through violent means has become a more prevalent offense in recent years. Topalli and Wright discuss how the sensational nature of this violent offense increases communities' overall general fear of crime when law-abiding citizens can become so randomly victimized. The authors focus on the decision-making factors that precede the actual criminal activity and the situated interaction between particular sorts of perceived opportunities and needs and desires of the offenders.

5

CONVICTED MURDERERS' ACCOUNTS OF THEIR CRIMES: A STUDY OF HOMICIDE IN SMALL COMMUNITIES

MELVIN C. RAY

RONALD L. SIMONS

The authors conducted interviews with 24 persons convicted of homicide in a rural midwestern state. The focus of the study was on the ways in which the 24 offenders explained the reasons for their crime. The results show that the vast majority of the offenders did not identify with a violent subculture. Only two of the respondents felt that their action was justified; the rest believed that taking another person's life was wrong, and they regretted what they had done. They attempted to lessen their violent action by either claiming self-defense or offering excuses that included being drunk, experiencing overbearing stress, and being depressed.

In the last few years an alternative approach to studying violence has developed. Some sociologists have adopted a situational approach whereby they explore the processual development and interpersonal dynamics of violent situations (Athens 1977, 1980; Dobash and Dobash 1984; Felson 1978, 1982; Felson and Steadman 1983; Hepburn 1973; Luckenbill 1977). Much of this research has employed an impression management explanation of violent encounters. Luckenbill (1977), for example, concludes that homicide was usually the result of what Goffman (1967) terms "character contests." He reports that violent transactions were characterized by a series of "moves" that were initiated by both the victim and the offender on the basis of the other's moves and, in some instances, the reaction of an audience. The first stage in Luckenbill's model includes the opening move. This behavior was generally

EDITOR'S NOTE: Copyright © 1986 by JAI Press. Reprinted by permission of University of California Press. Reprinted from *Symbolic Interaction*, *10*(1), Spring 1987, pp. 57–70.

committed by the victim and was interpreted by the offender as an assault to his/her face. The subsequent stages of the model describe the way in which the interaction escalated into violence as the parties reacted to each other in a manner calculated to prove oneself and save face.

Felson and Steadman (1983) also examined the sequential nature of situations that end in criminal homicide. Based on their study of 159 incidents of murder and/or assault, they modified Luckenbill's thesis on character contests by including the significance of retaliation in violent situations. They cited the results of several experimental studies showing that physical retaliation is often an important strategy for addressing "face saving" concerns.

In contrast to these impression management studies, Athens (1977, 1980) focuses on the definition of the situation, self-concept, and generalized other employed by violent actors. Based on his own research and a reanalysis of some of the data presented by Luckenbill, Athens (1985) contends that there is little evidence for the character contest explanation of interpersonal aggression. He states:

> A character contest presumes that people always commit violent criminal acts in order to display a strong character and maintain honor and face or to avoid displaying a weak character and losing honor and face. However, this is not the meaning which the perpetrators of violent criminal acts often attribute to their actions. (1985, pp. 425–426)

Similarly, in their study of situations involving domestic abuse, Dobash and Dobash (1984) failed to find support for the impression management explanation.

To an extent, the conflicting findings associated with the situational approach may be a result of differences in samples and methods. With regard to differences in samples, Luckenbill focused on convicted murderers; Felson and Steadman studied persons convicted of felonious assault, manslaughter, or murder; Athens included individuals convicted of homicide, aggravated assault, forcible and attempted rape, and robbery; and Dobash and Dobash only considered women who had experienced repeated domestic abuse. Rape, robbery, homicide, and spouse abuse are very different events.

One might therefore expect the interpersonal dynamics and motives associated with these various situations to also be dissimilar. Character contests might be endemic to one type of violent offense but not another.

Concerning differences in methodology, Luckenbill, and Felson and Steadman attempted to construct the behavioral sequencing of violent events through the use of court records, whereas Athens employed data obtained through interviews with perpetrators and Dobash and Dobash utilized information gleaned from interviews with victims. If one is concerned with the meaning that perpetrators of violence attach to their actions and to those of their victims, court records or reports from victims might be considered less valid and reliable sources than interviews with perpetrators themselves.

In the present study several individuals convicted of homicide or manslaughter were interviewed in an attempt to identify the meanings they imputed to the circumstances and events surrounding the commission of their offense. Of particular concern were the situated reasons or motives that the perpetrators perceived as guiding their violent actions.

Focus of the Study

Some years ago, Mills (1940) exhorted sociologists to give more attention to the study of "vocabulary of motives" associated with various types of situations. He noted:

> A satisfactory or adequate motive is one that satisfies the questioners of an act or program, whether it be the other's or the actor's. . . . The words which in a type situation will fulfill this function are circumscribed by the vocabulary of motives accepted for such situations. Motives are accepted justifications for present, future, past programs or acts. (1940, p. 906)

Particular vocabularies of motive are considered appropriate to particular types of classes of situations. When an actor engages in motive talk he or she describes a proposed line of action and the reasons for the action. If the proposed line of action and reasons are to be perceived as legitimate both by the actor and the others present, the motive talk must be perceived as consistent with the definition of the situation by the participants. Stated differently, routine social interaction requires that participants employ motives consistent with the vocabulary of motive assumed to be appropriate in situations of that type.

Groups often differ in the way they classify or define various situations. As a consequence, they may differ with regard to the vocabularies of motive they consider appropriate in a particular situation. As Mills observes:

> A labor leader says he performs a certain act because he wants to get higher standards of living for the workers. A businessman says that this is rationalization, or a lie; that it is really because he wants more money for himself from the workers. . . . What is reason for one man is rationalization for another. The variable is the accepted vocabulary of motives of each man's dominant group about whose opinion he cares. (1940, p. 913)

The present study attempted to classify the vocabularies of motive employed by persons convicted of homicide and manslaughter. Scott and Lyman's (1968) typology of accounts was used as an aid in analyzing the perpetrators' descriptions of their crimes. As Scott and Lyman contend: "The study of deviance and the study of accounts are intrinsically related, and a clarification of accounts will constitute a clarification of deviant phenomena" (1968, p. 62). A major focus of this study was identification of the range of accounts associated with situations involving homicide.

METHODS AND PROCEDURES

From October 1984 to October 1985, in-depth interviews were completed with all individuals entering the penal system of a midwestern state as a result of having been convicted of homicide or manslaughter. Given the small population and relatively low crime rate of the state this amounted to only 26 persons. Most of the respondents were residents of and had committed their homicides in communities ranging in size from 5 to 300,000. Hence, it should be emphasized at this point that the study focused on homicides in relatively small communities and the results should not be generalized to large metropolitan areas. All of the interviews took place at the state's orientation center as the inmates were entering prison to begin serving their sentences.

Six of the 24 respondents were female. Five were black and the remaining individuals were white. The black population in the state is a little over 1%. Eight of the offenders were married, 14 were single and had never been married, and 2 were divorcees. Educational level varied from 5 years of elementary schooling to 4 years of college. The average education was 10 years. Comparing the occupational scores to educational attainment, it appears that many of the individuals were underemployed at the time of their crime. Finally, consistent with previous studies of individuals convicted of homicide (Stanton 1969), most of the respondents had no previous felony record.

RESULTS

Consistent with previous findings (Wolfgang 1958), most of the cases involved the killing of someone the perpetrator knew. In 1 case the victim was a spouse, in 4 cases it was a relative, in 13 cases it was a friend or well-known acquaintance, while in only 6 cases was the victim a stranger. The location of the killing reflected this relationship between the victim and perpetrator. In 13 cases the homicide took place in the home of either the victim or the perpetrator. One of the murders took place in a bar, another at a party, and the remaining 9 were carried out in various other locations. In 15 cases a gun was used to kill the victim. A knife was employed in 6 cases. The remaining 3 homicides involved some other form of killing the victim.

Accounts Involving Justifications

Six individuals presented definitions of the situation that they felt justified their having killed the victim. In these accounts the respondents accepted responsibility for the homicide but felt that their actions were legitimate given the circumstances. In 2 cases the perpetrators argued that their victims deserved what they got. CJ was convicted of killing an individual with whom he had had a fight some weeks earlier. CJ had beaten the victim because he had killed a puppy belonging to CJ's sister. When the victim and some of his friends cornered CJ some time later with the intention of retaliating, CJ pulled a gun and shot him. CJ claimed the victim was "no good" and deserved what he got.

RO was convicted of killing his wife during an argument. In describing the homicide he stated that:

> Then Jill started to argue with me about why I didn't go over there . . . I got a cup of coffee . . . then I came out in the living room and then me and Jill was fighting right off the bat. And then that's when the violence came in and that's when I hit her and pulled her hair and threw her on the floor, then that's when the knife came in. And then she got up and I turned around and just cut her right around the neck and it went in too deep and killed her. . . . She knows me too well, what that does to me.

The accounts of both CJ and RO imply that violence is a legitimate tactic for dealing with interpersonal problems. In the course of describing the homicides they committed, both persons referred to fights they had had with other individuals. Their accounts might be taken as evidence of identification with a violent generalized other or subculture.

The 4 others who provided justification accounts cited self-defense as the reason for their violent actions. For instance, consider the case of GD, who allowed his brother-in-law, a man experiencing a number of personal problems, to move in with him. GD returned hom from church one night to be accosted by his brother-in-law who demanded that he give him some money.

> I had always given him money before. I was paying his rent, feeding him and everything else, but he was going too far. I wasn't working either. I was out of work too so the money was kind of scarce. . . . So he came in and asked for money. I told him I didn't have no money. He demanded money and he got louder and louder. He said if I didn't give it to him he'd kill me. I got concerned when he said he'd kill some people. So I tried to walk away from him and asked him to leave. And he got violent and violent; he got louder and

louder. And when I tried to escort him out the door he wouldn't go. He ran off to the bathroom to the right. He went in and got his knife out, and I saw him coming toward me. He kept right on coming so I had to shoot him. And then I called the police. He was only 3 feet away from me and I had to shoot him otherwise he was going to kill me. It was a split-second decision, you had to do it or he was going to kill me.

Another example is SK, a young female who stabbed another woman outside of a bar:

> I had been in [state] for about two weeks and was living with an aunt and uncle, and the next-door neighbors owned a bar and asked me to join them for a birthday party. I did and it's strange the way it all happened. These girls decided they didn't like me, kept giving me a lot of stares, wow the California girl, look at her . . . I made a big hit with the guys, O.K. I mean I guess it's pretty normal. I was a new girl from California. I dressed pretty good, minding my own business, drank my own beer, just sat and shot the shit with the bar maid. . . . Some guy came up and started talking to me and asked me if he could buy me a drink, and I said, "Sure," so he bought it and I got up and started playing pool and one of the girls comes up and told me that it was her guy. I said how the hell was I supposed to know this. And she went back and was talking about this, that I was supposedly flirting and I could just hear the little comments in the background. . . . I sat there for about 30 minutes putting up with this . . . I tried to leave but they didn't have a telephone. I asked them where the closest one was and they said across the street on the corner. So I want out and started to make the call and here comes this same girl and three of her friends. I just got nervous and reached in my purse and pulled out a pocket knife. I expected to show it to her so she would leave me alone. I never intended to stab her, but she came at me and we started wrestling. She went one direction and I went the other and then someone came up and said the girl had been stabbed.

The 4 persons who provided a self-defense justification showed emotional responses during their interviews suggesting that they felt regret about the death of the victim. They displayed tears or developed a husky voice when describing the events surrounding their violent act. In each case the death of the victim was defined as regrettable, but necessary given the circumstances. There was no evidence in the interviews that anyone of the 4 identified with a violent subculture. Although these individuals viewed their actions as constituting legitimate self-defense, the court had ruled in each case that more force was used than was required. Hence, they were convicted of either voluntary or involuntary manslaughter.

Accounts Involving Excuses

The vast majority of respondents, 18 of the sample of 24, provided accounts that they believed at least partially excused their violent action. They admitted that what they did was bad or inappropriate, but they provided reasons as to why they should not be held fully responsible. Classifying the accounts, 8 persons claimed they were so intoxicated on drugs or alcohol that either they did not know what they were doing or it was an accident, 4 individuals told a "sad story" concerning the stress they were under at the time of the act, in 3 cases the respondent cited a sad series of events and maintained he/she was intoxicated, and in three cases scapegoats were blamed for the violent act.

Intoxication

Respondents using this excuse claimed they didn't really understand what they were doing. In the extreme, they maintained that they had blacked out and had no recollection of the murder. PA, for example, is a 34-year-old mother of two convicted of killing her boyfriend. She described the events surrounding the homicide as follows:

George comes to the house every day to see my dog, and from there we go to the bar and drink. Then I come back and I might clean up his house or cook at my house and I send food over there. I wasn't mad at him. . . . I'd been drinking, smoking reefer, shooting heroin, and dropping pills. . . . The only thing I remember that I was in my own house sleeping on the couch, and the police came and got me. I don't remember going over to George's house. I must have blacked out. It could have been something in the drugs. PCP or whatever. I don't know. They say [I killed him] with a knife. They had pictures in the courtroom

and stuff. I didn't see the pictures until I was getting ready to have my trial when they was picking the jury. I asked my motive and said let me see the pictures, and he let me see the pictures. I couldn't believe that's what I had did to him.

PA, like other individuals who employed this type of excuse, expressed regret over the killing and maintained it never would have happened if she had not been intoxicated.

Sad Stories

Persons providing accounts of this type described, usually in great detail, a series of unfortunate events that they encountered prior to their act of violence. These events included unemployment, divorce, death of a loved one, serious financial loss, abuse from others, and so forth. The respondents contended that they were so emotionally distraught by these stressful events that they were not thinking and acting rationally. EL, a 48-year-old male convicted of killing a police officer, might be considered typical of persons citing sad stories.

It was a real emotional thing. It had to do with me and my wife. We was going through a, she had decided to get a divorce. When I met up with her that night at the bar I seen her embraced with another guy, and a heated argument started, and from there it proceeded on home. . . . She went home first and I followed her home. And when I got there the police was there. She had called the police on me and I'd never been arrested before. I just went into the house and loaded the gun, I guess for protection. I don't know to this day why I loaded or took the gun out. I just, there was something driving me. I had to get out of the house. As I proceeded out of the house toward my car, all of these lights came on. It was just like everything was closing in on me and just out of reaction I shot, I guess at this light. At the time I didn't know it was an officer. Then I found out later I killed this officer. I didn't even know it was an officer. I have no bitterness against the police. It was just like everything was closing in on me and, a, maybe it was just self-destruction on my part. Maybe I wanted them to open up on me. I don't know.

As is evident in this case, the individuals who provided sad story accounts maintained that they were not themselves at the time of the homicide. They contended their violent act was an anomaly stemming from the extreme duress they were experiencing.

Sad Stories Combined With Intoxication

Some respondents maintained that a series of stressful events had driven them to use intoxicants as a way of numbing their psychological pain. They contended that in the midst of this stress-motivated intoxication, they acted irrationally. In other words, they provided accounts that combined the two types of excuses cited previously. An example is HK, a young man convicted of shooting his girlfriend:

> Me and my girlfriend had broke up and I had been heavily depressed all day long for about 3 or 4 days. I was feeling badder and badder . . . and I was drinking pretty heavy that night and I was doing cocaine and smoking weed . . . I just wanted to kill myself. And I wrote a suicide note to my brother and I went over to my girlfriend's house . . . and had took a gun out of the cabinet and went over there. . . . I went in and there she was standing with a butcher knife and says "Come on motherfucker," like that. Then that's when I took the gun out and put it to my head. And then she panicked and ran into the bedroom, and I just went into the bedroom to talk to her. . . . I asked her if it was really over between us, and she went down on her knees and said, "Oh my God," or something like that. I don't believe this is happening. And then I went to pick her up to tell her that I loved her and then the gun went off somehow.

Individuals like HK claimed that they should not be held responsible for the terrible act they committed because they were out of their mind due to extreme stress and intoxication.

Scapegoat

Individuals who employed this excuse were convicted of homicides involving more than one perpetrator. The respondents maintained that they participated in the act because they were coerced or were afraid of the other participant in the crime. For instance, LV contended that she went along with her husband because she was afraid of what he would do if she resisted him. She stated:

I did everything he wanted me to do. I didn't really make decisions for myself. That's where my own stupidity came in. I let him get me into trouble which I should have known better. I should have put my foot down, but I was in fear of him 'cause he always threatened that if I ever left him or tried to divorce him he'd kill me which he tried to kill. . . . He put a knife to my throat one day. I wore marks, his fingermarks around my neck clear around to the other side from where he tried to choke me. . . . And I did get up enough nerve one time to leave him. I went clear out to Colorado, Idaho, he found me.

LV, like the other two individuals who claimed that they were scapegoats, expressed regret and guilt over the incident but claimed she was not the kind of person who would normally do such a thing. She argued that it was her inability to escape from her violent husband that caused her to go along with the act.

The individuals who employed excuses in accounting for their crimes did not appear to identify with a violent generalized other. Some of them lived in environments where violence occasionally occurred, but they did not try to justify violence as an approach to solving interpersonal problems. Indeed, several of the respondents who employed excuses manifested a fear or dread of violence. This fear was also apparent in two of the individuals who provided self-defense justifications. Almost all of the persons who cited excuses expressed deep regret concerning the victim's loss of life. Their remorse was evident in the teary eyes and husky voice that often accompanied their accounts.

Factors Involved in Violent Crimes

Handguns

In 15 of the 24 homicides considered in the study, the weapon employed was a gun, and in 12 of the 15 gun-related incidents, the weapon was a handgun. In case after case one was struck by the way the presence of a handgun made the homicide possible. The respondent often lived in an area or frequented places where he/she felt a gun was necessary for protection. In most instances it seemed that it was not a violent orientation but a fear of violence that led to the acquisition of the gun. The respondent usually kept the gun readily available in his/her home or car. Then, during a period of intense emotional distress and/or intoxication, or in the midst of a perceived threat from another, the individual went for the handgun and used it on someone.

Suicide Ideology

In 6 cases the respondent talked about wanting to kill him/herself at the time of the homicide. Four of the individuals described in some detail the suicide fantasies they were having just prior to the violent incident and one person had even written a suicide note. Two of the respondents reported that they were carrying a handgun at the time of the homicide because they intended to use it to kill themselves.

Drugs and Alcohol

Sixteen of the 24 respondents were heavy users of alcohol and/or other drugs. Most of these individuals described recurrent episodes of extreme intoxication. Thirteen of them had taken some type of intoxicant prior to their act of violence. In six cases the individual had combined large amounts of alcohol with other drugs.

Character Contents

Only two of the respondents offered accounts that included references to insulting, slanderous remarks made by the victim. RO, whose case was described earlier, claimed that he was frustrated by his wife's nagging and arguing. In the midst of his wife's denigrating remarks, he lost his temper and killer her with a kitchen knife.

The second case, that of SK, was also cited previously. Recall that she was convicted of stabbing another woman outside of a bar. In her description of the events surrounding the crime she stated:

> One of the girls came up and told me that it was her guy. . . . And she went back and I could just tell she was talking about this, that I was supposedly flirting and I could just hear the little comments in the background and I sat there for about 30 minutes and put up with this.

Contrary to the character contest or face-saving theory of homicide, SK did not attempt to counter the insults. Instead she went outside to a telephone booth in order to call for a ride. She was followed by the woman who had been insulting her. According to SK, the woman came at her and it was out of fear, not humiliation, that she stabbed the woman. Thus, this case is not consistent with the character contest explanation. The victim did verbally abuse the perpetrator. But, the victim's remarks served to scare and intimidate the respondent rather than to instill a desire for face-saving retaliation.

DISCUSSION AND CONCLUSIONS

Few of the respondents in this study provided evidence of identification with a violent subculture. Only two persons contended that their action was justified, that the victim got what he/she deserved. The remaining individuals took the position that killing another is wrong and regrettable. Having expressed regret, and sometimes guilt, over the incident, they attempted to reduce their culpability by either claiming self-defense or citing excuses concerning intoxication, psychological stress, and so forth.

Although few of the respondents might be considered members of a lower class, violent subculture, social class did seem to be an important predisposing structural factor. With some exceptions, the individuals tended to be underemployed and living on low incomes. As such, they were victims of the usual stresses associated with life in the lower social strata: poor housing, periodic unemployment, and financial worries. They manifested many of the recognized consequences of such living conditions: domestic disputes, alcohol and drug use, depression, and so on. As one listened to these "murderers" talk about their lives they elicited more pity than fear. Feelings of injustice and relative deprivation are themes that characterize many of their life stories. In most cases they were individuals from impoverished backgrounds living rather dead-end lives, who in the course of mounting life stress, struck out at someone, usually while intoxicated on drugs or alcohol. Frequently they were depressed, and in some cases suicidal, at the time of the act, and the person they attacked was usually a relative or acquaintance.

Only one of the cases is consistent with the character contest or impression management theory of homicide. This man killed his wife for persistently nagging him concerning certain of his personality traits. In another instance a respondent was verbally assaulted by the victim but the individual's motive for attacking the victim appeared to be unrelated to the insulting comments. The latter case underscores the importance of obtaining a perpetrator's definition of the situation when testing the impression management explanation. Witnesses may indicate that the participants in a violent altercation insulted and slandered each other. However, because the parties insulted each other prior to or in the course of a fight does not mean that the insults caused the conflict or contributed to its escalation. One must determine the meanings that the participants attached to each other's actions in order to determine the extent to which they were affected by the insults and were motivated to retaliate in order to save face. In the present study the respondents were more concerned with love relationships, drugs, money, and possessions than with face-saving. In their encounter with the victim they were concerned with justice, with getting or protecting what they felt they were entitled to, rather than with salvaging or promoting an esteemed identity.

Given the current controversy over handgun legislation, it is interesting to note the role that handguns played in these homicides. In half of the cases a handgun was used to kill the victim. In almost all of these incidents the gun was kept in the perpetrator's car or home for protection. However, in the midst of intense emotional distress and/or intoxication, the respondent employed the weapon to settle an interpersonal problem. In most of the cases it appears that the homicide would have been quite unlikely if a handgun had not been so readily available.

The results of this study should not be generalized to homicides in large metropolitan areas. It may be that violent subculture and impression management explanations are more applicable to violence in densely populated urban settings of the United States. However, based on a sample of persons convicted of homicide in a small, midwestern state, these factors do not appear to be important in the etiology of violent acts committed in rural areas and small cities.

REFERENCES

Athens, Lonnie. 1977. "Violent Crimes: A Symbolic Interactionist Study." *Symbolic Interaction* 1: 56–71.

——. 1980. *Violent Criminal Acts and Actors.* London: Routledge & Kegan Paul.

——. 1985. "Character contests and Violent Criminal Conduct: A Critique." *Sociological Quarterly* 26: 419–431.

Dobash, R. Emerson, and Russell P. Dobash. 1984. "The Nature and Antecedents of Violent Events." *British Journal of Criminology* 24: 269–288.

——. 1976. "Is There a 'Subculture of Violence' in the South?" *Journal of Criminal Law and Criminology* 66: 483–490.

Felson, Richard B. 1978. "Aggression as Impression Management." *Social Psychology Quarterly* 41: 205–213.

——. 1982. "Impression Management and the Escalation of Aggression and Violence." *Social Psychology Quarterly* 45: 245–254.

Felson, Richard B., and Henry J. Steadman. 1983. "Situational Factors in Disputes Leading to Criminal Violence." *Criminology* 21: 59–74.

Goffman, Erving. 1967. *Interaction Ritual: Essays on Face-to-Face Behavior.* Garden City, NY: Doubleday.

Hepburn, John R. 1973. "Violence Behavior in Interpersonal Relationships." *Sociological Quarterly* 14: 419–429.

Luckenbill, David. 1977. "Criminal Homicide as a Situated Transaction." *Social Problems* 25: 176–187.

Mills, C. Wright. 1940. "Situated Actions and Vocabularies of Motive." *American Sociological Review* 5: 904–913.

Scott, Marvin, and Stanford Lyman. 1968. "Accounts." *American Sociological Review* 33: 46–62.

Stanton, John W. 1969. "Murders on Parole." *Crime and Delinquency* 15: 149–155.

Stuart, Richard B. 1981. "Violence in Perspective." Pp. 3–30 in *Violent Behavior: Social Learning Approaches to Prediction, Management, and Treatment,* edited by Richard B. Stuart. New York: Brunner/Mazel.

Wolfgang, Marvin E. 1958. *Patterns in Criminal Homicide.* New York: Wiley.

6

Motives of Reward Among Men Who Rape

Robert Hale

This study examines the various ways that men think about their illegal activities and what they perceive as the goal of sexual assault on a woman. Hale interviewed and surveyed men incarcerated for rape in two prisons in a southern state. One hundred and thirty inmates made up the sample. The author looked at the different ways offenders think about rape and found that men who sexually assault do have specific goals in mind when they choose a female victim. The respondents often consider the rewards of their crime along with the risk of being caught. Such concerns about being detected are outweighed by their desire to injure the targeted victim. Also, Hale concludes that many sexual assaulters consider the effects that their violence will have on the emotional, physical, and psychological states of the women they rape.

Rape and those who commit rape have been the focus of a range of studies since the mid-1970s. The statistical and demographic characteristics of rape have been analyzed in an attempt to better understand the circumstances of this crime. The background characteristics of men who commit rape also have been explored in an attempt to develop a composite or profile. In addition, the motivations of men who commit rape have been examined to better understand what compels the rapist. . . .

Deterrence of criminal behavior is achieved as the individual balances beliefs of potential punishment against the anticipated risk of the criminal act. . . .

While studies of rape have a long history, Scully (1990) and Scully and Marolla (1984; 1985; 1995) have produced a progression of highly regarded works exploring the motives of reward among men who commit rape. In addition, Decker, Wright, and Logie (1992) have produced a model study of perceptual deterrence. They encourage future research that attempts to cross-validate perceptual deterrence across different types of offenses and offenders. This study will merge the ideas of rape and perceptual deterrence to a sample of men convicted for rape in an attempt to identify the anticipated gains (rewards) that motivate these men. . . .

EDITOR'S NOTE: From Hale, R., "Motives of reward among men who rape," in *American Journal of Criminal Justice, 22,* pp. 101-119. Copyright © 1997. Reprinted with permission.

Explanations of Rape

... Traditionally, the rapist is seen as compelled to rape, driven by either "uncontrollable urges" (Edwards, 1983) or a "disordered personality" (Scully and Marolla, 1984). Attributing rape to an uncontrollable impulse implies that normal restraints of self-control are reduced or erased by an innate sex drive. The major thesis is that sexual deprivation causes predisposed individuals to lose control of their behavior and force a victim into unwanted sexual relations (Symons, 1979).

The "disordered personality" approach tends to attribute rape to a mental illness or disease. Rape is thought to be a symptom of some deeply rooted abnormality, either cognitive or organic. Lanyon (1986) believes that all sexual offenders, not just rapists, are acting from the same sickness, although the sickness is not identified. Ellis and Hoffman (1990) state that a "mutant gene" may be the cause, but they do not rule out other genetic or hormonal influences. Scully and Marolla (1985, p. 298) point out that the "belief that rapists are or must be sick is amazingly persistent," and Lanyon (1986, p. 176) asserts that this belief "tends to be the view held by the judicial system, by social service agencies, and the general public." The further implication of either of these approaches is that thinking is not a part of the process of rape; this holds particular consequences for the etiology of the crime and rehabilitation of men who commit rape.

However, an abundance of research shows that thinking and choice are a part of the rapist's construction of the crime, and the learning approach is a common theme in the literature on rape. From this perspective a number of typologies of men who rape have been created (Burgess, 1991; Holmes, 1991; Knight and Prentky, 1987; Scully and Marolla, 1985, 1995). Scully (1990) and Weiner, et al. (1990) identify three categories of rape; Holmes identifies five categories of rape, the same number given by Knight and Prentky; while the Scully and Marolla study (1985) describes six categories of motive for rape. Although there is some overlap between the descriptions in these studies, there are also some mutually exclusive categories. From the studies noted above and other sources,

eight categories of reward for rape can be identified.

The first category originates in the work of Black (1983) and is expanded by Scully and Marolla (1985). The rapist commits his crime but views his act as a legitimate form of revenge. This view is referred to as "The Punishment Model" by Felson and Krohn (1990). In these rapes, men are seeking to hold accountable and punish a female for some action that the rapist subjectively perceives as an insult. The action may have been committed directly by the victim, or she may be a victim of "collective liability," in which the victim is a representative of a larger class of individuals (Black, 1983; Scully and Marolla, 1984).

The revenge motive is common when the victim is the primary source of frustration, whereas punishment is often the goal when the victim is held accountable for collective liability. In these rapes, the rapist views his action as an attempt to subordinate women for an attempt to challenge male dominance. This perception comes from the rapist's belief that he has the right to discipline and punish.

The second category reflects an anger within the rapist, often emerging from his perception that a woman is interested in having sex with him but later refuses to fulfill his desire. This category would be comparable to the "deniers" found by Scully (1990). When seeking this type of reward, the rape becomes a means to a specific end, which is sexual satiation. The rapist believes he can seize what he desires, reflecting his belief that even if a woman declines sexual advances, rape is a suitable means to what is desired (Groth and Birnbaum, 1979; Holmes, 1991).

The third category reflects the desire of the rapist that the sex act be performed totally for his satisfaction, without regard for the feelings or emotions of the victim. Thus, the rapist can avoid feelings of intimacy and caring while having his desires fulfilled. This approach would be applicable in some respects to the impersonal, anonymous sex that often occurs in casual "one-night stands" and between prostitutes and their customers (Felson and Krohn, 1990; Russell, 1988, 1995; Scully and Marolla, 1995).

The fourth category of reward includes rapists who commit their crimes for the excitement they receive. The sex is secondary to the rush of emotion that goes along with the act. Sexual deviants do not emerge at the most severe levels of deviance or illegality: Evidence abounds that chronic sexual deviants have a long history of involvement with "nuisance" sex behaviors before progressing into such crimes as rape (Holmes, 1991; Rosenfield, 1985). Some rapists have reported feelings of an intense "high" as they anticipate and then complete the rape (Knopp, 1984; Warren et al., 1991). Rapists who report this goal as a reward are seen as pushing the limits of excitement and personal enjoyment to the ultimate illegal extreme.

The fifth category includes those who rape for the adventure or challenge the act affords them. The gang rapist is typical of those included in this category. The challenges of forcing the victim to complete a sex act and of being able to perform sexually in front of other male participants and the danger inherent in attempting a rape are often cited as reasons for rape by those included in this category (Holmes, 1991). This category is similar to the proving of "masculinity" observed by Russell (1989), Kimmel (1996), and Messner (1989). The danger of the act seems to be the motivating factor for these men: The danger lies either in the risk of identification and capture or in the potential inability to perform in the situation. The thrill lies in the promotion of loyalty and brotherhood among those associated with the commission of the rape.

The sixth category reflects the notion that the underlying motivation for rape is domination of the victim, rather than the sex that accompanies the rape. The goal of this type of rapist is to ensure that women are frightened and intimidated sufficiently to not challenge the power and superiority of males (Deming and Eppy, 1981; Herman, 1995).

The seventh category has its basis in the evolutionary theory of rape, which proposes that rape originates from pressures of natural selection that require men to be more aware of sexual opportunities, to have sexual relations with a wide variety of sexual partners, and to use force when necessary to fulfill their sexual desires (Quinsey, 1984; Shields and Shields, 1983; Symons, 1979; Thornhill and Thornhill, 1983, 1987).

The final category was proposed by Abel and Blanchard (1974) and is detailed by Prentky and Burgess (1991). The reward for the rapist in this category is the physical fulfillment of a fantasy. The role of fantasy has been documented as a precursor to a number of violent crimes, including serial rape (MacCullough et al., 1983; Warren et al., 1991). The consensus is that the fantasy-world beliefs of many sadistic offenders become overt behavior when the offender feels compelled to live out his fantasy.

This study will measure the motives of reward among a sample of men incarcerated for committing rape; the men will be asked to rank order their motives for committing their crime. Then, based on the motive, the men will be asked whether a low or high likelihood of arrest would deter their committing the crime. From these results, the type of men who rape that would be deterred by the risk of incarceration should be apparent.

DATA AND METHODS

The subjects in this study were a sample of the population of men incarcerated for rape within two maximum-security state penitentiaries in the Deep South. . . .

Data were collected at two different times. First, subjects were briefed about the purpose of the study and then were asked to complete a questionnaire. This questionnaire gathered demographic information before presenting the eight categories of reward for committing rape. The men also consented to an interview, which was conducted at another time. During the interview, subjects were asked to provide details behind their motives. . . . Subjects were selected at random to provide the follow-up interviews. The questionnaire and interviews were administered during counseling sessions and were conducted by either the leader of the counseling session or the author. Men in this study were surveyed over a three-year period that ended in 1996.

. . . Only those convicted of rape against an adult female were included in the study. While definitions of what constitutes rape can vary, men in this study had been found guilty under statutes that required "forcible sexual intercourse" achieved "against her (the victim's) will." Those convicted of other sex offenses, such as nuisance sex offenses, sexual assault or molestation, offenses involving pornography, or pedophilia, were excluded from the analysis. Subjects must have been convicted of the charge of rape in order to be included in the final sample. This stipulation created a sample of 132 subjects. Two potential problems could arise from using this sample. One is that subjects might provide answers for which they would anticipate being rewarded in some way, perhaps with preference for early release. To bypass this issue of validity, subjects were informed that all answers would be anonymous and that prison counselors would not know who did or did not respond to the survey. It was also stressed that no inmate would receive preferential treatment for completing the survey, just as no respondent would be deprived of benefits for refusing to participate.

A second flaw is that the entire model could be seen as speculative since all subjects are incarcerated at the time of the survey and thus cannot fully answer concerning their motivation as if they were truly free to act on their perceptions of risk and reward. Surveying men who were not incarcerated might provide the better measure of cost and benefit, rather than posing artificial situations to men who are not free to act. . . .

The group of 132 rapists had an average age of 32.6, with a range from 17 to 67 years of age. The mean educational level of the sample was 12.5 years, with a range from seventh grade to post-college graduate education. At the time of their offenses, 72% had been or were married and another 13% had been living with a woman. The majority (57%) were employed in skilled or unskilled labor positions.

The group had a lengthy involvement of sex offenses. On average, they began their sex offenses (e.g., voyeurism, exhibitionism, and scatophilia) in their early teens (mean = 13.4), and they committed their first rape at age 17.4. Of the sample, 85% had a prior conviction for a sexual offense, and 52% had a prior rape conviction. Most reported being heterosexual (82%), 7% reported being homosexual, 5% reported being bisexual, and 8 of the men chose not to answer this question.

. . . The goal of this study is to examine how men incarcerated for rape view the rewards of their actions. Subjects in this study were asked to remember a rape they had committed or had thought about committing. Each respondent was then asked to fill in details describing the crime, such as the time of day and characteristics of the setting and of the victim; the goal was to make the crime situation as real as possible to the respondent. Subjects were then presented with eight categories of potential reward and were asked to select which reward would compel them to commit the rape.

Subjects were asked to rank order the eight categories of reward that would influence them to commit rape. . . . After the men had ordered their perceptions of the reward they anticipated, they were asked to consider how the risk of apprehension would affect the likelihood of committing rape. . . . The men were asked whether they would attempt to commit the rape they had visualized earlier within the context of the amount of reward anticipated and the risk of apprehension. Four responses were possible: (1) No; (2) Not Likely; (3) Likely; (4) Yes.

RESULTS

Table 6.1 provides four types of information. The second column (Number) details how often each category was selected as the primary goal for committing rape by the 132 respondents; the third column (Percent) lists what percentage of the men are reflected by the number in column two.

Twenty-eight men (21%) answered they would rape for the reward of "Revenge and Punishment." Twenty-three of the men (17%) committed rape "To exert control or power" while 19 (14%) committed rape to release "Anger."

The cumulative points given to each category were then calculated; these scores are derived from the addition of the scores given to each category by each of the 132 respondents. The highest possible score for any category in this column is 1,056, assuming each of the men gave that particular category the highest score of eight; as shown, "Anger" is the motive receiving the highest cumulative score, followed

Table 6.1   Frequency of Selection as Primary Goal for Committing Rape

Category	Number	Percent	Cumulative Points	Average Score
Revenge/Punishment	28	21	963	7.3
Anger	19	14	1003	7.6
Impersonal Sex	9	6	726	5.5
To Feel Good	4	3	950	7.2
Adventure/Danger	18	13	831	6.3
Control/Power	23	17	752	5.7
Masculinity	14	10	946	7.2
Fulfill a Fantasy	17	12	990	7.5

by those whose motive was "To fulfill or experience a fantasy" and those raping for "Revenge and Punishment."

Although the reward of "Revenge and Punishment" is the single highest category, the cumulative totals reveal that more men were raping out of "Anger" or "To fulfill or experience a fantasy." The category of "Anger" as a reward received a cumulative number of 1003 points, while the category "Fulfill a Fantasy" received a cumulative total of 990 points. The category of "Revenge and Punishment" received 963 points and was the third highest category in terms of cumulative points. . . .

When the rapist is seeking a reward of personal satisfaction and enjoyment, he is more likely to defy a "High Risk" of apprehension in order to commit the rape. Follow-up interviews with rapists in this category affirmed the hypothesis that the high risk of apprehension added to the satisfaction derived from the rape. According to "Vance," an inmate who fits into this category:

> The idea that she might later figure out who I was added to the thrill, at the time. . . . I had seen her around, and I'm sure that she had seen me. Plus, the place where I got her (in a car parked outside a shopping center) was pretty open, too. I guess I picked that spot because I knew there was more chance of getting caught. But I never figured it was going to happen.

For men in this category, the attempt to commit rape in spite of a high risk of apprehension added to the pleasure of completing the act.

Their satisfaction was actually the result of two separate situations.

In only two other categories of reward were the men willing to chance a high risk of apprehension, . . . when the reward was either the danger or challenge of the rape or to fulfill a fantasy. Research points out that these rapes are often committed in front of an audience, such as instances of gang rape (Holmes, 1991; Scully and Marolla, 1984; Scully, 1990). When witnesses are present, whether or not they are active participants in the rape, the risk of apprehension increases. It is assumed that those who are willing to commit rape in front of witnesses are aware of the increased risk of apprehension. As "Larry" pointed out:

> The guys that were with me that night, we had known each other forever, we had grown up together. They had stuff on me, and I had stuff on them. I knew they would not talk, 'cause I could turn on them, and they knew it.

Thus, the rapist who is seeking an adventure in spite of a high risk of apprehension appears to believe that his accomplices will not corroborate the crime.

Those who rape in order to experience a fantasy also appear to be unconcerned when the risk of apprehension is high. Fantasy is an integral part of not only "normal" sexual relations but also of deviant and criminal sexual acts (Burgess et al., 1986; Holmes, 1991; Prentky and Burgess, 1991). . . . Fantasy is defined as a series of thoughts that preoccupy the individual into the rehearsal of the script. As the fantasy evolves, it may include feelings, emotion, and dialogue (Burgess, 1991). . . .

[T]he fantasy of rape provides a secondary motive for rape for many men in this study. Research has asserted that fantasies of violence are often a precursor to violent behaviors (Cohen et al., 1971; MacCullough et al., 1983; Prentky et al., 1985; Warren et al., 1991). These fantasies are thought to be linked to the rapist's feelings of anger, of revenge, or of danger and adventure. The rapist is motivated by these emotions and has fantasized a script that allows for their resolution. This assumption was borne out through follow-up interviews with men who fit this classification. Typical of those who raped out of a motive of revenge was "Ken," who stated:

I had a dream of getting even to her for the divorce. Then she remarried and I lost track of her. Even though I didn't have no interest in finding her, I still hated her for what she had done. I wished of getting back at her, but the only thing I could do was to hurt somebody like her, who reminded me of her. It still wasn't enough, but I felt better after I did it (rape).

The revenge expressed by "Travis" had a longer history:

I never felt like my own mother ever cared for me. She was always off doing her own thing, she never seemed to have time for me. Never seemed to care. The only women I ever got involved with were those I could treat however I wanted. And it was never good. I used to sit around and think of ways of getting even, when I had the chance. I figure now it goes back to my mother. I probably knew it then, but I couldn't help myself. And I'll probably do the same if I ever get out of here.

While some who raped from a motive of revenge had over time put some thought into their actions, the anger-rapist was more impulsive. His act of rape stemmed from an immediate situation that, in the perception of the rapist, was an affront that demanded a response. For "Dan," the affront was the discovery his wife was having an affair:

When I heard this, I just lost my cool. You know, how they say a heat of passion thing, or something

like that? Everything just turned red and black in my head, like something went off. . . . I looked at her, and thought, "If you are giving it to someone else, then I'll take it any way I want to." I saw it as having sex, although I was forcing her to, and I knew she didn't want to. At least, not the way I was making her do it. . . . They called it rape later, but I didn't see it that way. . . . I'm not sure I see it that way even now.

Sometimes the anger-rape has nothing to do with an affront by another person. For "Doug," it was simply a combination of events, none of which were directly related to a female:

That night I was really pissed off about a lot of things. I mean, not really a lot of things, but some thing were really eating at me. I had been looking for a job, and that was not going well. Money was tight, and I was getting hassled over the bills. I was just mad at everything in general, nothing seemed to be going my way. And then, out in the bar, I saw this girl who seemed to have it all together, who seemed to have a lot. I decided right then I was going to show her before the night was over what it was like to lose something . . . what it felt like to hurt.

DISCUSSION

This study provides support for the idea that rape can occur for a variety of motives. Among these motives, violence is certainly an issue to consider. While definitions of violence vary, emotions of danger, hatred, and revenge, along with beliefs of punishment and privilege, are repeatedly used as descriptions. In this study, violence emerges as a leading factor in rape. More men rape out of revenge, or to punish the victim, than for any other single motive, while anger is a latent motive for many men who commit rape.

Given the amount of research detailing rape as a crime of violence, these findings are not surprising. The data add support to the contention that violence is an integral part of the link between men and rape. . . . Evidently men who rape from a motive of revenge or anger are not considering, or affected by, the danger of committing the act. As mentioned earlier, many of the men whose primary motive was the element of danger were involved in gang rapes or were seeking personal satisfaction; revenge and anger were not part of the incident. Similarly, men who raped with the motives of anger or revenge were not concerned with the danger, but were more concerned with correcting a perceived "Wrong" (Katz, 1988). . . .

The link between power, anger, and revenge as factors in rape has long been noted, but this study fails to provide support for it. They key variable seems to be "power," as anger and revenge have been discussed as important within this study. It is plausible that men in this study did not want to admit (either to themselves or the researcher) that they were seeking power, given that power rapists tend toward insecurity with doubts about their masculinity (Holmes, 1991; Maletzky, 1991; Scully and Marolla, 1995). It may also be that the anger and revenge felt by these men simply overshadowed the need for power, although all three sensations were present.

While the analyses in this study show that violence is a factor influencing many men to commit rape, violence is present in every case. Certainly rape is a violent act, and the control of a victim often requires that violence be used by the perpetrator. However, there are motives for rape that do not include violence. When men are committing rape "to feel good," to experience a new "adventure," or to "fulfill a fantasy," violence does not have to be a part of the script. It is entirely possible that "Larry" and "Vance" committed their crimes without intending to use violence. Of course, this explanation is certainly open to debate, depending on the definition of violence and the definition of the situation; however, from the perspective of the perpetrator, there are instances when rape does occur without violence.

Responses in this study imply that the majority of rapists are not suffering from psychoses or other serious psychiatric disorders, as those from the "disordered personality" perspective would believe. Certainly, these men have a distorted perception of the role of sex and of females. This is not, however, a full sign of mental illness or other cognitive impairment. It is more a reflection of poor judgment, which is more typical of faulty social or learning skills. . . .

REFERENCES

Abel, G. G. and Blanchard, E. B. (1974). "The role of fantasy in the treatment of sexual deviation." *Archives of General Psychiatry, 30,* 467–475.

Black, D. (1983). "Crime as social control." *American Sociological Review, 48,* 34–45.

Burgess, A. W. (Ed.). 1991. *Rape and sexual assault: A research handbook.* New York: Garland.

Burgess, A. W., Hartman, C. R., Ressler, R. K., Douglas, J., and McCormack, A. (1986). "Sexual homicide: A motivational model." *Journal of Interpersonal Violence, 1,* 251–272.

Decker, S., Wright, R., and Logie, R. (1992). "Perceptual deterrence among active residential burglars: A research note." *Criminology, 31,* 135–147.

Deming, M. B. and Eppy, A. (1981). "The sociology of rape." *Sociology and Social Research, 64,* 357–380.

Edwards, S. (1983). "Sexuality, sexual offenses, and conception of victims in the criminal justice process." *Victimology: An International Journal, 8,* 113–128.

Ellis, L. and Hoffman, H. (Eds.). (1990). *Crime in biological, social, and moral contexts.* New York: Praeger.

Felson, R. B. and Krohn, M. (1990). "Motives for rape." *Journal of Research in Crime and Delinquency, 27*(3), 222–242.

Groth, A. N. and Birnbaum, J. (1979). *Men who rape.* New York: Plenum Press.

Herman, J. L. (1995). "Considering sex offenders: A model of addiction. In P. Searles and R. Berger (Eds.), *Rape and society* (pp. 74–98). Boulder, CO: Westview Press.

Holmes, R. (1991). *Sex crimes.* Newbury Park, CA: Sage Publications.

Holmes, R., DeBurger, J., and Holmes, S. T. (1988). "Inside the mind of the serial murderer." *American Journal of Criminal Justice, 13*(1), 1–9.

Katz, J. (1988). *Seductions of crime.* New York: Basic Books.

Kimmel, M. (1996). *Manhood in America: A cultural history.* New York: Free Press.

Knight, R. A. and Prentky, R. A. (1987). "The developmental antecedents and adult adaptations of rapist subtypes." *Criminal Justice and Behavior, 14*(4), 403–426.

Knopp, F. (1984). *Retraining adult sex offenders: Methods and models.* Syracuse, NY: Safe Society Press.

Lanyon, R. I. (1986). "Theory and treatment in child molestation." *Journal of Consulting and Clinical Psychology, 54*, 176–182.

MacCullough, M., Snowden, P., Wood, J., and Mills, H. (1983). "Sadistic fantasy, sadistic behavior, and offending." *British Journal of Psychiatry, 143*, 20–29.

Maletzky, B. (1991). *Treating the sexual offender.* Newbury Park, CA: Sage Publications.

Messner, M. (1989). "When bodies are weapons: Masculinity and violence in sport." *International Review of the Sociology of Sport, 25*(3), 203–220.

Prentky, R. A. and Burgess, A. W. (1991). "Hypothetical biological substrates of a fantasy-based drive mechanism for repetitive sexual aggression." In A. W. Burgess (Ed.), *Rape and sexual assault* (pp. 235–256). New York: Garland.

Prentky, R. A., Cohen, M. L., and Seghorn, T. K. (1985). "Development of a rational taxonomy for the classification of sexual offenders: Rapists. *Bulletin of the American Academy of Psychiatry and the Law, 13*, 39–70.

Quinsey, V. L. (1984). "Sexual aggression: Studies of offenders against women." In D. Weisstub (Ed.), *Law and mental health: International perspectives* (Vol. 1, pp. 84–121). New York: Pergamon.

Rosenfield, A. (1985, April). "Sex offenders: Men who molest, treating the deviant." *Psychology Today,* 8–10.

Russell, D. E. H. (1988). Pornography and rape: A causal model. *Political Psychology, 9*, 41–73.

Russell, D. E. H. (1989). "Sexism, violence, and the nuclear mentality." *Exposing nuclear phallocies* (pp. 63–74). New York: Pergamon.

Russell, D. E. H. (1995). "White man wants a black piece." In P. Searles and R. Berger (Eds.), *Rape and society* (pp. 129–138). Boulder, CO: Westview Press.

Scully, D. (1984). "Convicted rapists' vocabulary of motives: Excuses and justifications." *Social Problems, 31*(5), 530–544.

Scully, D. (1990). *Understanding sexual violence: A study of convicted rapists.* Boston: Unwin Hyman.

Scully, D. and Marolla, J. (1985). "Riding the bull at Gilley's: Convicted rapists describe the rewards of rape." *Social Problems, 32*(3), 251–263.

Scully, D. and Marolla, J. (1995). "Riding the bull at Gilley's: convicted rapists describe the rewards of rape." In P. Searles and R. Berger (Eds.), *Rape and society* (pp. 58–73). Boulder, CO: Westview Press.

Shields, W. M. and Shields, L. M. (1983). "Forcible rape: An evolutionary perspective." *Ethology and Sociobiology, 4*, 115–136.

Symons, D. 1979. *The evolution of human sexuality.* New York: Oxford University Press.

Thornhill, R. and Thornhill, N. W. (1983). "Human rape: An evolutionary analysis." *Ethology and Sociobiology, 4*, 137–173.

Warren, J. I., Hazelwood, R. R., and Reboussin, R. (1991). "Serial rape: The offender and his rape career." In A. Burgess (Ed.), *Rape and sexual assault* (pp. 275–311). New York: Garland.

Weiner, N. A., Zahn, M. A., and Sagi, R. J. (1990). *Violence: Patterns, causes, and public policy.* New York: Harcourt Brace Jovanovich.

7

Stick-up, Street Culture, and Offender Motivation

Bruce A. Jacobs

Richard Wright

Jacobs and Wright conducted 86 interviews with active armed robbers. They looked at the foreground conditions, that is, what robbers were thinking before they committed their crime, rather than background characteristics, which are usually emphasized in the etiology of criminality. The authors focused on the motivating factors or decision-making process in the real-life settings of these active offenders. Most actual robbers decide to commit this criminal act without much thoughtful reasoning. Their participation in street culture and crime causes them not to think of options for their illegal behavior to such an extent that robbery almost seems to be deterministic. According to the authors, armed robbery participants are overwhelmed by their emotional, financial, and drug problems and perceive robbery as their only alternative.

In this article we attend to exploring the decision-making processes of active armed robbers in real-life settings and circumstances. Our aim is to understand how and why these offenders move from an unmotivated state to one in which they are determined to commit robbery. We argue that while the decision to commit robbery stems most directly from a perceived need for fast cash, this decision is activated, mediated, and channeled by participation in street culture. Street culture, and its constituent conduct norms, represents an essential intervening variable linking criminal motivation to background risk factors and subjective foreground conditions.

Methods

The study is based on in-depth interviews with a sample of 86 currently active robbers recruited from the streets of St. Louis, Missouri. Respondents ranged in age from 16 to 51. All but 3 were African-American; 14 were female. All respondents had taken part in armed robberies, but many also had committed strong-arm

EDITOR'S NOTE: From Jacobs, B. & Wright, R., "Stick-up, street culture, and offender motivation," in *Criminology, 37,* pp. 149-174. Reprinted with permission from the American Society of Criminology.

attacks. Respondents did not offend at equal rates, but all (1) had committed a robbery within the recent past (typically within the past month), (2) defined themselves as currently active, and (3) were regarded as active by other offenders. Sixty-one of the offenders admitted to having committed 10 or more lifetime robberies. Included in this group were 31 offenders who estimated having done at least 50 robberies. Seventy-three of the offenders said that they typically robbed individuals on the street or in other public settings, 10 reported that they usually targeted commercial establishments, and 3 claimed that they committed street and commercial robberies in roughly equal proportions.

MONEY, MOTIVATION, AND STREET CULTURE

Fast Cash

With few exceptions, the decision to commit a robbery arises in the face of what offenders perceive to be a pressing need for fast cash (see also Conklin, 1972; Feeney, 1986; Gabor et al., 1987; Tunnell, 1992). Eighty of 81 offenders who spoke directly to the issue of motivation said that they did robberies simply because they needed money. Many lurched from one financial crisis to the next, the frequency with which they committed robbery being governed largely by the amount of money—or lack of it—in their pockets:

> [The idea of committing a robbery] comes into your mind when your pockets are low; it speaks very loudly when you need things and you are not able to get what you need. It's not a want, it's things that you need, . . . things that if you don't have the money, you have the artillery to go and get it. That's the first thing on my mind; concentrate on how I can get some more money.
>
> I don't think there is any one factor that precipitates the commission of a crime, . . . I think it's just the conditions. I think the primary factor is being without. Rent is coming up. A few months ago, the landlord was gonna put us out, rent due, you know. Can't get no money no way else; ask family and friends, you might try a few other ways of getting the money and, as a last

resort, I can go get some money [by committing a robbery].

Many offenders appeared to give little thought to the offense until they found themselves unable to meet current expenses.

> [I commit a robbery] about every few months. There's no set pattern, but I guess it's really based on the need. If there is a period of time where there is no need of money . . . , then it's not necessary to go out and rob. It's not like I do [robberies] for fun.

The above claims conjure up an image of reluctant criminals doing the best they can to survive in circumstances not of their own making. In one sense, this image is not so far off the mark. Of the 59 offenders who specified a particular use for the proceeds of their crimes, 19 claimed that they needed the cash for basic necessities, such as food or shelter. For them, robbery allegedly was a matter of day-to-day survival. At the same time, the notion that these offenders were driven by conditions entirely beyond their control strains credulity. Reports of "opportunistic" robberies confirm this, that is, offenses motivated by serendipity rather than basic human need:

> If I had $5,000, I wouldn't do [a robbery] like tomorrow. But [i]f I got $5,000 today and I seen you walkin' down the street and you look like you got some money in your pocket, I'm gonna take a chance and see. It's just natural. . . . If you see an opportunity, you take that opportunity. . . . It doesn't matter if I have $5,000 in my pocket, if I see you walkin' and no one else around and it look like you done went in the store and bought somethin' and pulled some money out of your pocket and me or one of my partners has peeped this, we gonna approach you. That's just the way it goes.

Need and opportunity, however, cannot be considered outside the open-ended quest for excitement and sensory stimulation that shaped much of the offenders' daily activities. Perhaps the most central of pursuits in street culture, "life as party" revolves around "the enjoyment of 'good times' with minimal concern for obligations and commitments that are external to the . . . immediate social setting" (Shover and Honaker, 1992:283). Gambling, hard drug use, and heavy drinking were the behaviors of choice:

> I [have] a gambling problem and I . . . lose so much so I [have] to do something to [get the cash to] win my money back. So I go out and rob somebody. That be the main reason I rob someone.
>
> I like to mix and I like to get high. You can't get high broke. You really can't get high just standing there, you got to move. And in order to move, you got to have some money . . . Got to have some money, want to get high.

While the offenders often referred to such activities as partying, there is a danger in accepting their comments at face value. Many gambled, used drugs, and drank alcohol as if there were no tomorrow; they pursued these activities with an intensity and grim determination that suggested something far more serious was at stake. Illicit street action is no party, at least not in the conventional sense of the term. Offenders typically demonstrate little or no inclination to exercise personal restraint. Why should they? Instant gratification and hedonistic sensation seeking are quite functional for those seeking pleasure in what may objectively be viewed as a largely pleasureless world.

The offenders are easily seduced by life as party, at least in part because they view their future prospects as bleak and see little point in long-range planning. As such, there is no mileage to be gained by deferred gratification:

> I really don't dwell on [the future]. One day I might not wake up. I don't even think about what's important to me. What's important to me is getting mine [now].

The offenders' general lack of social stability and absence of conventional sources of support only fueled such a mindset. The majority called the streets home for extended periods of time; a significant number of offenders claimed to seldom sleep at the same address for more than a few nights in a row (see also Fleisher, 1995). Moving from place to place as the mood struck them, these offenders essentially were urban nomads in a perpetual search for good

times. The volatile streets and alleyways that criss-crossed St. Louis's crime-ridden central city neighborhoods provided their conduit (see also Stein and McCall, 1994):

> I guess I'm just a street person, a roamer. I like to be out in the street . . . Now I'm staying with a cousin . . . That's where I live, but I'm very rarely there. I'm usually in the street. If somebody say they got something up . . . I go and we do whatever. I might spend the night at their house or I got a couple of girls I know [and] I might spend the night at their house. I'm home about two weeks out of a month.

Keeping Up Appearances

The open-ended pursuit of sensory stimulation was but one way these offenders enacted the imperatives of street culture. No less important was the fetishized consumption of personal, nonessential, status-enhancing items. Shover and Honaker (1992:283) have argued that the unchecked pursuit of such items—like anomic participation in illicit street action—emerges directly from conduct norms of street culture. The code of the streets (Anderson, 1990) calls for the bold display of the latest status symbol clothing and accessories, a look that loudly proclaims the wearer to be someone who has overcome, if only temporarily, the financial difficulties faced by others on the street corner (see e.g., Katz, 1988). To be seen as "with it," one must flaunt the material trappings of success. The quest is both symbolic and real; such purchases serve as self-enclosed and highly efficient referent systems that assert one's essential character (Shover, 1996) in no uncertain terms.

> You ever notice that some people want to be like other people . . . ? They might want to dress like this person, like dope dealers and stuff like that. They go out there [on the street corner] in diamond jewelry and stuff. "Man, I wish I was like him!" You got to make some kind of money [to look like that], so you want to make a quick hustle.

The functionality of offenders' purchases was tangential, perhaps irrelevant. The overriding goal was to project an image of "cool transcendence," (Katz, 1988) that, in the minds of offenders, knighted them members of a mythic street aristocracy. As Anderson (1990:103–104) notes, the search for self-aggrandizement takes on a powerful logic of its own and, in the end, becomes all-consuming. Given the day-to-day desperation that dominates most of these offenders' lives, it is easy to appreciate why they are anxious to show off whenever the opportunity presents itself (particularly after making a lucrative score). Of course, it would be misleading to suggest that our respondents differed markedly from their law-abiding neighbors in wanting to wear flashy clothes or expensive accessory items. Nor were all of the offenders' purchases ostentatious. On occasion, some offenders would use funds for haircuts, manicures, and other mundane purchases. What set these offenders apart from "normal citizens" was their willingness to spend large amounts of cash on luxury items to the detriment of more pressing financial concerns.

Obviously, the relentless pursuit of high living quickly becomes expensive. Offenders seldom had enough cash in their pockets to sustain this lifestyle for long. Even when they did make the occasional "big score," their disdain for long-range planning and desire to live for the moment encouraged spending with reckless abandon. That money earned illegally holds "less intrinsic value" than cash secured through legitimate work only fueled their spendthrift ways (Walters, 1990:147). The way money is obtained, after all, is a "powerful determinant of how it is defined, husbanded, and spent" (Shover, 1996:104). Some researchers have gone so far as to suggest that through carefree spending, persistent criminals seek to establish the very conditions that drive them back to crime (Katz, 1988). Whether offenders spend money in a deliberate attempt to create these conditions is open to question; the respondents in our sample gave no indication of doing so. No matter, offenders were under almost constant pressure to generate funds. To the extent that robbery alleviated this stress, it nurtured a tendency for them to view the offense as a reliable method for dealing with similar pressures in the future. A self-enclosed cycle of reinforcing behavior was thereby triggered (see also Lemert, 1953).

Why Robbery?

The decision to commit robbery, then, is motivated by a perceived need for cash. Why does this need express itself as robbery? Presumably the offenders have other means of obtaining money. Why do they choose robbery over legal work? Why do they decide to commit robbery rather than borrow money from friends or relatives? Most important, why do they select robbery to the exclusion of other income-generating crimes?

Legal Work

That the decision to commit robbery typically emerges in the course of illicit street action suggests that legitimate employment is not a realistic solution. Typically, the offenders' need for cash is so pressing and immediate that legal work, as a viable money-making strategy, is untenable: Payment and effort are separated in space and time and these offenders will not, or cannot, wait. Moreover, the jobs realistically available to them— almost all of whom were unskilled and poorly educated—pay wages that fall far short of the funds required to support a cash-intensive lifestyle:

> Education-wise, I fell late on the education. I just think it's too late for that. They say it's never too late, but I'm too far gone for that . . . I've thought about [getting a job], but I'm too far gone I guess . . . I done seen more money come out of [doing stick-ups] than I see working.

Legitimate employment also was perceived to be overly restrictive. Working a normal job requires one to take orders, conform to a schedule, minimize informal peer interaction, show up sober and alert, and limit one's freedom of movement for a given period of time. For many in our sample, this was unfathomable; it cramped the hedonistic, street-focused lifestyle they chose to live:

> I'm a firm believer, man, God didn't put me down on this earth to suffer for no reason. I'm just a firm believer in that. I believe I can have a good time every day, each and every day of my life, and that's what I'm trying to do. I never held a job.

The longest job I ever had was about nine months . . . at St. Louis Car; that's probably the longest job I ever had, outside of working in the joint. But I mean on the streets, man, I just don't believe in [work]. There is enough shit on this earth right here for everybody, nobody should have to be suffering. You shouldn't have to suffer and work like no dog for it, I'm just a firm believer in that. I'll go out there and try to take what I believe I got comin' [because] ain't nobody gonna walk up . . . and give it to me. [I commit robberies] because I'm broke and need money; it's just what I'm gonna do. I'm not going to work! That's out! I'm through [with work]. I done had 25 or 30 jobs in my little lifetime [and] that's out. I can't do it! I'm not going to!

The "conspicuous display of independence" is a bedrock value on which street-corner culture rests (Shover and Honaker, 1992:284): To be seen as cool one must do as one pleases. This ethos clearly conflicts with the demands of legitimate employment. Indeed, robbery appealed to a number of offenders precisely because it allowed them to flaunt their independence and escape from the rigors of legal work.

This is not to say that every offender summarily dismissed the prospect of gainful employment. Twenty-five of the 75 unemployed respondents claimed they would stop robbing if someone gave them a "good job"—the emphasis being on good:

My desire is to be gainfully employed in the right kind of job . . . If I had a union job making $16 or $17 [an hour], something that I could really take care of my family with, I think that I could become cool with that. Years ago I worked at one of the [local] car factories; I really wanted to be in there. It was the kind of job I'd been looking for. Unfortunately, as soon as I got in there they had a big layoff.

Others alleged that, while a job may not eliminate their offending altogether, it might well slow them down:

[If a job were to stop me from committing robberies], it would have to be a straight up good paying job. I ain't talkin' about no $6 an hour . . . I'm talkin' like $10 to $11 an hour, something like

that. But as far as $5 or $6 an hour, no! I would have to get like $10 or $11 an hour, full-time. Now something like that, I would probably quit doing it [robbery]. I would be working, making money, I don't think I would do it [robbery] no more . . . I don't think I would quit [offending] altogether. It would probably slow down and then eventually I'll stop. I think [my offending] would slow down.

While such claims may or may not be sincere, it is unlikely they will ever be challenged. Attractive employment opportunities are limited for all inner-city residents and particularly for individuals like those in our sample. Drastic changes in the post World War II economy—deindustrialization and the loss of manufacturing jobs, the increased demand for advanced education and high skills, rapid suburbanization and out-migration of middle class residents (Sampson et al., 1997)—have left them behind, twisting in the wind. The lack of legal income options speaks to larger societal patterns in which major changes in the U.S. economy have reduced the number of available good-paying jobs and created an economic underclass with unprecedented levels of unemployment and few options—beyond income-generating crime—to exercise (Wilson, 1987). Governmental directives, such as changes in requirements and reductions in public transfer payments, decidedly reduce the income of already marginalized persons in inner-city communities (Johnson and Dunlap, 1997)—those at highest risk for predatory crime. This only intensifies their economic and social isolation (Sampson et al., 1997), makes their overall plight worse, and their predisposition to criminality stronger.

Most offenders realized this and, with varying degrees of bitterness, resigned themselves to being out of work:

I fill out [job] applications daily. Somebody [always] says, "This is bad that you got tattoos all over looking for a job." In a way, that's discrimination. How do they know I can't do the job? I could probably do your job just as well as you, but I got [these jailhouse] tattoos on me. That's discriminating. Am I right? That's why most people rob and steal because, say another black male came in like me [for a job], same haircut, same everything. I'm dressed like this, tennis shoes, shorts and tank top. He has on [a] Stacy Adams pair of slacks and a button-up shirt with a tie. He will get the job before I will. That's being racist in a way. I can do the job just as well as he can. He just dresses a little bit better than me.

Clearly, these offenders were not poster children for the local chamber of commerce or small business association. By and large, they were crudely mannered and poorly schooled in the arts of impression management and customer relations. Most lacked the cultural capital (Bourdieu, 1977) necessary for the conduct of legitimate business. They were not "nice" in the conventional sense of the term; to be nice is to signal weakness in a world where only the strong survive.

Even if the offenders were able to land a high-paying job, it is doubtful they would keep it for long. The relentless pursuit of street action—especially hard drug use—has a powerful tendency to undermine any commitment to conventional activities (Shover and Honaker, 1992). Life as party ensnares street-culture participants, enticing them to neglect the demands of legitimate employment in favor of enjoying the moment. Though functional in lightening the burdensome present, gambling, drinking, and drugging—for those on the street—become the proverbial "padlock on the exit door" (Davis, 1995) and fertilize the foreground in which the decision to rob becomes rooted.

Borrowing

In theory, the offenders could have borrowed cash from a friend or relative rather than resorting to crime. In practice, this was not feasible. Unemployed, unskilled, and uneducated persons caught in the throes of chronically self-defeating behavior cannot, and often do not, expect to solve their fiscal troubles by borrowing. Borrowing is a short-term solution, and loans granted must be repaid. This in itself could trigger robberies. As one offender explained, "I have people that will loan me money, [but] they will loan me money because of the work [robbery] that I do; they know they gonna get their money [back] one way or another." Asking for money also was perceived

by a number of offenders to be emasculating. Given their belief that men should be self-sufficient, the mere prospect of borrowing was repugnant:

> I don't like always asking my girl for nothing because I want to let her keep her own money . . . I'm gonna go out here and get some money.

The possibility of borrowing may be moot for the vast majority of offenders anyway. Most had long ago exhausted the patience and goodwill of helpful others; not even their closest friends or family members were willing to proffer additional cash:

> I can't borrow the money. Who gonna loan me some money? Ain't nobody gonna loan me no money. Shit, [I use] drugs and they know [that] and I rob and everything else. Ain't nobody gonna loan me no money. If they give you some money, they just give it to you; they know you ain't giving it back.

When confronted with an immediate need for money, then, the offenders perceived themselves as having little hope of securing cash quickly and legally. But this does not explain why the respondents decided to do robbery rather than some other crime. Most of them had committed a wide range of income-generating offenses in the past, and some continued to be quite versatile. Why, then, robbery?

For many, this question was irrelevant; robbery was their "main line" and alternative crimes were not considered when a pressing need for cash arose:

> I have never been able to steal, even when I was little and they would tell me just to be the watch-out man . . . Shit, I watch out, everybody gets busted. I can't steal, but give me a pistol and I'll go get some money. . . . [Robbery is] just something I just got attached to.

When these offenders did commit another form of income-generating crime, it typically was prompted by the chance discovery of an especially vulnerable target rather than being part of their typical modus operandi:

I do [commit other sorts of offenses] but that ain't, I might do a burglary, but I'm jumping out of my field. See, I'm scared when I do a burglary [or] something like that. I feel comfortable robbing . . . , but I see something they call "real sweet," like a burglary where the door is open and ain't nobody there or something like that, well . . .

Many of the offenders who expressed a strong preference for robbery had come to the offense through burglary, drug selling, or both. They claimed that robbery had several advantages over these other crimes. Robbery took much less time than breaking into buildings or dealing drugs. Not only could the offense be committed more quickly, it also typically netted cash rather than goods. Unlike burglary, there was no need for the booty "to be cut, melted down, recast or sold," nor for obligatory dealings with "treacherous middlemen, insurance adjustors, and wiseguy fences" (Pileggi, 1985:203). Why not bypass all such hassles and simply steal cash (Shover, 1996:63).

> Robbery is the quickest money. Robbery is the most money you gonna get fast . . . Burglary, you gonna have to sell the merchandise and get the money. Drugs, you gonna have to deal with too many people, [a] bunch of people. You gonna sell a $50 or $100 bag to him, a $50 or $100 bag to him, it takes too long. But if you find where the cash money is and just go take it, you get it all in one wad. No problem. I've tried burglary, I've tried drug selling . . . the money is too slow.

Some of the offenders who favored robbery over other crimes maintained that it was safer than burglary or dope dealing:

> I feel more safer doing a robbery because doing a burglary, I got a fear of breaking into somebody's house not knowing who might be up in there. I got that fear about house burglary . . . On robbery I can select my victims, I can select my place of business. I can watch and see who all work in there or I can rob a person and pull them around in the alley or push them up in a doorway and rob them. You don't got [that] fear of who . . . in that bedroom or somewhere in another part of the house.
> [I]f I'm out there selling dope somebody gonna come and, I'm not the only one out there robbing

you know, so somebody like me, they'll come and rob me . . . I'm robbin' cause the dope dealers is the ones getting robbed and killed you know.

A couple of offenders reported steering clear of dope selling because their strong craving for drugs made it too difficult for them to resist their own merchandise. Being one's own best customer is a sure formula for disaster (Waldorf, 1993), something the following respondent seemed to understand well:

> A dope fiend can't be selling dope because he be his best customer. I couldn't sell dope [nowadays]. I could sell a little weed or something cause I don't smoke too much of it. But selling rock [cocaine] or heroin, I couldn't do that cause I mess around and smoke it myself. [I would] smoke it all up!

Others claimed that robbery was more attractive than other offenses because it presented less of a potential threat to their freedom:

> If you sell drugs, it's easy to get locked up selling drugs; plus, you can get killed selling drugs. You get killed more faster doing that.
> Robbery you got a better chance of surviving and getting away than doing other crimes . . . You go break in a house, [the police] get the fingerprints, you might lose a shoe, you know how they got all that technology stuff. So I don't break in houses . . . I leave that to some other guy.

Without doubt, some of the offenders were prepared to commit crimes other than robbery; in dire straits one cannot afford to be choosy. More often than not, robbery emerged as the "most proximate and performable" (Lofland, 1969:61) offense available. The universe of money-making crimes from which these offenders realistically could pick was limited. By and large, they did not hold jobs that would allow them to violate even a low-level position of financial trust. Nor did they possess the technical know-how to commit lucrative commercial break-ins, or the interpersonal skills needed to perpetrate successful frauds. Even street-corner dope dealing was unavailable to many; most lacked the financial wherewithal to purchase baseline inventories—inventories many offenders would undoubtedly have smoked up.

The bottom line is that the offenders, when faced with a pressing need for cash, tend to resort to robbery because they know of no other course of action, legal or illegal, that offers as quick and easy a way out of their financial difficulties. As Lofland (1969:50) notes, most people under pressure have a tendency to become fixated on removing the perceived cause of that pressure "as quickly as possible." Desperate to sustain a cash-intensive lifestyle, these offenders were loathe to consider unfamiliar, complicated, or long-term solutions (Lofland, 1969:50–54). With minimal calculation and "high" hopes, they turned to robbery, a trusted companion they could count on when the pressure was on. For those who can stomach the potential violence, robbery seems so much more attractive than other forms of income-generating crime. Contemplating alternative offenses becomes increasingly difficult to do. This is the insight that separates persistent robbers from their street-corner peers:

> [Robbery] is just easy. I ain't got to sell no dope or nothing, I can just take the money. Just take it, I don't need to sell no dope or work . . . I don't want to sell dope, I don't want to work. I don't feel like I need to work for nothing. If I want something, I'm gonna get it and take it. I'm gonna take what I want . . . If I don't have money, I like to go and get it. I ain't got time [for other offenses]; the way I get mine is by the gun. I don't have time to be waiting on people to come up to me buying dope all day . . . I don't have time for that so I just go and get my money.

DISCUSSION

The overall picture that emerges from our research is that of offenders caught up in a cycle of expensive, self-indulgent habits (e.g., gambling, drug use, and heavy drinking) that feed on themselves and constantly call for more of the same (Lemert, 1953). It would be a mistake to conclude that these offenders are being driven to crime by genuine financial hardship; few of them are doing robberies to buy the proverbial loaf of bread to feed their children. Yet, most of their crimes are economically motivated. The offenders perceive themselves as needing money and robbery is a response to that perception.

Being a street robber is a way of behaving, a way of thinking, an approach to life (see e.g., Fleisher, 1995:253). Stopping such criminals exogenously—in the absence of lengthy incapacitation—is not likely to be successful. Getting offenders to "go straight" is analogous to telling a lawful citizen to "relinquish his history, companions, thoughts, feelings, and fears, and replace them with [something] else" (Fleisher, 1995:240). Self-directed going-straight talk on the part of offenders more often than not is insincere—akin to young children talking about what they're going to be when they grow up: "Young storytellers and . . . criminals . . . don't care about the [reality]; the pleasure comes in saying the words, the verbal ritual itself brings pleasure" (Fleisher, 1995:259). Gifting offenders money, in the hopes they will reduce or stop their offending (Farrington, 1993), is similarly misguided. It is but twisted enabling and only likely to set off another round of illicit action that plunges offenders deeper into the abyss of desperation that drives them back to their next crime.

REFERENCES

Agar, Michael
　1973　Ripping and Running: A Formal Ethnography of Urban Heroin Addicts. New York: Seminar Press.

Akers, Ronald L.
　1985　Deviant Behavior: A Social Learning Approach. 3d ed. Belmont, Calif.: Wadsworth.

Anderson, Elijah
　1990　Streetwise. Chicago: University of Chicago Press.

Ball, John C.
　1967　The reliability and validity of interview data obtained from 59 narcotic drug addicts. American Journal of Sociology 72:650–654.

Baron, Stephen and Timothy Hartnagel
　1997　Attributions, affect, and crime: Street youths' reactions to unemployment. Criminology 35:409–434.

Becker, Howard S.
  1963   Outsiders. New York: Free Press.

Bennett, Trevor and Richard Wright
  1984   Burglars on Burglary: Prevention and
          the Offender. Aldershot: Gower.

Berk, Richard A. and Joseph M. Adams
  1970   Establishing rapport with deviant
          groups. Social Problems 18:102–117.

Bottoms, Anthony and Paul Wiles
  1992   Explanations of crime and place. In
          David Evans, Nigel Fyfe, and Derek
          Herbert (eds.), Policing and Place:
          Essays in Environmental Criminology.
          London: Routledge.

Bourdieu, Pierre
  1977   Outline of a Theory of Practice.
          Cambridge: Cambridge University
          Press.

Braithwaite, John
  1989   Crime, Shame, and Reintegration.
          Cambridge: Cambridge University
          Press.

Briar, Scott and Irving Piliavin
  1965   Delinquency, situational inducements,
          and commitment to conformity. Social
          Problems 13:35–45.

Chaiken, Jan M. and Marcia R. Chaiken
  1982   Varieties of Criminal Behavior. Santa
          Monica, Calif.: Rand Corporation.

Chambliss, William J. and Robert Seidman
  1971   Law, Order, and Power. Reading,
          Mass.: Addison-Wesley.

Cloward, Richard A. and Lloyd E. Ohlin
  1960   Delinquency and Opportunity. New
          York: Free Press.

Cohen, Albert K.
  1955   Delinquent Boys. New York: Free
          Press.

Colvin, Mark and John Pauly
  1983   A critique of criminology: Toward an
          integrated structural-Marxist theory of
          delinquent production. American
          Journal of Sociology 89:513–551.

Conklin, John
  1972   Robbery. Philadelphia: JB Lippincott.

Cornish, Derek B. and Ronald V. Clarke (eds.)
  1986   The Reasoning Criminal: Rational
          Choice Perspectives on Offending.
          New York: Springer-Verlag.

Davis, Peter
  1995   Interview, "If You Came This Way."
          All Things Considered, National
          Public Radio, October 12.

Ekland-Olson, Sheldon, John Lieb, and Louis
  Zurcher
  1984   The paradoxical impact of criminal
          sanctions: Some microstructural find-
          ings. Law & Society Review 18:
          159–178.

Elliott, Delbert S., David Huizinga, and Suzanne
  S. Ageton
  1985   Explaining Delinquency and Drug
          Use. Beverly Hills, Calif.: Sage.

Empey, LaMar T. and Mark C. Stafford
  1991   American Delinquency. 3d ed.
          Belmont, Calif.: Wadsworth.

Farrington, David P.
  1993   Motivations for conduct disorder and
          deliquency. Development and Psy-
          chopathology 5:225–241.

Feeney, Floyd
  1986   Robbers as decision-makers. In Derek
          B. Cornish and Ronald V. Clarke
          (eds.), The Reasoning Criminal:
          Rational Choice Perspectives on
          Offending. New York: Springer-
          Verlag.

Felson, Marcus
  1987   Routine activities in the developing
          metropolis. Criminology 25:911–931

Fleisher, Mark S.
  1995   Beggars and Thieves: Lives of Urban
          Street Criminals. Madison: University
          of Wisconsin Press.

Gabor, Thomas, Micheline Baril, Maurice Cusson,
  Daniel Elie, Marc LeBlanc, and Andre
  Normandeau
  1987   Armed Robbery: Cops, Robbers, and
          Victims. Springfield, Ill.: Charles C
          Thomas.

Johnson, Bruce D. and Eloise Dunlap
  1997   Crack Selling in New York City.
          Paper presented at the 49th Annual
          Meeting of the American Society of
          Criminology, San Diego.

Katz, Jack
  1988   Seductions of Crime: Moral and
          Sensual Attractions in Doing Evil.
          New York: Basic Books.

Lemert, Edwin
  1953   An isolation and closure theory of
          naive check forgery. Journal of
          Criminal Law, Criminology, and
          Police Science 44:296–307.

Lofland, John
  1969   Deviance and Identity. Englewood
          Cliffs, N.J.: Prentice-Hall.

Pileggi, Nicholas
  1985   Wiseguy. New York: Simon &
          Schuster.

Sampson, Robert J., Stephen W. Raudenbush, and
  Felton Earls
  1997   Neighborhoods and violent crime: A
          multilevel study of collective efficacy.
          Science 277:918–924.

Shover, Neal
  1996   Great Pretenders: Pursuits and Careers
          of Persistent Thieves. Boulder, Colo.:
          Westview.

Shover, Neal and David Honaker
  1992   The socially-bounded decision mak-
          ing of persistent property offenders.
          Howard Journal of Criminal Justice
          31:276–293.

Stein, Michael and George McCall
  1994   Home ranges and daily rounds:
          Uncovering community among urban
          nomads. Research in Community
          Sociology 1:77–94.

Tunnell, Kenneth D.
  1992   Choosing Crime: The Criminal
          Calculus of Property Offenders.
          Chicago: Nelson-Hall.

Waldorf, Dan
  1993   Don't be your own best customer:
          Drug use of San Francisco gang drug
          sellers. Crime, Law, and Social
          Change 19:1–15.

Walters, Glenn
  1990   The Criminal Lifestyle. Newbury
          Park, Calif.: Sage.

Wilson, William J.
  1987   The Truly Disadvantaged. Chicago:
          University of Chicago Press.

8

Dubs and Dees, Beats and Rims: Carjackers and Urban Violence

Volkan Topalli

Richard Wright

In order to study carjacking as a violent crime, Topalli and Wright conducted a field study with 28 active carjackers. They focused their research on factors (opportunities, risks, rewards) that those offenders think about when contemplating carrying out their crimes. Similar to the study on armed robbers, the authors were interested in carjackers' motivations, planning, execution, and postcrime activities. The authors analyzed the links between carjackers' lifestyles and the contextual situation in which offending decisions were made. They conclude that the decision to carjack is based on two important factors: perceived situational inducements and opportunity. Either one of these variables is sufficient to cause an offender to carjack. The degree of involvement in street crime culture also plays an important role in determining if an individual will commit a carjacking by lowering one's resistance to temptation.

With the exception of homicide, probably no offense is more symbolic of contemporary urban violence than carjacking. Carjacking, the taking of a motor vehicle by force or threat of force, has attained almost mythical status in the annals of urban violence and has played an undeniable role in fueling the fear of crime that keeps urban residents off of their own streets. What is more, carjacking has increased dramatically in recent years. According to a recent study (BJS 1999), an average of 49,000 carjackings were attempted each year between 1992–96, with about half of those attempts being successful. This is up from an average of 35,000 attempted and completed carjackings between 1987–92—a 40 percent increase.

Although carjacking has been practiced for decades, the offense first made national headlines in 1992 when a badly botched carjacking in suburban Washington DC ended in homicide. Pamela Basu was dropping her 22-month-old daughter at pre-school when two men commandeered her BMW at a stop sign. In full view of neighborhood residents, municipal workers, and a school bus driver, the two men tossed her daughter (still strapped to her car seat) from the vehicle and attempted to drive off with Basu's arm tangled in the car seat belt. She was dragged over a mile to her death. This incident focused a nation-wide spotlight on carjacking and legislative action soon followed with the passing of the Anti Car Theft Act of 1992. Carjacking was made a federal crime punishable by up to a 25 year term in prison or—if the victim is killed—by death.

Like other forms of robbery, carjacking bridges property and violent crimes. Although a manifestly violent activity, it appears often to retain elements of planning and calculation typically associated with instrumental property crimes such as burglary. Unlike most robberies, however, carjacking apparently is directed at an object rather than a subject.

Most of the research on carjackings is based on official police reports or large pre-existing data sets such as the National Crime Victimization Survey. From this research, we know that carjackings are highly concentrated in space and time, occurring in limited areas and at particular hours (Friday and Wellford 1994). These studies also indicate that carjackers tend to target individuals comparable to themselves across demographic characteristics such as race, gender, and age (Friday and Wellford 1994; Armstrong 1994). We know that weapons are used in 66–78 percent of carjackings, and that weapon usage increases the chance that an offense will be successful (BJS 1999; Donahue, McLaughlin, and Damm 1994; Fisher 1995; Rand 1994). Finally, these studies suggest that carjacking is often a violent offense; approximately 24–38 percent of victims are injured during carjackings (BJS 1999; Fisher 1995; Rand 1994).

Despite these studies, much about carjacking remains poorly understood. By their very large-scale nature, such studies are incapable of providing insight into the interaction between motivational and situational characteristics that govern carjacking at the individual level. What

is more, they overrepresent incidents in which the offenders and victims are strangers. Recent literature on the nature of acquaintance robbery (e.g., Felson, Baumer, and Messner 2000) and drug robbery (see Jacobs, Topalli, and Wright 2000, and Topalli, Wright, and Fornango 2002) suggests that this limitation may represent a crucial gap in our understanding of the social and perceptual dynamics associated with carjacking. If, for example, offenders target victims who they know or "know of," the chance of serious injury or death may increase because within-offense resistance and post-offense retaliation both are more likely.

We conducted a field-based study of active carjackers, focusing on the situational and interactional factors (opportunities, risks, rewards) that carjackers take into account when contemplating and carrying out their crimes. Drawing on a tried and tested research strategy (Jacobs, Topalli, and Wright 2000; Jacobs 1999; Wright and Decker 1997, 1994), we recruited 28 active offenders (with three asked back to participate in follow-ups) from the streets of St. Louis, Missouri and interviewed them at length about their day-to-day activities, focusing on the motivations, planning, execution, and aftermath of carjackings. This methodological strategy allowed us to examine the perceptual links between offenders' lifestyles and the immediate situational context in which decisions to offend emerge, illuminating the contextual cues that mediate the carjacking decision. Interviews focused on two broad issues: (1) motivation to carjack and vehicle/victim target selection, and (2) aftermath of carjacking offenses (including vehicle disposal, formal and informal sanction risk management, use of cash, etc.). The issue of how carjacking occurs (i.e., offense enactment) is covered across the discussion of the these two broader themes, because enactment represents a behavioral bridge that unites them. Thus, the procedural characteristics of carjacking naturally emanated from discourse regarding motivation, target selection, and aftermath.

MOTIVATION AND TARGET SELECTION

In the area of motivation, our interviews focused on the situational and interactional factors that underlie the decision to commit a carjacking, and the transition from unmotivated states to those in which offenses are being contemplated. On its face carjacking seems risky. Why risk a personal confrontation with the vehicle owner when one could steal a parked car off the street? Respondents felt that car theft was more dangerous because they never knew if the vehicle's owner or law enforcement might surprise them.

Low-Down: I done did that a couple of times too, but that ain't nothing I really want to do 'cause I might get in a car [parked on] the street and the motherfucker [the owner of the vehicle] might be sitting there and then it [might not] be running [any] ways. I done got caught like that before, got locked up, so I don't do that no more. I can't risk no motherfucking life just to get in to a car and then the car don't start. That's a waste of time. I would rather catch somebody at a light [or] a restaurant drive-thru or something like that.

Throughout the interviews, two global factors emerged as governing motivation, planning, and target selection: the nature of a given carjacking *opportunity* (that is, its potential risks and rewards) and the level of *situational inducements* (such as peer pressure, need for cash or drugs, or revenge). When these factors, in some combination, reached a critical minimal level, the decision to carjack became certain.

INTERNAL AND EXTERNAL PRESSURES: SITUATIONAL INDUCEMENTS AND CARJACKING

Many of the offenders we spoke to indicated that their carjackings were guided by the power of immediate situational inducements. Such inducements could be internal (e.g., money, drugs, the avoidance of drug withdrawal, need to display a certain status level, desire for revenge, jealousy) or external (objective or subjective strains, such as pressure from family members to put food on the table, the need to have a vehicle for use in a subsequent crime). Situational inducements could be intensely compelling, pressing offenders to engage in

carjacking even under unfavorable circumstances, where the risk of arrest, injury, or death was high or the potential reward was low. Here, the individuals' increased desperation caused them to target a vehicle or victim they would not otherwise consider (such as a substandard car, or one occupied by several passengers), initiate an offense at a time or location that was inherently more hazardous (e.g., day-time, at a busy intersection), or attempt a carjacking with no planning whatsoever.

Internal situational inducements usually were linked to the immediate need for cash. Most street offenders (including carjackers) are notoriously poor planners. They lead cash-intensive lifestyles in which money is spent as quickly as it is obtained (due to routine drug use, street gambling, acquisition of the latest fashions, heavy partying; see e.g., Jacobs 2000; Wright and Decker 1994, 1997; Shover 1996). As a result, they rapidly run out of money, creating pressing fiscal crises, which then produce other internal situational inducements such as the need to feed oneself or to avoid drug withdrawal.

The sale of stolen vehicles and parts can be a lucrative endeavor. Experienced carjackers sometimes stripped the vehicles themselves (in an abandoned alley or remote lot) and sold the items on the street corner or delivered them to a chop shop owner with whom they had a working relationship. Of particular value were "portable" after-market items, such as gold or silver plated rims, hub caps, and expensive stereo components. Across our 28 respondents, profits from carjacking per offense ranged anywhere from $200 to $5,000, with the average running at $1,750. The cash obtained from carjacking served to alleviate ever emergent financial needs.

Little Rag, a diminutive teen-aged gangbanger, indicated that without cash the prospect of heroin withdrawal loomed ahead.

INT: So, why did you do that? Why did you jack that car?
Little Rag: For real? 'Cause it's the high, it's the way I live. I was broke. I was fiending [needed drugs]. I had to get off my scene real quick [wanted to get back on my feet]. I sold crack but I'd fallen off [ran out of money] and I had to go and get another lick [tempting crime target] or

something to get back on the top. I blow it on weed, clothes, shoes, shit like that. Yeah, I truly fuck money up.

The need for drugs was a frequent topic in our discussions with carjackers. Even the youngest offenders had built up such tolerances to drugs like heroin and crack that they required fixes on a daily basis. Many were involved in drug dealing and had fallen into a well known trap; using their own supply. Whether they sold for themselves or in the service of someone else, the need for cash to replenish the supply or feed the habit was a powerful internal motivator. L-Dawg, a young drug dealer from the north side of St. Louis also had developed a strong addiction to heroin. Only two days before his interview with us, he had taken a car from a man leaving a local night club.

L-Dawg: I didn't have no money and I was sick and due some heroin so I knew I had to do something. I was at my auntie's house [and] my stomach started cramping. I just had to kill this sickness, 'cause I can't stay sick. If I'd stayed sick I would [have to] do something worse. The worse I get sick, to me, the worse I'm going to do. That's how I feel. If I've got to wait on it a long time, the worse the crime may be. If it hadn't been him then I probably would have done a robbery. One way or another I was going to get me some money to take me off this sickness. I just seen him and I got it.

External situational inducements could be just as compelling. Pressures from friends, family, other criminal acquaintances, or even the threat of injury or death were capable of pushing offenders to carjacking. For example, C-Low described an incident that occurred while he was with a friend in New Orleans. The two were waiting in the reception room of a neighborhood dentist when a group of men hostile to C-Low's friend walked into the office.

C-Low: They knew him. I didn't know them. It was something about some fake dope. I think it was some heroin. He got caught. We weren't strapped [armed] at the time. We booked out. We left. We just left 'cause he know this person's gonna be strapped, and I didn't know this. So my partner was like, "Man, just burn out man, just leave." So we was leaving and they was coming up behind us [and started] popping [shooting] at us just like that, popping at my partner, just started shooting at him, so my partner he was wounded. We had no car or nothing so we were running through and the guy was popping at us. So, there was a lady getting out of her car, and he stole it. We had to take her car because we had no ride. She worked at the [dentist's]. She like a nurse or something. It was a nice little brand new car. Brand new, not the kind you sort of sport off in like. She saw I was running. She heard the gun shots. I know she heard them, but she didn't see the guy that was shooting at us though. She had the keys in her hand. She was getting out her car, locking her door, yeah. She had her purse and everything. [My partner] just came on her blind side, just grabbed her, hit her. She just looked like she was shocked, she was in a state of shock. She was really scared. And [we] took her car and we left. We could've got her purse and everything, but we were just trying to get away from the scene 'cause we had no strap and they were all shooting at us. We just burnt on out of there. Got away. But then [later that day] he got caught though . . . somebody snitched on him and they told them [the police] that he had the car. He gave me the car but he got caught for it, they couldn't find the car 'cause I'd taken it to the chop shop. I sold the car for like twenty-seven hundred bucks and about 2 ounces of weed.

Similarly, Nicole, a seasoned car thief and sometimes carjacker, described a harrowing spur-of-the-moment episode. She and a friend had been following a young couple from the drive-in, casually discussing the prospect of robbing them, when her partner suddenly stopped their vehicle, jumped out and initiated a carjacking without warning. Nicole was instantly drawn into abetting her partner in the commission of the offense.

Nicole: My partner just jumped out of the car. He jumped out of the car and right then when I seen him with the gun I [realized] what was happening. I had to move. Once he got the guy out of the car he told me, "Come get the car." The girl was already out of the car screaming, "Please don't kill me, please don't kill me!" She was afraid because [she could see] I was high. You do things [when] you high. She's running so I'm in the car waiting on him. He's saying, "Run bitch and don't look back." She just started running . . . across the parking lot. [At] the same time he made the guy get up and run, "Nigger you do something, you look back, I'm gonna kill your motherfucking ass." As he got up and ran he shot him any ways.

RISKS AND REWARDS: HOW OPPORTUNITY DRIVES CARJACKING

Need was not the only factor implicated in carjacking. Some carjackers indicated that they were influenced by the appeal of targets that represented effortless or unique opportunities (e.g., isolated or weak victims, vehicles with exceptionally desirable options). Here, risks were so low or potential rewards so great that, even in the absence of substantive internal or external situational inducements, they decided to commit a carjacking. Such opportunities were simply too good to pass up.

Po-Po (short for "Piss Off the POlice") described just such an opportunity-driven incident. She and her brother had spent the day successfully pickpocketing individuals at Union Station, a St. Louis mall complex. On their way out, she noticed an easy target, an isolated woman in the parking lot, preoccupied with the lock on her car door.

Po-Po: It was a fancy little car. I don't know too much about names of cars, I just know what I like. A little sporty little car like a Mercedes Benz like car. It was black and it was shiny and it looked good. I just had to joy ride it. She was a white lady. It looked like she worked for [a news station] or somebody. We just already pick pocket[ed] people down at Union Station, but fuck. So we just walked down stairs and [I] said, "You want to steal a car? Come on dude, let me get this car." I didn't have a gun on me. I just made her think I had a gun. I had a stick and I just ran up there to her and told her, "Don't move, don't breathe, don't do nothing. Give me the keys and ease your ass away from the car." I said, "You make one sound I'm going to blow your mother-fucker head off and I'm not playing with you!" I said, "Just go on around the car, just scoot on around the car." Threw the keys to my little

brother and told him go on and open the door. And she stood around there at the building like she waiting on the bus until we zoomed off. We got away real slow and easy.

Likewise, Kow, an older carjacker and sometimes street robber, was on his way to a friend's house to complete a potentially lucrative drug deal when he happened on an easy situation—a man sitting in a parked car, talking to someone on a pay phone at 2 am.

INT: What drew you to this guy? What were you doing? Why did you decide to do this guy?

Kow: Man, it ain't be no, "What you be doing?," [it's] just the thought that cross your mind be like, you need whatever it is you see, so you get it, you just get it. I was going to do something totally different [a drug deal] but along the way something totally different popped up so I just take it as it comes. I was like, "Whew! Get that!" I don't know man, your mind is a hell of thing. On our way to this other thing. It just something that just hit you, you know what I'm saying? Plus, [he looked like] a bitch. I don't know, it's just something, he look like a bitch, just like we could whip him, like a bitch, you know what I'm saying? Easy.

Not all irresistible opportunities were driven solely by the prospect of monetary gain. In a city the size of St. Louis, offenders run into one another all the time, at restaurants, malls, movie theaters and night-clubs. As a result, individuals with shared histories often encounter unique chances for retaliation or personal satisfaction. Goldie emphasized how such opportunities could pop up at a moment's notice. While cruising the north side of St. Louis, he spotted an individual who had sexually assaulted one of his girlfriends.

Goldie: I did it on the humbug [spontaneously]. I peeped this dude, [saw that] he [was] pulling up at the liquor store. I'm tripping [excited] you know what I'm saying, [as I'm] walking there [towards the target]. You know, peep him out, you dig? He [was] reaching in the door to open the door. His handle outside must have been broke cause he had to reach in [the window] to open the door. And I just came around you know what I'm saying. [I] put it [the gun] to his head, "You want to give me them keys, brother?" He's like, "No, I'm not given' you these keys." I'm like, "You gonna give me them keys, brother. It's as simple as that!" Man, he's like, "Take these, motherfucker, fuck you and this car. Fuck you." I'm like, "Man, just go on and get your ass home." [Then I] kicked him in his ass, you know what I'm saying, and I was like, "Fuck that, as a matter of fact get on your knees. Get on your knees, motherfucker . . ." Then I seen this old lady right, that I know from around this neighborhood. I was like "Fuck!," jumped on in the car [and] rolled by. I wanted to hit him but she was just standing there, just looking. That's the only thing what made me don't shoot him, know what I'm saying? 'Cause he's fucking with one of my little gals. 'Cause he fucked one of my little gals. Well, she was saying that he didn't really fuck her, you know, he took the pussy, you know what I'm saying? He got killed the next week so I didn't have to worry about him. Motherfuckers said they found him dead in the basement in a vacant house.

ALERT AND MOTIVATED OPPORTUNISM

Offenses motivated purely by either irresistible opportunities or overwhelming situational inducements are relatively rare. Most carjackings occur between these extremes, where situational inducements merge with potential opportunities to create circumstances ripe for offending. What follows are descriptions of offenses spurred by the combination of internal or external situational inducements and acceptable (or near acceptable) levels of risk and reward. The degree to which a given situation was comprised of rewards and risks on the one hand and internal and external pressures on the other varied, but when the combination reached a certain critical level, a carjacking resulted.

Offenders often described situations in which inducements were present, but *not* pressing, where they had *some* money or *some* drugs on them, but realized that the supply of either or both was limited and would soon run out. In such cases, the carjackers engaged in a state of what Bennett and Wright (1984; see also, Feeney 1986) refer to as *alert opportunism*. In other words, offenders are not desperate, but

they anticipate need in the near term and become increasingly open to opportunities that may present themselves during the course of their day-to-day activities. Here, would-be carjackers prowled neighborhoods, monitoring their surroundings for good opportunities, allowing potential victims to present themselves.

Corleone, a sixteen-year-old with over a dozen carjackings under his belt, had been committing such offenses with his cousin since the age of 13. The two were walking the streets of St. Louis one afternoon looking for opportunities for quick cash when they saw a man walking out of a barbershop toward his parked car, keys in hand. Motivated by the obliviousness of their prey and the lightness of their wallets, they decided to take his car.

Corleone: It was down in the city on St. Louis Avenue. We was just walking around, you know. We just look for things to happen you see just to get money. We just walk around and just see something that's gonna make us money. We just happened to be going to the Chinaman [a restaurant] to get something to eat. [We had] about five or six dollars in our pocket which ain't nothing. It was this man driving a blue Cutlass. It had some chrome wheels on there. He just drove up and we was going to the Chinaman and . . . my cousin was like, "Look at that car, man, that's tough [nice]. I'm getting that. I want that." [I was] like, "Straight up, you want to do it?" He was like, "Yeah." He was all G[ood] for it. Then he [the victim] came out the barbershop. It was kind of crowded and we just did what we had to do. There was this little spot where [my cousin] stash[es] his money, drugs and all that type of stuff and then he got the gun [from the stash]. He got around the corner. He say, "Hey, hey." I asked him for a cigarette so he went to the passenger's side [of the vehicle to get one]. I ran on the driver's side with a gun. Put it to his head and told him to get out the car.

INT: Did you know that you were going to do carjacking or . . . ?

Corleone: No. Not necessarily. But since that was what came up, that's what we did.

No matter how alert one is, however, good opportunities do not always present themselves. Over the course of time, situational inducements mount (that is, supplies of money and drugs inevitably dry up), and the option of waiting for ideal opportunities correspondingly diminishes. Such conditions cause offenders to move from a "passive" state of alert opportunism to an "active" state of what could be referred to as *motivated opportunism* (creating opportunities where none previously existed or modifying existing non-optimal opportunities to make them less hazardous or more rewarding). Here, attention and openness to possibilities expands to allow offenders to tolerate more risk. Situations that previously seemed unsuitable start to look better.

Binge, a 45-year-old veteran offender who had engaged in burglary, robbery, and carjacking for over 20 years, discussed his most recent decision to get a car on a wintry January day. He had been carrying a weapon (a 9mm Glock) since that morning, looking to commit a home invasion. After prowling the streets for hours and encountering few reasonable prospects, he happened on an easy opportunity—a man sitting parked in a car, its engine running, at a Metrolink (trolley) station.

Binge: Well, I was out hustling, trying to get me a little money and I was walking around. I was cold. I was frustrated. I couldn't get in [any] house[s] or nothing, so I say [to myself], "Well I'll try and get me a little car, and you know, just jam off the heat and shit that he [the vehicle's owner] got," you know? I was strapped [carrying a firearm] and all that, you know and I was worried about the police catching me, trying to pull my pistol off, and I see this guy. Well, he was at the Metrolink you know, nodding [falling asleep] in his car. So, I went up to the window. I just think that I just peeped it on [happened on the situation]. I was at the Metrolink you know, I was standing at the bus station trying to keep warm and so I just walked around with no houses to rob, and I seen this dude you know sitting in his car, you know, with the car running. And I said "Ah man, if I can get a wag at this [take advantage of this opportunity]," you know. It wasn't just an idea to keep warm or nothing like that. I was cold and worried, and it just crossed my mind and I thought I can get away with it, and I just did it. I'd do anything man, I'd do anything. If I want something and I see I can get away with it I'm gonna do it. That's what

I'm saying. That night I saw an opportunity and I took it, you know. It just occurred to me.

Just as compelling were instances where third parties placed demands on offenders. A number of our respondents indicated that they engaged in carjacking to fulfill specific orders or requests from chop shop owners or other individuals interested in a particular make and model of car or certain valuable car parts. The desire to fulfill such orders quickly created conditions ripe for motivated opportunism.

Goldie, for instance, was experiencing strong internal situational pressures (the need for cash) and external pressures (the demand for a particular vehicle by some of his criminal associates) combined with a moderately favorable opportunity (inside information on the driver of the wanted vehicle and its location):

Goldie: He's from my neighborhood. He's called Mucho. He's from the same neighborhood but like two streets over. Them two streets don't come over on our street. You know, we not allowed to over on they street. It was a nice car. The paint, the sound system in it, and the rims. [It] had some beats, rims. Rims cost about $3,000, some chrome Daytons. 100 spokes platinums.

INT: OK. That's a lot of money to be putting on a car. What does he do for a living?

Goldie: I don't know. I don't ask. What they told me was they wanted this car and they are going to give me a certain amount of money.

INT: You say they told you they wanted this car. You mean they told you they wanted his car or they wanted a car like that?

Goldie: His car. His car. They want [Mucho's] car. They said, "I need one of these, can you get it for me?" And they knew this guy. So now, I need that car. That car.

Low-Down also specialized in taking orders from chop shops:

Low-Down: What I do, I basically have me a customer before I even go do it. I ask a few guys that I know that fix up cars, you know what I'm saying. I ask them what they need then I take the car. But see, I basically really got a customer. I'm talking about this guy over in East St. Louis. Me and Bob, we real cool. He buy 'em cause he break 'em down, the whole car down and he got an autobody shop. He sell parts. He'll take the car and strip it down to the nitty gritty and sell the parts. He get more money out of selling it part by part than selling the car. And, before I get it I already set the price.

He also had a drug habit:

Low-Down: The main reason basically why I did it was I be messing with heroin, you know what I'm saying. I be using buttons [heroin housed in pill-form]. I be snorting some, but I be snorting too much, you know? I got a habit for snorting cause I be snorting too much at a time, that's how I call it a habit. I probably drop about 5 or 6 [buttons] down first [thing in the morning]. [So] I was basically really sick and my daughter needed shoes and shit like that and my girlfriend was pressuring me about getting her some shoes. She had been pressing me about two or three days. Baby food and stuff like that. But the money I had, I had been trying to satisfy my habit with it. Basically I just thought it was a good thing to do. It was a good opportunity.

AFTERMATH

The second portion of our interviews with car-jackers dealt with the aftermath of carjackings. Here, we were concerned with basic questions: What did they do with the vehicles? What did they do with their money? Given the propensity of many carjackers to target other offenders, how did they manage the threat of retaliation? The majority of our respondents immediately disposed of the vehicle, liquidating it for cash. As Corleone put it, "there's a possibility that report[ed] the car stolen and while I'm driving around the police [could] pull me over. I ain't got time to hop out [of the vehicle] and run with no gun. I just want to get the money that I wanted."

Although most of our respondents immediately delivered the vehicle to a chop shop or dismantled it themselves, a fair number of them chose to drive the vehicle around first, showing it off or "flossing" to other neighborhood residents and associates. Despite the possibility that the vehicle's owner or the police might catch up with them, they chose to floss.

INT: What do you like to do after a carjacking?

Binge: Well, what I like to do is just like to, see my friends. They don't give a damn either, I just go pick them up and ride around, smoke a little bit [of] weed, and get some gals, and to par-tying or something like that you know. I know it's taking a chance but, you know like I say, they don't give a damn.

INT: Is that what you did with the car that you took off the guy at the Metrolink?

Binge: Yeah. I was just riding around listening to the music, picked up a couple of friends of mine. We rode around. I told them it was a stolen car. It was a nice little car too. Black with a kind of rag top with the three windows on the side. The front and back ones had a little mirror and another window right in the roof. Oh yeah, oh yeah—[it had] nice sounds. I was chilling man, I was chill-ing, you know? I was driving along with the music playing up loud. Ha ha. You know I wasn't even worried. I was just feeling good. 'Cause I'm not used to driving that much you know 'cause I don't have a car you know. That's why when I do a car-jacking I just play it off to the tee, run all the gas off, keep the sounds up as loud as I can, keep the heat on, you know just abuse the car you know. That's all about carjacking like that.

C-Low described his desire to floss as having to do with the ability to gain status in his neighborhood.

C-Low: Put it this way, you got people you know that's driving around. We just wanna know how it feels. We're young and we ain't doing shit else. So they [people from the neighbourhood] see you driving the car, they gonna say, "Hey, there's C-Low!" and such and such. That makes us feel good 'cause we're riding, and then when we're done riding we wreck the car or give it to some-body else and let him ride. We took the car and drove around the hood, flossing everything. And then we wrecked it on purpose. We ran it into a ditch. I don't know, we were fucked up high, we were high man, just wild! Wrecked the thing.

But even for offenders like these, the prospect of getting caught and losing profits eventually began to outweigh the benefits of showing off.

INT: So how long did you drive around in the [Chevrolet] Suburban before you stripped it?

Loco: Oh, we was rolling that. We drove for a good thirty minutes, then I said that I want[ed] to get up out of it because they might report it stolen. We was [still driving] right there [near] the scene [of the crime] and they [the police] would have probably tried to flag me [pull me over]. And if they tried to flag me, I would [have to] have taken them through a high-speed chase. Fuck that.

Sleezee-E informed us (as did other respondents) that disposing of the vehicle quickly was the key to getting away with a carjacking. Indeed, almost all respondents were aware of the police department's "hot sheet" for stolen vehicles (although their estimations of how long it took for a vehicle to show up on the hot-sheet varied greatly, from as soon as the vehicle was reported stolen to 24 hours or even longer afterwards).

Sleezee-E: [People think that] the cops will wait 24 hours just to see what you are going to do with the car. Because some idiots, when they jack a car, they just drive it around and then they leave it someplace. I don't do that. That's how you get caught. Driving it around. You take that car right to the chop shop and let them cut that sucker up.

Once the vehicle was stripped, most carjackers disposed of the vehicle by destroying it somehow.

Littlerag: 'Cause it was hot man! It was too hot. All I [wanted] was to take the rims, take the beats, the equalizer, the detachable face. Got all that off, then I just pour gas on it and burnt that motherfucker up. I had fingerprints [on it]. I didn't have no gloves on. I had my own hands on the steering wheel. I left my fingerprints.

Nicole and a boyfriend chose a less conventional method of getting rid of their stolen vehicle.

Nicole: We got rid of the car first. We drove the car two blocks and went back down a ways to the park. We drove the car up there, we parked right there and sat for about ten minutes, made sure how many cars come down this street before we can push it over there. It's a pond, like it's a lake out there with ducks and geeses in it.

CASH FOR CARS: LIFE AS PARTY

While a few respondents reported that they used the proceeds from carjacking to pay for necessities or bills, the overwhelming majority indicated that they blew their cash indiscriminately on drugs, women, and gambling.

We had interviewed Tone on a number of previous occasions for his involvement in strong-arm drug robbery. Although robbery was his preferred crime, he engaged in carjacking occasionally (about once every two months) when easy opportunities presented themselves. During his most recent offense, he and three of his associates took a Cadillac from a neighborhood drug dealer and made six thousand dollars. When we asked what he did with his portion, he indicated that he, " . . . spent that shit in like, two days."

INT: You can go through fifteen hundred dollars in two days?

Tone: Shit, it probably wasn't even two days, it probably was a day, shit.

INT: What did you spend fifteen hundred on?

Tone: It ain't shit that you really want. Just got the money to blow so fuck it, blow it. Whatever, it don't even matter. Whatever you see you get, fuck it. Spend that shit. It wasn't yours from the getty-up, you know what I'm saying? You didn't have it from the jump so. . . . Can't act like you careful with it, it wasn't yours to care for. Easy come, easy go. The easy it came, it go even easier. Fuck that, fuck all that. I ain't trying to think about keeping nothing. You can get it again.

INT: So what does money mean to you?

Tone: What money mean? Shit, money just some shit everybody need that's all. I mean, it ain't jack shit.

INT: Ok, so it's not really important to you?

Tone: Fuck no. Cause I told you, easy come, easy go.

Mo had taken a Monte Carlo from two men residing in another neighborhood. He had planned the offense over the course of a month and finally, posing as a street window cleaner, carjacked them as they exited a local restaurant. The vehicle's after-market items netted five thousand dollars in cash.

INT: I'm just kind of curious how you spend like five thousand dollars!

Mo: Just get high, get high. I just blow money. Money is not something that is going to achieve for nobody, you know what I'm saying? So everyday, there's not a promise that there'll be another [day] so I just spend it, you know what I'm saying? It ain't mine, you know what I'm saying, I just got it, it's just in my possession. This is mine now, so I'm gonna do what I've got to do. It's a lot of fun. At a job you've got to work a lot for it, you know what I'm saying? You got to punch the clock, do what somebody else tells you. I ain't got time for that. Oh yeah, there ain't nothing like gettin' high on five thousand dollars!

Binge and Loco confirmed that the proceeds from their illegal activities went to support this form of conspicuous consumption.

Binge: I just blowed it man. With the money me and my girlfriend went and did a bit of shopping, stuff for Christmas. But, the money I got from his wallet? I just blowed that, drinking and smoking marijuana.

For Corleone, the motivation to carjack was directly related to his desire to manage the impressions of others in his social milieu. His remarks served as a poignant comment on sociocultural and peer pressures experienced by many inner city youths. The purpose of carjacking was to obtain the money he needed to purchase clothing and items that would improve his stature in the neighborhood.

Corleone: [$1500 is] a lot. [I bought] shoes, shoes, everything you need. Guys be styling around our neighborhood. The brand you wear, shoes cost $150 in my size. Air Jordans, everybody want those. Everybody have them. I see everybody wearing those in the neighborhood. I mean come on, let's go get a car. I'm getting those, too.

INT: How many pairs of sneakers have you got?

Corleone: Millions. I got, I got, I got a lot of shoes. Clothes, gotta get jackets.

INT: Well, why do you have to look good, what's so big about looking good?

Corleone: It's for the projects, man. You can't be dressing like no bum. I mean you can't, you can't go ask for no job looking like anything.

INT: So you're saying like if you don't look good, you can't get girls, if you don't have the nicest shoes?

Corleone: You can't. Not nowadays, not where I'm from. You try to walk up to a girl, boy, you got on some raggedy tore up, cut up shoes they're gonna spit on you or something. Look at you like you crazy. Let's say you walking with me. I got on creased up pants, nice shoes, nice shirt and you looking like a bum. Got on old jeans. And that dude, that dude, he clean as a motherfucker and you look like a bum.

INT: So you're competing with each other, too?

Corleone: Something like that. Something like a popularity contest.

INT: Well, you know, you can look nice and clean and not have to spend one hundred and fifty bucks on shoes, you know.

Corleone: It's just this thing, it's a black thing. You ain't going [to] understand, you don't come from the projects.

THE HAZARDS OF CARJACKING: RETALIATION AND THE SPREAD OF VIOLENCE

Interestingly, a sizeable proportion of our respondents purposely targeted people who themselves were involved in crime. Such individuals make excellent targets. Their participation in street culture encourages the acquisition of vehicles most prized by carjackers (those with valuable, if often gaudy, after-market items). And, because they are involved in a number of illegal activities (such as drug selling), they cannot go to the police. As Mr. Dee put it;

Mr. Dee: You can't go to no police when you selling drugs to buy that car with your drug money. So, I wasn't really worried about that. If he would have went to the police he would have went to jail automatically 'cuz they would have been like, "Where'd you get this thousand dollar car from?" He put about $4,000 into the car. So, he ain't got no job, he ain't doin it like that bro. He'd be goin' to the police station lookin' like a fool tellin' his story. I [could] see if he's workin' or something . . . and slinging. It'd be different 'cause he could show them his check stub from work.

However, there is a considerable danger associated with targeting such individuals because, unable to report the robbery of illegal goods to the police, they have a strong incentive to engage in retaliation—those who fail to do so risk being perceived as soft or easy (see e.g., Topalli, Wright, and Fornango 2002). This introduces the possibility that incidents of carjacking likely are substantially higher than officially reported.

When asked about the possibility of retaliation many of the carjackers, displaying typical street offender bravado, indicated that they had no fear. The need to see oneself as capable and tough was essential to respondents. Such self-beliefs served to create a sense of invulnerability that allowed carjackers to continue to engage in a crime considered by many to be hazardous. As Playboy put it, "It [can't be] a fear thing. If you're gonna be scared then you shouldn't even go through with stuff like that [carjacking people]."

Likewise, Big-Mix expressed an almost complete disregard for the consequences of his actions. His comments confirm the short-term thinking characteristic of many street offenders, "I don't give a damn. I don't care what happens really. I don't care. That's how it always is. Whether they kill us or whether we kill them, same damn shit. Whatever. I don't fucking care." Pacman, a younger carjacker who worked exclusively with his brother-in-law, indicated that thinking about the possible negative consequences was detrimental to one's ability to execute an offense. When asked if he was worried about retaliation, he was dismissive.

Pacman: Yeah, you be pretty pissed. But like I say, I'm not looking over my back, you know what I'm saying? Because, I wouldn't be here for sure. I couldn't [keep carjacking]. I definitely wouldn't last man. I wouldn't have lasted as long as I lasted. Because it would be too many motherfuckers [that I've victimized], you know what I'm saying, [for me to look] over my shoulder all the time. When I look what the fuck could I do anyhow? I could get a few of them, but it would take a lot of motherfucking looking over my shoulder. I try to avoid that altogether. I'm going to avoid all that.

Other carjackers relied on hypervigilance (obsessive attention to one's surroundings and to the behavior of others), or anonymity maintenance (e.g., targeting strangers, not talking about the crime, using of disguises, carjacking in areas away from one's home ground; see Jacobs, Topalli, and Wright 2000; Jacobs 2001) to minimize the possibility of pay-back. Sexy-Diva, a female carjacker who worked with Sleezee-E, often spent hours with potential victims at night clubs before taking their cars, "I just disguises myself. I change my hair . . . my clothes. I change whatever location I was at. And then I don't even go to that area no more. They can't find me. No way, no how."

Nukie sacrificed a great deal of his day-to-day freedom by engaging in behaviors designed to anticipate and neutralize the threat of retaliation.

Nukie: That's why I don't go out. If I go somewhere to get me a beer, if I'm gonna get me some bud [marijuana] or something, I stay in the hood. I don't go to the clubs. There's too many people going there at night, you know what I'm saying? I don't need to be spotted like that. That's why I keep on the DL [down-low, out of sight]. You see, I stay in the hood. [If] I be riding [in a car] while I'm riding I might have my cats [friends] with me. You know, no motherfucker's gonna try to fuck with us like that. Yeah, I be with some motherfucker most of the time. If we're [going] to do something, go get blowed [high]—see, we get blowed everyday—I be with people, shit.

Pookie chose to employ similar preemptive tactics, but also emphasized the need to be proactive when dealing with the threat of retaliation, predicated on the philosophy that, "the best defense is a good offense."

Pookie: Well, you know the best thing [to deal] with retaliation like this here, you know, in order for you to get some action you got to bring some action. If I see you coming at me and you don't look right, then this is another story here. If you doing it like you're reaching for something, I'm gonna tear the top of your head off real quick, you know. I'm gonna be near you, where you're at because they ain't nothing but some punk-ass tires and rims that I took from you, that's all it is. What you gotta understand is that you worked hard for it, and I just came along and just took them, you know. You go back and get yourself another set son, 'cause if I like them then I'm gonna take them again.

In the end, there were no guarantees. No matter how many steps a carjacker took to prevent retaliation, the possibility of payback remained. As self-confidence bred the perception of security, so too did it breed over-confidence. This was true in the case of Goldie, whose motivation to carjack a known drug dealer named Mucho was described above. His attempt did not go as planned.

Goldie: He was going to put up a fight trying to spin off with [the car], I jumped in and threw it in park so now I'm tussling with him, "Give me this motherfucker!" He's trying to speed off. He got like in the middle of the intersection. I dropped my gun on the seat and he grabbed me like around here [the neck], trying to hold me down in the car, and throw it back in Drive, with me in the car, you dig? You know, I'm like no, I ain't going for that shit. I had my feet up on the gear [shift], you know what I'm saying? He ain't tripping off the gun. He trying to hold me, "Nigger motherfucker, you ain't going to get this car! Punk-ass nigger! What the fuck wrong with you? What the fuck do you want my car for?" [I said], "Look boy, I don't want that punk ass shit dude! I'm getting this car. This is mine. Fuck you!" The gun flew on the passenger's seat. So I grabbed the gun and put it to his throat, "So what you gonna to do? Is you gonna die or give up this car?" [He replied], "Motherfucker, you're going to have to do what you are going to have to do." He don't want to give up his car, right? So I cocked it one time, you know, just to let him know I wasn't playing, you dig? But I ain't shoot him on his head, put it on his thigh. Boom! Shot him on his leg. He got to screaming and shit hollering, you know what I'm saying, "You shot me! You shot me! You shot me!" like a motherfucker gonna hear him or

something. Cars just steady drive past and shit, you know what I'm saying. By this time I opened up the door, "Fuck you!" Forced his ass on up out of there. He laying on the ground talking about, "This motherfucker shot me! Help, help!" Hollering for help and shit. But before I drove off I backed up, ran over him I think on the ankles like. While he was laying on the street, after I shot him. Ran over his bottom of his feet or whatever you know what I'm saying. Oh, yeah. I felt that. Yeah. Boom, boom. "Aaaah!" scream. I hear bones break, like all this down here was just crushed. I didn't give a fuck though. Sped off. Went and flossed for a minute.

INT: I don't know—two streets over and he sounds like he's pretty scandalous. You're not worried about him coming up on you for this?

Goldie: No. I pretty much left him not walking. And he don't know who I am. [Later on] I heard about that. [people were saying]," Motherfucker Mucho, he got knocked [attacked], motherfucker tried to knock him, took his car, you know what I'm saying, on the block." I'm like, "Yeah, I heard about that. You know what I'm saying. I wonder who did this shit." You know what I'm saying?

Three months later, we spoke to Goldie from his hospital bed. Mucho had tracked him down and shot him in the back and stomach as he crossed the street to buy some marijuana.

Goldie: I call them a bad day . . . I got shot. I saw him [Mucho] drive by but I didn't think he seen me. He caught up to me later. [I got shot] in the abdomen (pointing at his stomach) . . . here's where they sewed me up. I had twenty staples.

INT: How did it go down?

Goldie: [I wanted to] stop on the North[side] and get me a bag of grass, grab me a bag of weed or something. So, [we were] going around to the set [the dealer's home turf] and I'm getting out, I see [Mucho's] car parked this time. He wasn't in it. I'm thinking in my mind like you know, "That's that puss ass." So I'm like, "Damn I'm having bad vibes already." So I instantly just turned around like, "Fuck it. I'll go somewhere else to get some grass." I'm walking [back] to [my] car and hear a gunshot. Jump in the car. You know . . . you [don't] feel it for a minute. [Then,] my side just start hurting, hurting bad you know what I'm saying? I'm like damn. Looked down, I'm in a puddle of blood, you know. She freaking out and screaming, "You shot! You shot!" and shit. [She] jumped out the car like she almost should be done with me, you know what I'm saying? So I had to immediately take myself to the hospital. [They] stuffed this tube all the way down my dick all the way to my stomach . . . fucking with my side, pushing all of it aside. [I was there] about a good week. I done lost about fifteen or twenty pounds. That probably wouldn't have happened if I wouldn't have to go do that. Wanted some more grass. At the wrong spot at the wrong time.

Goldie made it clear during the interview that he felt the need to counter-retaliate to protect himself from future attack by maintaining a tough reputation, a valuable mechanism of deterrence.

> **INT:** You don't feel like you all are even now? You shot him—he shot you. Why go after him?
>
> **Goldie:** It's [about] retaliation. When I feel good is when he taken care of . . . and I don't have to worry about him no more. I mean my little BG's [Baby Gangsters, younger criminal protégées] look up to me. Me getting shot and not going and do [something about, it they would say,] "Ah, [Goldie's] a bitch. Aw, he's a fag." Now down there [in the neighborhood], when they hit you, you hit them back. You know, if someone shoot you, you gotta shoot them back. That's how it is down there or you'll be a bitch. Everybody will shoot you up, whoop your ass. Know what I'm saying? Treat you like a punk. It's just I got to do what I have to do, you know what I'm saying.

Many carjackers echoed such sentiments, indicating a common belief in the importance of following unwritten rules of conduct and behavior related to street offending, especially when they refer to matters of honor or reputation (see Anderson 1999; Katz 1988).

CONCLUSION

This chapter has demonstrated that the decision to commit a carjacking is governed by two things: perceived situational inducements and perceived opportunity (see Hepburn 1984; Lofland 1969). Situational inducements involve immediate pressures on the would-be offender to act. They can be internal (e.g., the need for money or desire for revenge) or external (e.g., the peer pressure of co-offenders). Opportunities refer to risks and rewards tied to a particular crime target in its particular environmental setting.

Carjackings occur when perceived situational inducements and a perceived opportunity, alone or in combination, reach a critical level, thereby triggering that criminogenic moment when an individual commits to the offense. It is important to reiterate that either a perceived opportunity or perceived situational inducement on its own may be sufficient to entice an individual to commit a carjacking (see Hepburn 1984). Numerous examples of this have been detailed throughout the first part of the chapter. It is also important to note that background and foreground factors (such as membership in a criminogenic street culture) can increase the chance that a carjacker will go after a vehicle by lowering his/her capacity to resist the temptation to offend.

More often, carjackings were motivated through the combined influence of opportunity and inducement. The carjackers' responses indicate that offenses triggered by *pure* opportunity or *pure* need are relatively rare. Most carjackings occur between the extremes, where opportunities and situational inducements overlap.

Owing to their precarious day-to-day existence—conditioned by risk factors such as persistent poverty, and exacerbated by "boom and bust" cycles of free-spending when money *is* available—carjackers are always under some degree of pressure and thus are encouraged to maintain a general openness to offending. During a "boom" period, carjackers anticipate future needs, but are not desperate to offend. This encourages Bennett and Wright's (1984) previously described notion of alert opportunism—a general willingness to offend if a particularly good opportunity presents itself.

But as time passes and no acceptable opportunities emerge, situational pressures to offend begin to mount in the face of diminishing resources. Approaching "bust" periods increasingly promote an active willingness to offend, driven by heightened situational inducements. Dormant or anticipated needs become pressing ones, moving carjackers from a state of alert opportunism to a state of motivated opportunism. As they continue to become more situationally desperate, their openness to offending expands to include opportunities perceived to have greater risk or lower reward (see Lofland 1969). Targets that previously seemed unsuitable become increasingly attractive and permissible. The logical outcome is a carjacking triggered almost exclusively by pressing needs.

It is also possible for carjackers to move from a state of motivated opportunism to the lower state of alert opportunism, especially where the decision to commit such an offense is driven by a desire for revenge. Retaliatory urges tend to be high initially, and then to dissipate over time. This is not to say however, that an offended party has necessarily forgiven the offending party. They may simply be getting on with their lives, even as they keep their eyes open for the object of their wrath.

Although infrequent when compared to strong-arm robbery or drug robbery, carjacking's proportional impact on the spread of violence is probably more significant than has been suspected. When offenders themselves are targeted carjacking, like other forms of violent crime, can produce retaliatory behavior patterns that serve to perpetuate and proliferate cycles of violence on the streets. In addition, their sensationalist nature increases the public's general fear of crime when law-abiding citizens are victimized. In either case, the preceding evidence and discussion indicate that carjacking is a unique and dynamic form of crime that probably deserves its own categorization (separate from robbery or auto theft) or, at the very least, further study and attention by those interested in criminal decision-making.

REFERENCES

Armstrong, L. (1994). Carjacking, District of Columbia (September, 1992–December, 1993; NCJRS, abstracts data base).

Anderson, E. (1990). *Streetwise*. Chicago: University of Chicago Press.

Bennett, T. & Wright, R. (1984). *Burglars on burglary: Prevention and the offender*. Brookfield, VT: Gower.

Bureau of Justice Statistics (1994). *Carjacking: National Crime Victimization Survey* (Crime Data Brief).

Bureau of Justice Statistics (1999). *Carjacking: National Crime Victimization Survey* (Crime Data Brief).

Donahue, M., McLaughlin, C. & Damm, L. (1994). Accounting for carjackings: An analysis of police records in a Southeastern city (NCJRS, abstracts data base).

Feeney, F. (1986). Robbers as decision-makers. In R. Clarke and D. Cornish (Eds.), *The reasoning criminal*. New York: Springer-Verlag.

Felson, R., Baumer, E. & Messner, S. (in press). Acquaintance robbery. *Justice Quarterly*.

Fisher, R. (1995). Carjackers: A study of forcible motor vehicle thieves among new commitments (National Institute of Justice/NCJRS abstracts data base).

Friday, S. & Wellford, C. (1994). Carjacking: A descriptive analysis of carjacking in four states, preliminary report (National Institute of Justice/ NCJRS abstracts data base).

Jacobs, B. A. (1999). *Dealing crack: The social world of streetcorner selling*. Boston: Northeastern University Press.

Jacobs, B. A., Topalli, V. & Wright, R. (2000). Managing retaliation: The case of drug robbery. *Criminology, 38*, 171–198.

Jacobs, B. A. & Wright, R. (1999). Stick-up, street culture, and offender motivation. *Criminology, 37*, 149–173.

Katz, J. (1988). *Seductions of crime: Moral and sensual attractions in doing evil*. New York: Basic Books.

Loftin, C. (1985). Assaultive violence as a contagious social process. *Bulletin of the New York Academy of Medicine, 62*, 550–55.

Rand, M. (1994). Carjacking (Bureau of Justice Statistics/NCJRS abstracts data base).

Rosenfeld, R. & Decker, S. H. (1996). Consent to search and seize: Evaluating an innovative youth firearm suppression program. *Law and Contemporary Problems, 59*, 197–219.

Shover, N. (1996). *Great pretenders: Pursuits and careers of persistent thieves*. Boulder, CO: Westview.

Topalli, V., Wright, R. & Fornango, R. (2002). Drug dealers, robbery and retaliation: Vulnerability, deterrence, and the contagion of violence. *The British Journal of Criminology, 42,* (in press).

Uniform Crime Reports (1997). Federal Bureau of Investigation, Washington, DC.

Wright, R. & Decker, S. (1994). *Burglars on the job.* Boston: Northeastern University Press.

Wright, R. & Decker, S. (1997). *Armed robbers in action: Stickups and street culture.* Boston: Northeastern University Press.

Wright, R. & Bennett, T. (1990). Exploring the offender's perspective: Observing and interviewing criminals. In K. Kempf, *Measurement issues in criminology.* New York: Springer-Verlag.

Part III

Sex Crimes

Although I have placed forcible rape in the prior category of violent crimes, where it clearly belongs, there are a host of other sex offenses that are illegal in nature but not thought of as violent acts. In this particular section, the articles discuss certain criminal activity involving offenses of a sexual nature that most often are not well understood by the public and, furthermore, do not gain media attention on a frequent basis. Most of the following sexual illegalities are out of public view, with the occasional exception of street prostitution, and are not considered that serious by the justice system. Although these types of crimes are not life threatening to their victims, they nevertheless often result in long-term harm to those victims who experience them. In one case you will learn how sex offenders themselves experience constant harm in their attempts to live law-abiding lives once returned to the community from prison.

In their study of female street prostitutes, Romenesko and Miller conducted topical life histories of a small research population of "street hustlers" in a midwestern city. They enter a life of prostitution as a result of a perceived lack of legitimate employment opportunities and are attracted to the economic potential that they believe this type of work offers. In truth, once in the life women become part of a pimp's pseudo-family, which comprises the man they work for and all the other women he controls. The authors vividly describe the organizational hierarchy of the pseudo-family along with status positions and rivalries that exist among the women. As female prostitutes age, they become marginal to the pimp and are often disowned when younger and more attractive hustlers join the pimp and his family of prostitutes. Romenesko and Miller found that once disregarded, aging street hustlers experience a double jeopardy by a capitalistic-patriarchal structure that exists in conventional society and is even more profound in the illegal world of street prostitution.

Sexual intimacy between therapists and clients is explicitly recognized as one of the most serious violations of the professional-client relationship, subject to both regulatory and administrative penalties and, in several states, criminal sanctions. In this article, Pogrebin, Poole, and Martinez examine the written accounts submitted to a state's mental health grievance board by psychotherapists who have had complaints of sexual misconduct filed against them by former clients. The authors employed Scott and Lyman's classic formulation of accounts and Goffman's notion of the apology as conceptual guides in organizing the vocabularies of motive used by the group of therapists to explain their untoward, often illegal behavior.

It has become common practice in most states to place stringent conditions on the release of convicted sex offenders into the community. Zevitz and Farkas studied the effects of community notification on released sex offenders and discuss the results of making this group of parolees more visible to the public. The authors focus on the social and psychological effects of community notification in their attempts at offender reintegration in those communities where notification is practiced. By interviewing a population of released sex offenders, they found that community notification may have a negative impact on sex offenders' reintegration. The article further points out the obstacles these offenders face as a direct result of public notification and conclude that stable housing and employment could mitigate the negative impact that they face.

The final selection in this section presents an offense that many may have experienced, but that has rarely been studied. Obscene telephone callers assault women by manipulating or violating conventions of conversational interaction that most people in conventional society abide by. Warner, in analyzing this illegal phenomenon, claims that women create images of their offenders. The images are of male callers they consider threatening and unpredictable, and of strangers who are socially distant to themselves and their social world. The obscene calls force female victims to question their moral character, and often they blame themselves for becoming a recipient of such immoral behavior. Warner found that assessments by official agencies, family members, and friends of the victim often challenge their sense of self when these groups make moral judgments that in turn put victimized women on the defensive about their own decisions and behavior in handling these obscene phone encounters.

9

THE SECOND STEP IN DOUBLE JEOPARDY: APPROPRIATING THE LABOR OF FEMALE STREET HUSTLERS

KIM ROMENESKO

ELEANOR M. MILLER

In this article Romenesko and Miller obtained "topical" life histories of 14 female street "hustlers" ranging in ages from 18 to 35. The majority of the women study participants belonged to a street institution called a pseudofamily composed of a man and the women who worked hustling for him. Women hustlers have had few if any legitimate opportunities for employment and see the economic potential of the street, along with membership in the pimp's pseudofamily, as a refuge from their sad past life. In reality, the whole economic and social existence of the pseudofamily existence actually serves to repress further, rather than improve the women hustlers' opportunities for social-economic mobility. The authors note that life in the pimp's family is characterized by competition among female family members for status and attention from the men they work for. As these hustlers age and their arrest records lengthen, they are cast aside and become marginal when younger, more attractive women join the pseudofamily and carry more favor with the male head of the group.

This article is about Milwaukee women who make their living from street hustling. We attempt to document the operation of a patriarchal structure within the world of women's illicit work that parallels that which exists in the licit world of women's work and, ironically, further marginalizes females who hustle the streets of American cities primarily

EDITOR'S NOTE: From Romenesko, K. & Miller, E., "The second step in double jeopardy: Appropriating the labor of female hustlers," in *Crime and Delinquency, 35,* pp. 109-135. Reprinted with permission from Sage Publications, Inc.

because of preexisting economic marginalization. The socioeconomic status of these women is, as a result of their experiences in this alternative labor market, even further reduced with respect to the licit market because of the criminal involvement, frequent drug and alcohol dependency, psychological and physical abuse, and ill-health connected with the work of street hustling. A particularly tragic element of this process resides in the fact that for many, the world of the streets appears an attractive array of alternative work opportunities when compared to their experience of "straight" work. The impoverishment, dependency, and, ultimately, redundancy of women, particularly women of color, who work in "women's jobs" in the licit world of work *and* the impoverishment, dependency, and redundancy they experience in the illicit work world, we argue, constitute a situation of "double jeopardy."

That there are few viable opportunities for poorly educated, unskilled women like themselves in the licit job market is very clear to women who eventually become female street hustlers. That a parallel situation exists in the illicit job market, however, escapes the view of new female recruits to street hustling because that view is clouded by the appeal of *feeling,* many for the first time, personally desirable, agentive, and productive as *women* workers. Furthermore, we will attempt to demonstrate that the central mechanism by which female street hustlers are made dependent, and financially, socially, and physically insecure, the mechanism by which they, themselves, are commodified is the "pseudo-family"—a patriarchal unit made up of a "man" and the women who work for him.

Since nearly all of the literature written about street-level prostitutes (which we prefer to call female street hustlers because of the clear diversity in their everyday work that includes a broad array of street hustles), indicates that a major reason women enter street life is because of the perceived financial rewards available to them (Miller, 1986; Brown, 1979; James and Meyerding, 1976; Laner, 1974), we became interested in learning how participation in street life in general, and the pseudo-family in particular, actually affects women's financial situation and, more broadly, their life chances. It is the embeddedness of women's hustling work in the social network that is the "pseudo-family," then, that is the focus of this article.

METHODS

In-depth interviews with fourteen women who claimed to have participated in, or have knowledge of, pseudo-families were conducted during the spring of 1986. The majority of the women interviewed were institutionalized at the time either at the Milwaukee House of Correction or at a residential drug treatment facility in Milwaukee.

Interviews were voluntary and a payment of $10.00 was made to all respondents upon completion of the interview. Interview schedules followed the "topical life history" method. That is, emphasis was placed not only upon subject behaviors across a range of more or less discrete topical areas, but across time as well. Interviews lasted between forty-five minutes and two hours, with the average interview taking approximately eighty minutes to complete. Interviews were tape recorded and later transcribed. The fourteen respondents whose interviews are included in this analysis ranged in age from 18 to 38 with an average age of 25.5. Nine women were black, three white; one described herself as Puerto Rican/white, and one as Mexican/Indian. The average length of time these women spent in school is 10.7 years.

BACKGROUND

Given the inaccessibility of lucrative legal work to most women in our society and the failure of our welfare system to maintain women and their children at a livable standard (Sidel, 1986), participation in the illegal activities of street life is a route that a certain proportion of women see as an occupational alternative. As James and Meyerding (1976, p. 178) say:

> Money-making options are still quite limited for women in this society, especially for unskilled or low-skilled women. Recognition of this basic sex inequality in the economic structure helps us see prostitution as a viable occupational choice, rather

than as a symptom of the immorality or "deviance" of individual women.

One of our respondents, Elsie, described how and why she became involved in street life.

I got married when I was eighteen. My husband, okay, when I got married, that was my biggest mistake 'cuz he never was around. I had all these little babies. When I was eighteen I had two [babies], when I was nineteen I had three. He wouldn't work, he wouldn't help support them, and there I'd sit in this little apartment with nothin'—no TV, nothin'—no milk for your babies and, you know, they gotta have milk and stuff. You know, I was so tired of runnin' to my people askin' 'em for, you know, favors: "Would you go buy the babies some milk?" You know, I was embarrassed doin' that. So then that's when I learned to take a check that wasn't mine and go cash it. You know that's what really drove me into it. For their sake, not to benefit myself—not at first it wasn't.

The women of this study, we argue, look to street life to obtain the status and economic stability that is largely unachievable for the children of the poor and near poor, be they male or female. The commodification of female sexuality offers them entree to the streets, often initially as prostitutes. As this analysis will show, however, street women are doubly jeopardized: Involvement in illicit street life is not a viable avenue to financial stability for women because of the institutionalized obstacles that exist in the underworld—obstacles as difficult to circumvent as those in "straight" society.

Work History

Most of the fourteen female street hustlers included in this analysis worked at straight jobs at some point in their lives. (The few that did not, became participants in street hustling in their early teens.) Straight work usually took place before the women's entrance into street life; however, in a few cases, women tried straight jobs in an attempt to leave street life. While some have not given up on the idea that there is some kind of straight work that they could do as the basis of a career, most admitted

that they could not go back to boring, low-paying work with inflexible hours. They believe that "the life" (short for street life, also called "the fast life") will pay off sometime in the near future. They think that things will improve, money will be saved, and they will retire from "the life" in comfort. If our analysis is correct, they are probably mistaken.

The types of straight jobs our respondents performed were low-status, low-paying, often part-time jobs that are heavily dominated by females. Examples of the types of jobs the women worked at are fast food attendant or cook, box checker at a department store, housekeeper, dietary aide, beautician, bakery shop clerk, assembly line worker for a manufacturing company, child care aide, hot dog stand attendant, cashier, waitress, hostess, receptionist, go-go dancer, and secretary.

Some women who tried to make it in straight society have been frustrated by their experience and have, essentially, given up. Brandy, a young black woman who participated in the study, had her first straight job when she was 18 years old. She performed clerical duties in Chicago but was eventually laid off. She said:

I got laid off cuz the plant closed down and opened up in Oak Park and I didn't have a way out to Oak Park from the South Side of Chicago. After that, believe it or not, I was working at [a] hotel on Michigan Avenue in Chicago, Illinois, as a maid. After I took two tests, they gave me a promotion—I was supervisor of house-cleaning. . . . That's why it's weird for me to be goin' through the things I'm goin' through now [Brandy was incarcerated at the Milwaukee House of Correction at the time of the interview]. I'm not a bad person at all.

How long were you supervisor of house-cleaning?

Brandy: About two and a half years [until she was 21 years old].

What did you do then?

Brandy: Stayed at home and collect unemployment . . . and was lookin' for another job, but I never found one.

Other women, although they had no problem finding jobs, were not satisfied with the pay. Once women get a taste for making "fast money" on the street, it is difficult to go back. Tina describes this attraction to the street in the following comments:

What [type of work] were you doing?

Tina: Oh, I did a lot of things. . . . I was workin' at Marc's Bigboy as a waitress. I've been a waiter, I've been a host, I've been a private secretary, I've been a cashier. I've got a lot of job skills, also.

You've had all these straight jobs and you've got lots of skills. What's the attraction to the streets?

Tina: The money, the fast money. That sums it up.

Elsie, in the passage below, relates not only how difficult it is to live on the wages paid by many employers, but also how sexism worked (and works) to keep women marginal to the job market. She is describing the first straight job that she had as a dietary aide and why she became disaffected.

Elsie: I got the job 'cuz I really needed it. In fact, I had just turned 18 so they was payin' me $1.30 an hour. This was like in '67. Yea, '67 cuz my son was about a year old. They was payin' me $1.30 an hour. . . . See, I was stayin' at home. My mother, she was keepin' my baby. I'd get my check, I'd give her half and I'd keep half. Then, the shop was in walkin' distance, you know, from her house where we was stayin' and we'd get our lunch and stuff free cuz I was workin' in the kitchen. But still, that wasn't enough money for that hard work.

How long did you work there?

Elsie: I worked there nine months. . . . I would have stayed there but I was pregnant again. I was tryin' to keep it from them 'cuz back then you could only work 'til you was five or six months pregnant. And so I tried to keep it from 'em. I worked until I was nine months 'cuz, you know, I never got real big. So I was tellin' 'em I was like four months and I was really eight—so I could keep my job.

Finally, a few women reported simply being intolerant of supervised nine-to-five work and ending up getting fired from their jobs or quitting. It is interesting to note that the women who

fall into this category tended to have the least skills.

What did you do then?

Rita: Worked in a hospital in the laundry department and hated it because it was hot, it was hard work, and I hated my boss.

How much did it pay?

Rita: Minimum wage.

How long did you work there?

Rita: Three months.

Why did you leave?

Rita: I've always gotten fired. But, okay, you could say that I've always gotten fired, but before they could fire me, I know when they're going to fire me, [so] I always quit. Before that "You're fired" comes out, I always quit.

What other kinds of straight jobs have you had?

Rita: Okay, my first job was in the hospital. Nurses aide was my last—I got fired there. A security guard for the telephone company.

What was the reason behind quitting most of them?

Rita: I just am not a nine-to-five person. . . . Most of my jobs were during the day. I had to be there in the morning. I didn't have a car; transportation by bus, I didn't like that. So, it was like, I don't like to take buses. I like being my own boss. Work when I want to work, make good money.

For the uneducated and largely unskilled women of this study—whose experience with straight work was unfulfilling or whose jobs were unstable due to a volatile and sexist market place—participation in street life, with its promise of money, excitement, and independence, seemed the answer to their problems.

STREET LIFE

According to the women of this study, a prerequisite to working as a street hustler is that a woman must have a male sponsor, a "man," to act as a "keep-away" from other "men" who vie for a living on the street. She must turn her earnings over to this "man" in order to be considered "his woman" and to enjoy the "man's" protection. The "man," besides providing protection from other "men," gives his women material necessities and gifts and, most important, sex and love (Merry, 1980).

The following, from an interview with a respondent whom we have named "Chris," illustrates why street women need "men" to work. When asked why she worked and turned her earnings over to a "man" instead of working for herself and keeping her money, she said:

> **Chris:** You can't really work the streets for yourself unless you got a man—not for a long length of time . . . 'cause the other pimps are not going to like it because you don't have anybody to represent you. They'll rob you, they'll hit you in the head if you don't have nobody to take up for you. Yea, it happens. They give you a hassle . . . [The men] will say, "Hey baby, what's your name? Where your man at? You got a man?"
>
> So you can't hustle on your own?
>
> **Chris:** Not really, no. You can, you know, but not for long.

Women who do hustle on their own are called "outlaw" women, a term that clearly indicates that their solo activities are proscribed. Women who work as outlaws lose any protection that they had formerly been granted under the "law" of the street and are open targets for "men" who wish to harass them, take their money, or exploit them sexually.

Street life is male dominated and so structured that "men" reap the profits of women's labor. A "man" demands from his women not only money, but respect as well. Showing respect for "men" means total obedience and complete dedication to them. Mary reports that in the company of "men" she had to "talk mainly to the women—try not to look at the men if possible at all—try not to have conversation with them." Rita, when asked about the rules of the street, said, "Just basic, obey. Do what he wants to do. Don't disrespect him. . . . I could not disrespect him in any verbal or physical way. I never attempted to hit him back. Never." And, in the same vein, Tina said that when her "man" had others over to socialize, the women of the family were relegated to the role of servant. "We couldn't speak to them when we wasn't spoken to, and we could not foul up on their orders. And you cannot disrespect them."

Clearly, "men" are the rulers of the underworld. Hartmann's (1984, p. 177) definition of patriarchy is particularly apropos of "street-level" male domination.

> We can usefully define patriarchy as a set of social relations between men, which have a material base, and which, though hierarchical, establish or create interdependence and solidarity among men that enable them to dominate women. Though patriarchy is hierarchical and men of different classes, races, or ethnic groups have different places in the patriarchy, they also are united in their shared relationship of dominance over their women; they are dependent on each other to maintain that domination. . . . In the hierarchy of patriarchy, all men, whatever their rank in the patriarchy, are bought off by being able to control at least some women.

"Men," as the quotations below illustrate, and as Milner and Milner (1972) have written with regard to pimps, maintain a strong coalition among themselves allowing them to dominate women. As the number of women a "man" has working for him increases, the time he can spend supervising each female's activities decreases. It becomes very important that "men" be in contact with one another to assure the social and economic control of street women. With watchful "men" keeping tabs on their own and other "men's" women, women know that they must always work hard and follow the rules.

Rose gives an example, below, of one of the many ways in which "men" protect one another. Rose's sister, who is also involved in street life, was badly beaten by her "man" and decided to press charges against him. She decided to drop the charges because:

> **Rose:** She got scared 'cuz he kept threatening her . . . His friends [were] callin' her and tellin' her if she didn't drop the charges they was gonna do somethin' to her. So she dropped the charges.

It is interesting to note that when working women are caught talking with "men" who are not their own, it is considered to be at the women's initiative and they are reprimanded. In other words, "men" can speak with whom they please, when they please, but women are not allowed to speak, or even look, at other "men." Ann gives another example of this element of the cohesiveness of "men": "If another pimp see you 'out of pocket' [breaking a rule] or bent over talkin' to another pimp or somethin' in their car, they will tell your man, 'Yea, she was talkin' to some dude in a Cadillac—he try and '"knock your bitch.'" That the word they use."

What does it mean?

Ann: He tryin' to come up with her. He tryin' to have her for himself. That what they mean. That what they be sayin' all the time: "He tryin' to 'knock your bitch,' man. You better put her 'in check.'" In other words, you better, how should I say it? "In check" mean, tell her what's happenin', keep her under control.

Or, if you be walkin' around on the stroll and stuff and somebody might see you sittin' down on the bench and they say, "Man, your ho [whore] is lazy! All the cars were passin' by—that girl didn't even catch none of 'em." 'Cuz they [the "men"] know you and stuff.

The mechanisms that keep women oppressed at the street level are parallel to those of the broader culture. The wage structure in the United States is such that many women cannot financially endure without men. On the street, women also need men to survive, because they are not allowed to earn money in their absence. In addition, women of the street must give all of the money they earn to their "man." A "man," in turn, uses the money to take care of his needs and the needs of his women—although, as we shall see, money is usually not distributed unconditionally but, rather, is based upon a system of rewards and punishments that are dependent upon the behavior of women.

THE PSEUDO-FAMILY

The pseudo-family is a familylike institution made up of a "man" and the women who work for him. Its female members refer to each other as "wives-in-law" and to the man for whom they work as "my man." By focusing on the pseudo-family, we try to isolate and expose a street-level mechanism for female oppression, and to explain why the "wife-in-law" is unlikely to revolt or leave her oppressive situation. Specifically, we attempt to describe how the pseudo-family, while seeming to offer love, money, and stability to the women who participate in it, is structured so that women, in fact, gain little. Our thesis, in other words, is that the security of the pseudo-family for the female street hustler is largely illusory.

The structure of the family is hierarchical and the "man" holds the top position. He collects all of the money that women earn and makes decisions about how it is spent. He is also the disciplinarian of the family: When a rule is broken, he is the person who decides upon and metes out the punishment. The "man" also has the final say about who is recruited into his family. Despite the fact that his women may oppose the idea of additional "wives-in-law," there is typically little they can do about it. If women refuse to accept their "wives-in-law" or try to make life difficult for them, they are likely to be thought "disrespectful" and, as such, will be punished. Next in the hierarchy of the pseudo-family is the position of "bottom woman." When a "man" and woman start out working only as an entrepreneurial couple who, at some later date, recruit another woman into the family to become the first woman's "wife-in-law," the most senior woman is accorded the privileged position of "bottom woman." "Bottom women" help the "man" take care of business by making sure that the household is in good working order. "Bottom women" are also expected to keep tabs on their "wives-in-law" and to smooth out any differences that may occur between them.

The "man" of the family places a special trust in the woman that occupies the position of "bottom woman." It is, therefore, an enviable position and one to which "wives-in-law" aspire. To safeguard the position, then, and to control the jealousy that her "wives-in-law" may feel toward her because of her higher status, the "bottom woman" must convince her "wives-in-law"—each individually—that they, in fact, are the "man's" favorite.

Dee, a 26-year-old, self-described Puerto Rican/white woman, has been involved in street life since she was 15. She has an extensive background as "bottom woman" and described how she kept her families (of up to six "wives-in-law") together.

Were you always bottom woman?

Dee: Any man I had, yea. Because it was the mind, the mental thing. If you can avoid jealousy and all that and keep a family together, and me being from keepin' families together, it was an easy thing—right up my alley. [You'd have to] make them feel they were number one and you were just a peon. [It] kept them around, kept them bringin' in $400.00 and $500.00 dollars a night.

Did you get along with your wives-in-law?

Dee: I would force myself to get along with them. . . . I would be protection for them so, consequently, they wouldn't really fight with me, per se, but with each other. So that I would say that I'm . . . [your] . . . friend and her friend and be the mediator between the two. I never really had any problems unless one of them violated some type of code.

To gain the trust of the "bottom woman" and to prove that she is held in high esteem, the "man" often passes on privileged information to the "bottom woman" about her "wives-in-law." Having this information increases the "bottom woman's" feelings of power and prestige and instills confidence in her and her position, motivating the "bottom woman" to do a better job for the "man." Although the information she receives could be used to flaunt the fact that she is her "man's" favorite, the "bottom woman" must remain discreet if she is to retain her position and keep the family together.

A female street hustler can also become a "bottom woman" by working her way up through the ranks of the family. Depending upon the situation, becoming a "bottom woman" in this way can take anywhere from a week to a number of years. Generally, young white women have the best chance of promotion because they make the most money (due to the racism of customers and to the fact that they can work in more and "better" locations than black women without attracting attention from the police), are without lengthy criminal records when first recruited to the streets, and tend to be more obedient. At least one of our respondents said:

Rose: ["Men"] go for young womens and for the white womens. . . . They figure the young womens will get out there and maybe catch a case occasionally until she catch on [but] she won't go to jail . . . and then they think that white women can work anywhere, make more money . . . without catching no heat from the police.

Although there is limited advancement for women, they can never rise to a status equal to the "man." The rules of "the game" explicitly state that "men" are to dominate women, and women, as subservient creatures, are to respect and appreciate that fact (Milner and Milner, 1972). As there are opportunities for promotions within the family, and into safer and more prestigious hustles outside of the family, women cling to the false hope that "their day will come." In addition, the competition among women for these positions, and for their "man's" affection, creates such divisiveness that it is unlikely that women will conspire to fight for significant changes in the street or pseudo-family structure. Thus "men" are assured that the status quo is maintained.

Male Control

The maintenance of male dominance at the street level is dependent upon "men's" ability to control women. The control of women by "men" is made possible, first of all, by the very fact that women are not allowed involvement in the life without "men" as their sponsors. As "men" are well aware, however, women will not accept them and their rule as against the rule and control of another "man" unless given adequate enticements. These take the form of love, money, and the accompanying sense of security. Once a woman has become attached to the "man," he is then in a position to control her more effectively. Lastly, when all else fails, "men" will resort to physical violence in order to maintain control.

Getting women into positions where they can be dominated begins during the "turning out" process—the process whereby a "man" teaches a woman how to become a street hustler. The "turning out" process, as Miller (1978, p. 142) notes:

[I]nvolves a variety of subtle techniques of social control and persuasion on the part of the pimp. . . . [It] involves several steps in which the pimp initially attracts the woman through his sexual and economic appeal and later changes her mind about the propriety of prostitution and the proper relationship between men and women. . . . The critical factor in the beginning of the relationship is the establishment of control over the relationship by the pimp.

Elsie, whom we heard from earlier, is a 38-year-old black woman. She first met her "man" when she was 28 years old; he was 35. Since about the age of 19, Elsie had been intermittently involved in forgery and other petty street crimes. She was still on the fringe of the life, however, and therefore had never worked with a "man." Elsie described for us how, and under what circumstances, she was turned out by her "man."

Okay, I had met this guy . . . I knew about stuff like that but I had never indulged in it (prostitution). I'd done, you know, check forgin' and stealin' but that's it. Me and him was talking—he was dressed up all nice and nice lookin' and he had a big nice Continental outside. So I'm lookin,' you know? So I asked him, I said, "What kind of work do you do?" He said, "I sell insurance. . . ." He said, "You married?" I said, "No, I'm separated but livin' by myself." He said, "Well, could I call you or come by your house, could you give me your number?" I said, "Yea." So I gave him the number. He called me a couple of times and I said, "Well, come back to my house!" I was all excited—I had the house all spic and span and he came over. Right about that time I was having a few problems 'cuz I owed some back bills and stuff—my phone was gettin' ready to be cut off and stuff. I didn't have no stereo or nothin'. So he come by there that night, looked around. I still didn't really know what was happening, you know. I told him I was in financial trouble, they were gettin' ready to cut my phone off—I need a TV and stereo and stuff. He said, "Well, I'm goin' to pay your phone bill and everything for you." And he did! He paid the phone bill the next day, he brought me a stereo, he did all that. He bought me a new outfit.

One of my girlfriends . . . she said, "Elsie, you're too naive." She said, "You don't know

what he's supposed to be?" I said, "No." She said, "I think he calls himself a pimp." I said, "He's nice, look what he did." She said, "You know, that's the way they do. He trying to set you up." I said, "Girl, go away. I know what I'm doin." I was older than she was, you know, and she's trying to tell me. And sure enough, that's what it was.

How did you find out?

Elsie: He told me about it, okay. When I found out, I found out on the telephone. I said, "Yea, I heard you was, you know." He said, "No, who told you that?" I said, "I just heard it 'cuz a pimp is noticed on the streets." And he used the name Peachy, that was his street name. So I said, "If that's what you is, stay away from me, 'cuz I don't even want to get involved with nobody like that." He said, "No, it ain't like that with me and you, we can just be friends." I said, "Well, I don't even want nothin' to do with you if it's like that. I got four kids, I ain't got time for nothin' like you." He said, "Elsie, it ain't like that. If you don't wanna do nothin' like that." He said, "I admit, I do do that. I got some ladies that, you know, give me money and stuff, but you ain't gonna have to be like that." He said we could just be friends, that we could have a relationship different from that, you know.

Did you believe him?

Elsie: Yea, I did! 'Cuz I liked him! So I fell for it. And then the next thing I knew, he came by and said, "Elsie, I got somethin' set up for you."

After Elsie had serviced her first trick, her "man" came to pick her up. She said:

I got in the car and I was sittin' up, lookin' all quiet and stuff. He said, "Did you get the money?" I said, "Yea." He said, "Where is it?" I said, "I got it." He said, "Give it here." I said, "All of it?" He said, "Yea, and then if you want some I'll give it to you." That's the part that hurted. I gave it to him. [I thought], "I'll never do that no more." But then, after that, I got smart, not really smart but I got a little wiser. Okay, when I would do somethin' like that, I would stash me some money.

Rita, a 26-year-old white woman, first met her "man" when she was 18. However, as she describes it, "I was 18 when I met my man, but I was 19 when I first put money in his hand." Rita said that for some months she had had a girlfriend/boyfriend relationship with her "man." He had a 30-year-old white woman working for him but, according to Rita:

> He wanted a white girl he could turn out himself. Apparently, she [his first woman] had been turned out by somebody else. But I was his turn out completely.

Her "man," like Elsie's, made sure that there was a strong emotional attachment before any money changed hands. Rita said:

> He made me fall in love with him before I put any type of money in his hand. I was working three months and still dating him and not paying him. But he knew that he was getting me.
>
> So he was turning you out but you weren't giving him any money?
>
> **Rita:** No I was not. He knew I was falling in love with him, okay? Finally, that day came where we sat in the car. We had just gone out. I was sitting in the car with him, he had brought me home from going out—we'd gone out drinking. I was sittin' in the car with him and he turns to me and says, "Well, you know, we've been going out . . . and you know what I'm about and my lady's gettin' kind of mad that ain't nothin' happening and you are working. It's either, you gotta start giving me some money or I can't see you no more." And I did not want that, I fell in love with the man. So it was like, okay, here you go buddy, am I yours now?

Rita also said that her "man" would not bail her out of jail during that initial period when she was not giving him any money. She said that the reason he wouldn't bail her out was that

> He was just tryin' to show me, "Look, if you go to jail, I'm there and I'm yours and you're mine, I'll get you out. But you're alone and I can't get you out. That's going against what this is all about."

As these excerpts illustrate, the establishment of an emotional tie between the "man" and the woman is an extremely important element of the "turning out" process. Even though a woman discovers relatively early that a "man's" affection is conditional (depending upon her payment, respect, and obedience to him), she believes the payoff—being "taken care of" financially, socially, and emotionally—to be worth it. When "wives-in-law" enter the picture, however, and the woman finds that she must "share" her "man" with others, jealousy and conflict arise, creating instability within the family.

GETTING DISENTANGLED FROM THE PSEUDO-FAMILY

Changes in the composition of the pseudo-family are frequent and occur for various reasons. The intervention of the criminal justice system, the inability of "men" to control adequately the jealously of their women, the heavy drug use of street women and their men, and the decreased marketability of women as they age are all factors that affect family stability.

Criminal Justice System

Because of the highly visible hustles in which street women engage, frequent contact with criminal justice authorities is inevitable. When a woman is arrested and incarcerated, a new woman is often recruited into the pseudo-family to replace the lost earnings of the incarcerated woman. When the incarcerated woman is released and returns to her family, the family must adapt to the new amalgam of personalities. Often, however, the adaptation is unsuccessful and the result is that one of the women may leave the family in search of a better situation. The woman who leaves is replaced by another, creating new stresses within the pseudo-family, and the whole process is repeated.

Also, while incarcerated, women may be recruited into new families. One of the women interviewed—a veteran of the game—chose her new "man" by telephone from the House of Correction. She was recruited by one of his women who was also serving time. Another woman reported that she was considering leaving her "man" of nine years for a new "man" whom she met and corresponded with in the House of Correction via smuggled letters. Women who are unhappy with their current family situation, then, or women who are "between" men, have opportunities while incarcerated to join new families.

Drug Use

Excessive drug use by street hustlers and their "men" also erodes family stability. "Men" have a responsibility to their women to take care of business—to pay bills and post bond, dispense clothing and pocket money, and generally behave as a responsive (if not caring) family member—and women, in return, must provide "men" with money and respect. As Joan says, a real pimp "gonna take care of her 'cuz she's takin' care of him." In other words, there are mutual obligations between "men" and women. When "men" use hard drugs excessively, they are unable to hold up their end of the bargain adequately. Similarly, women who are heavy users are unable to maintain good work habits. Both addicted "men" and women drain the financial resources of the family. In addition, "men" become dependent on their women to support their habits and thereby lose the respect of other "men" and women of the street.

> **Dee:** They ["men" in Milwaukee] would take and allow a woman to hold the fact that they're givin' 'em money over their heads. Where, in Chicago or New York, they would get rid of the woman— 'cuz women come too easy. The women here are the pimps. And if they're payin' the man they're payin' 'em not because of some strategic level he's at, it's because they want someone to protect them. But he's shootin' so much dope nine times out of ten, that the woman runs it [the family]. She says she's not going to work—he might get down on his hands and knees and beg her 'cuz he's such a punk he needs to go pop from the dopeman. . . . They're shootin' too much dope, they're doin' too much dope. Dope has become a big problem as far as the players and the females. That's taken a lot of good men who have taken the money that women have made 'em, put it in businesses, and smoked up the business. So the drugs alone has killed a lot of the men. It's a vicious circle—it all goes back to the dopeman.

Aging Out

As women age, their value as street hustlers declines and they are often discarded by "men" who favor younger, naive women who have yet to establish criminal records. As a woman becomes both increasingly cynical and less attractive (compared to her younger competitors), she is less assimilable into the pseudo-family. Jody, for example, said that she and her "man" permitted a "wife-in-law" of two years to remain in jail (pending $500.00 bail) because, "we wanted to leave the state and she was old and I had learned everything from her so she was no longer needed." When Rita, another respondent, entered a family, her "wife-in-law" left because, as Rita reported, "she knew that I was prettier than her, younger. She was 30 years old. I [was] 19 and freshly turned out."

In addition, older women are more inclined to consider their own needs rather than the needs of their man. Toni, a 29-year-old, after working as a street hustler for eleven years, seeks a secure life with a "good" man. As she says:

> You become more and more experienced and you have to do serious time and you have so much time to think about what you want. If you're going to stay in this life, [you begin to think] about what you want out of it instead of what he wants out of it. It becomes more of a priority. Now, I want someone who does not mess around with drugs. . . . Who has enough money of his own, has enough things going for himself, to where he doesn't need me. That's what I want now.

Since older women hustlers are less financially successful than younger women, and often are chary of exploitative "men," they are frequently spurned by "men."

Instability

Though women are not allowed control over their earnings, they often receive costly gifts from their "men." The process of leaving "men," however, is such that women frequently lose the material wealth that they have acquired during the relationship. Since the decision to leave a "man" is usually made in anger—after a beating, or after being left to "sit" in jail, or in a fit of jealousy—women commonly leave their families without their treasured material possessions. As Toni says, "I've had to leave without my stuff a few times. Then you gotta start all over each time. But there's been times when

I've got to keep my stuff, too." In addition, women who have accomplished some occupational mobility while with a "man" must start at the bottom of the occupational hierarchy when they choose a new one.

The more frequently women change men, then, the less likely their chance of accumulating any material wealth for themselves.

SUMMARY AND CONCLUSIONS

The pseudo-family, in addition to addressing the emotional and sexual needs of its members, is explicitly organized to realize and exploit the "profitability" that inheres in its female members—that is, their sexuality. But if female street hustlers are both labor and capital "rolled into one," it is difficult to imagine how the "man" can interject himself into the lives of women who, logically, are self-sufficient—in a purely entrepreneurial sense. In fact, the women who join pseudo-families "barter" their way through their family lives, endlessly exchanging their resources as women for, variously, the affections (and recognition) of the "man" and the material goods necessary to survival. There is a twofold answer to the question, therefore, of why women relinquish control of their natural assets in the sexual market of the street. One is that they bring to the street emotional and financial vulnerabilities and so fall prey to "men" prepared to exploit them; the second is that the "sexual" street scene represents a deeply entrenched, patriarchal structure that quickly and effectively punishes independent female hustlers.

REFERENCES

Hartmann, H. I. 1984. "The Unhappy Marriage of Marxism and Feminism: Towards a More Progressive Union." Pp. 172–189 in *Feminist Frameworks,* edited by A. M. Jaggar and P. S. Rothenberg. New York: McGraw-Hill.

James, J. and J. Meyerding. 1976. "Motivations for Entrance into Prostitution." Pp. 177–205 in *The Female Offender,* edited by L. Cites. Washington, DC: Heath.

Laner, M. R. 1974. "Prostitution as an Illegal Vocation: A Sociological Overview." Pp. 406–418 in *Deviant Behavior: Occupational and Organizational Bases* edited by C. Bryant. Chicago: Rand McNally.

Merry, S. 1980. "Manipulating Anonymity: Streetwalkers, Strategies for Safety in the City." *Ethnos* 45:157–175.

Miller, E. M. 1986. *Street Woman.* Philadelphia: Temple University Press.

——— 1988. "'Some People Calls It Crime': Hustling, the Illegal Work of Underclass Women." Pp. 190–232 in *The Worth of Women's Work,* edited by A. Statham, E. Miller, and H. Mauksch. Albany: SUNY Press.

Miller, G. 1978. *The World of Deviant Work.* Englewood Cliffs, NJ: Prentice-Hall.

Milner, C. and R. Milner. 1972. *Black Players: The Secret World of Black Pimps.* Boston: Little, Brown.

10

ACCOUNTS OF PROFESSIONAL MISDEEDS: THE SEXUAL EXPLOITATION OF CLIENTS BY PSYCHOTHERAPISTS

MARK R. POGREBIN, ERIC D. POOLE, AND AMOS MARTINEZ

Sexual relations between psychotherapists and their clients are explicitly recognized as one of the most serious violations of the professional–client relationship and have a long history of prohibition. If caught, therapists who are involved in such transgressions can expect strong sanctions from both criminal justice and regulatory bodies. Psychotherapists accused of sexual intimacy with clients have a vested interest in accounting for their deviant and often criminal behavior. Their career, income, and professional reputation are very dependent on a believable explanation for such actions. Pogrebin, Poole, and Martinez examined the written accounts submitted to the Colorado State Mental Health Grievance Board by therapists who have had complaints of sexual misconduct filed against them by former clients. The authors found that discredited therapists attempt to project a normal self and attempt to view their wrongful behavior as atypical and not representative of their true selves.

Sexual intimacy between therapists and clients is explicitly recognized as one of the most serious violations of the professional-client relationship, subject to both regulatory and administrative penalties and, in several states, criminal sanctions. In this paper we examine the written accounts submitted to the Colorado State Grievance Board by psychotherapists who have had complaints of sexual misconduct filed against them by former clients. We employ Scott and Lyman's classic formulation of accounts and Goffman's notion of the apology as conceptual guides in organizing the vocabularies of motive used by our group of therapists to explain their untoward behavior.

Intimate sexual relationships between mental health therapists and their clients have been increasingly reported in recent years (Akamatsu 1987). In a survey of over 1400 psychiatrists, Gartell, Herman, Olarte, Feldstein, and Localio (1987) found that 65% reported having treated a patient who admitted to sexual involvement with a previous therapist.[1] National self-report surveys indicate that approximately 10% of psychotherapists admit having had at least one sexual encounter with a client (Gartell, Herman, Olarte, Feldstein, and Localio 1986; Pope, Keith-Spiegel, and Tabachnick 1986). It is suggested that these surveys most likely underestimate the extent of actual sexual involvement with clients because some offending psychotherapists either fail to respond to the survey or fail to report their sexual indiscretions (Gartell et al. 1987). Regardless of the true prevalence rates, many mental health professional associations explicitly condemn sexual relations between a therapist and client. Such relationships represent a breach of canons of professional ethics and are subject to disciplinary action by specific licensing or regulatory bodies.

PSYCHOLOGICAL IMPACT ON CLIENT

Individuals who seek treatment for emotional or mental health problems assume a dependency role in a professional-client relationship in which direction and control are exerted by the therapist. The client's most intimate secrets, desires, and fears are revealed to the therapist. Therapeutic communication relies on the development of trust between client and therapist. In order to be successful, therapy requires the individual in treatment to abandon the psychic defenses that shield his or her genuine self from scrutiny (Pope and Bouhoutsos 1986). The lowering of these defenses in a therapeutic relationship increases the client's emotional vulnerability. Because the potential for manipulation or exploitation of the client is heightened in such relationships, Benetin and Wilder (1989) argue that the therapist must assume a higher degree of professional responsibility to ensure that personal trust is not abused.

As Finkelhor (1984) points out, the therapeutic relationship is fundamentally asymmetrical; thus, the controlling presumption is that a client's volition under conditions of therapeutic dependency must always be considered problematic. The client cannot be considered capable of freely consenting to enter into a sexual relationship with a therapist. The therapist's sexual exploitation of a client represents an obvious violation of trust, destroying any therapeutic relationship that has been established. The client often experiences intense feelings of betrayal and anguish at having been victimized by the very person who had been trusted to help (Pope and Bouhoutsos 1986). Sexual exploitation by a therapist can result in clients' suffering emotional instability, conflicts in interpersonal relationships, and disruptions in work performance (Benetin and Wilder 1989).

THE COLORADO STATE GRIEVANCE BOARD

Historically in Colorado, grievances against licensed mental health providers were handled by two separate licensing boards: the Board of Psychologist Examiners and the Board of Social Work Examiners. During the past 20 years, the state witnessed a proliferation of practitioners in the unregulated field of psychotherapy. Individuals trained in traditional professional fields of psychology, counseling, and social work may call themselves psychotherapists, but anyone else with (or without) training in any field may refer to their practice as psychotherapy. In short, psychotherapists are not subject to mandatory licensing requirements in Colorado. Largely through the lobbying efforts of licensed mental health practitioners, the state legislature was persuaded to address some of the problems associated with the operation of a decentralized grievance process that failed to regulate unlicensed practitioners of psychotherapy. The result was the passage of the Mental Health Occupations Act, creating on July 1, 1988 the State Grievance Board within the Colorado Department of Regulatory Agencies.

The State Grievance Board has the responsibility to process complaints and undertake disciplinary proceedings against the four categories of licensed therapists and against unlicensed psychotherapists. Upon the filing of a complaint, the eight-member board (comprising four licensed therapists and four public members) initiates the following action:

1. The named therapist receives written notice and is given 20 days to respond in writing;

2. When deemed appropriate by the board, the complainant may review the therapist's response and is given 10 days to submit further information or explanation; and

3. The board reviews the available information and renders a decision about the complaint.

If the board determines that disciplinary action against a licensed therapist is warranted, the board is increased by an augmenting panel of three members, each of whom is a licensed practitioner in the same field as the psychotherapist subject to sanctioning. The board can issue a letter of admonition, place restrictions on the license or the practice, require the therapist to submit to a mental or physical examination, or seek an injunction in a state district court to limit or to stop the practice of psychotherapy. When the complaint involves an unlicensed practitioner, injunctive action is the board's only disciplinary remedy.

The governing state statute further mandates that psychotherapists provide their clients with a disclosure statement concerning their credentials (e.g., degrees and licenses) and specific client rights and information (e.g., second opinion and legal confidentiality, as well as therapeutic methods and techniques and fee structure, if requested). In the Model Disclosure Statement developed by the State Grievance Board, the impropriety of sexual relations is specifically noted:

In a professional relationship (such as our [client and therapist]), sexual intimacy between a therapist and a client is never appropriate. If sexual intimacy occurs, it should be reported to the State Grievance Board.

During 1988 the state legislature also enacted a statute making sexual contact between therapist and client a criminal offense (Colorado Revised Statutes 18-3-405.5, Supplement 1988).

Since 1988, sexual intimacy between therapists and clients has been explicitly and formally recognized as one of the most serious violations of the professional-client relationship, subject to both regulatory or administrative and criminal penalties. Yet, between August 1, 1988 and June 30, 1990, 10% ($n = 33$) of the 324 complaints filed with the State Grievance Board involved allegations of sexual misconduct. Given the implications that these sexual improprieties raise for both the client as victim and the therapist as offender, we wish to examine the written accounts submitted to the board by psychotherapists who have had complaints of sexual misconduct filed against them.

THEORETICAL PERSPECTIVE

As Mills (1940, p. 904) observes, the "imputation and avowal of motives by actors are social phenomena to be explained. The differing reasons men give for their actions are not themselves without reasons." Mills draws a sharp distinction between cause and explanation or account. He focuses not on the reasons for the actions of individuals but on the reasons individuals give for their actions. Mills (1940, p. 906) views motive as "a complex of meaning, which appears to the actor himself or to the observer to be an adequate ground for his conduct." Yet, there may be another dimension: the individual's perception of how the motive will appear to others.

Mills argues that such motives express themselves in special vocabularies: first, they must satisfactorily answer questions concerning both social and lingual conduct; second, they must be accepted accounts for past, present, or future behavior. According to Scott and Lyman (1968), accounts are socially approved vocabularies that serve as explanatory mechanisms for deviance. These linguistic devices attempt to shape others' attribution about the actor's intent or motivation, turning it away from imputations that are harmful (e.g., personal devaluation, stigma, or imposition of negative sanctions).

It is no doubt true that, in many instances, being able to effectively present accounts will lessen the degree of one's moral responsibility. Moral responsibility is rarely a present-or-absent attribution. Just as there are degrees of deviation from expected conduct norms, there are probably types and degrees of accountability, as well as acceptability, to various audiences with respect to the accounts that individuals offer. "The variable is the accepted vocabulary of motives of each man's dominant group about whose opinion he cares" (Mills 1940, p. 906).

It is easy for most people to draw from a repertoire of accounts in explaining their untoward acts. This is not to suggest that their reasons are either sincere or insincere. Nor does it deny the validity of their claims; they may well have committed the disapproved behavior for the very reasons that are given. The important thing here is that they require an appropriate vocabulary of motive to guide their presentation of self. In the present study we seek to identify the meanings therapists imputed to the circumstances and events surrounding their sexual relations with clients. Of particular interest are the situated reasons or motives these individuals offer in accounting for their actions.

METHOD

To the 33 complaints of sexual misconduct filed from August 1988 through June 1990, 30 written responses from psychotherapists were submitted to the State Grievance Board.[2] Twenty-four therapists admitted to sexual involvement with clients; six denied the allegations. In the present study we examine the statements of the 24 therapists who provided accounts for their sexual relations with clients.[3] Twenty-one therapists are men; three are women.

The analytical method utilized in reviewing therapists' accounts was content analysis, which "translates frequency of occurrence of certain symbols into summary judgments and comparisons of content of the discourse" (Starosta 1984, p. 185). Content analytical techniques provide the means to document, classify, and interpret the communication of meaning, allowing for inferential judgments from objective identification of the characteristics of messages (Holsti 1969).

The 24 written responses ranged in length from 2 to 25 pages. Each response was assessed and classified according to the types of explanations invoked by therapists in accounting for their acknowledged sexual relations with clients. We employed Scott and Lyman's (1968) classic formulation of accounts (i.e., excuses and justifications) and Goffman's (1971) notion of the apology as conceptual guides in organizing the vocabularies of motive used by our group of therapists to explain their untoward behavior. Our efforts build upon the work of previous sociologists who have utilized the concept of accounts to analyze the vocabularies of motive of convicted rapists (Scully and Marolla 1984) and convicted murderers (Ray and Simons 1987) in prison interviews with researchers, as well as the vocabularies of criminal defendants in presentence interviews with probation officers (Spencer 1983) and of white-collar defendants in presentence investigation reports (Rothman and Gandossy 1982). We developed the following classification scheme consistent with the controlling themes identified in the written accounts:[4]

1. Excuse: an account in which an individual admits that an act was wrong or inappropriate, while providing a socially approved vocabulary for mitigating or relieving personal responsibility.
   (a) Appeal of defeasibility: an excuse in which an individual seeks to absolve himself or herself of responsibility by claiming to have acted on the basis of either lack of information or misinformation.
   (b) Scapegoating: an excuse in which an individual attempts to shift responsibility by asserting that his or her behavior was a response to the actions or attitudes of others.

2. Justification: an account in which the individual acknowledges the wrongfulness of the type or category of an act but seeks to have the specific instance in question defined as an exception.
   (a) Sad tale: a highly selective portrayal of distressing biographical facts through which the individual explains his or her present act as the product of extenuating conditions.

(b) Denial of injury: a justification in which the individual asserts that his or her act was permissible under the particular occasion since no one was harmed or the consequences were trivial (or even beneficial).

3. Apology: an account in which the individual acknowledges the wrongfulness of the act and accepts personal responsibility but seeks to portray his or her act as the product of a past self that has since been disavowed.

Excerpts from the therapists' written accounts, presented in our findings below, have been selected to illustrate the defining elements of each of the above types of accounts. Care has been taken to avoid disclosing any information that could be used to identify the source of the statements. All names are fictitious to ensure against any potential violation of anonymity.

FINDINGS

Accounts are "linguistic device[s] employed whenever an action is subjected to valuative inquiry" (Scott and Lyman 1968, p. 46). An important function of accounts is to mitigate blameworthiness by representing one's behavior in such a way as to reduce personal accountability. This involves offering accounts aimed at altering the prevailing conception of what the instant activity is, as well as one's role in the activity. Excuses, justifications, and apologies all display a common goal: giving a "good account" of oneself.

Excuses

Appeal of Defeasibility

In an appeal of defeasibility one accounts for one's behavior by denying any intention to cause the admitted harm, or by claiming a failure to foresee the unfortunate consequences of one's act, or both. As Lyman and Scott (1989, pp. 136–37) explain:

The appeal of defeasibility invokes a division in the relation between action and intent, suggesting that the latter was malfunctioning with respect to knowledge, voluntariness, or state of complete consciousness.

In the following account, the therapist claims ignorance of professional rules of conduct governing relations with clients:

I did not know that seeing clients socially outside of therapy violated hospital policy. . . . [I]f I realized it was strictly forbidden, I would have acted differently.

The next case involves a female therapist who had engaged in a long-term sexual relationship with a female client. The therapist couches her account in terms of failing to be informed by her clinical supervisors that the relationship was improper:

Both Drs. Smith and Jones had total access to and knowledge of how I terminated with her [the client] and continued our evolving relationship. Neither of them in any way inferred that I had done anything unethical or illegal. I do not understand how I can be held accountable for my actions. There were no guidelines provided by the mental health center around this issue. Both Drs. Smith and Jones knew of and approved of my relationship with her.

Other appeals of defeasibility incorporate elements of defective insight and reasoning, or just poor judgment, in an effort to deny intent. An appropriate vocabulary of motive is necessarily involved in the presentation of such appeals. For example, Scully and Marolla (1984, pp. 540–41) report that convicted rapists attempted to

. . . negotiate a non-rapist identity by painting an image of themselves as a "nice guy." Admitters projected the image of someone who had made a serious mistake but, in every other respect, was a decent person.

The deviant actor makes a bid to be seen as a person who has many of the same positive social attributes possessed by others. This individual presents the basic problem simply: "Everybody makes mistakes"; "It could happen to anyone"; or "We all do stupid things." Such fairly standard, socially approved phrases or ideas are used to sensitize others to their own mistakes, thereby reminding them of their own vulnerability and limiting their opportunity to draw lines between themselves and the individual deviant. The basic message is that the deviant act is not indicative of one's essential character (Goffman 1963). This message is supplemented by an effort to present information about the "untainted" aspects of self. In these presentational cues, deviant actors seek to bring about a softening of the moral breach in which they are involved and relieve themselves of culpability.

In the following example, a therapist admits that she simply misinterpreted her own feelings and did not consciously intend to become sexually involved with her client:

> It was after a short period of time that I first experienced any sexual feelings toward her. I did excuse the feelings I had as something which I never would act on. Unfortunately, I did not understand what was happening at the time.

Similarly, another therapist seeks to diminish culpability by attributing his sexual indiscretion to a misreading of his client's emotional needs:

> I experienced her expressions of affection as caring gestures of our spiritual bond, not lust. And I had no reason to suspect otherwise from her, since I had been so clear about my aversion to romantic involvement. We had sexual intercourse only once after termination. I am not promiscuous, neither sexually abusive nor seductive.

Another variant of the appeal of defeasibility involves a claim that the inappropriate behavior was an unforeseen outcome of the therapeutic process itself. This denial of responsibility requires articulating one's position in the professional argot of psychotherapy. Such professionals are able to provide rather complex and compelling accounts of themselves, attempting to convince an audience of peers of the "real" meaning or "correct" interpretation of their behavior. As Lofland (1969, p. 179) posits,

> . . . since they are likely to share in the universe of understandings and cultural ideology of expert imputors, they are more likely to be aware of what kinds of reasons or explanations such imputors will buy.

The therapist in the next account focuses on the unique problems arising in the professional-client relationship that contributed to the sexual misconduct:

> The two inappropriate interactions occurred when she was a practicing psychotherapist and I was seeing her as a client, supervisee, and socially. I believe that my unresolved counter-transference and her transference greatly contributed to the events.

In related accounts, therapists provide a professional assessment or opinion that the therapeutic techniques utilized in treating their clients got out of hand. This approach is shown in the following:

> The initial development of a change in the relationship centered around my empathetic feelings that touched on unresolved feelings of loss in my own life. One aspect of the treatment centered around a lifetime of severe feelings of abandonment and rejection that the client felt from her family. This worked powerful feelings within me and I responded by overidentifying with the client, becoming emotionally vulnerable and feeling inappropriate responsibility to ease the client's pain.

Some of these professional accounts provide lengthy and detailed descriptions of various treatment techniques utilized because of the ineffectiveness of prior intervention attempts. These therapists stress the multifarious nature of the problems encountered in treatment that warranted the use of more complex and often more risky types of treatment. The following case shows the compromising position in which the therapist placed himself in attempting to foster the client's amenability to treatment:

> Because we were at an impasse in therapy I adjusted the treatment to overcome resistance. I employed several tactics, one of which was to share more of my personal life with her; another was to see her outside the usual office setting.

A slight variation of this defeasibility claim involves what Scott and Lyman (1968, p. 48) call the "gravity disclaimer," where the actor recognizes the potential risks involved in the pursuit of a particular course of action but suggests that their probability could not be predetermined.

> When she came in she was very down, to the point that she was staring at the floor. I felt she was not being reached in a cognitive way, so I tried to reach her using a sensory approach. I was trying to communicate to her: caring, love, acceptance, compassion and so on. Unfortunately, with the sensory approach there is a fine line not to be crossed, and I crossed it.

The appeal of defeasibility is a form of excuse that links knowledge and intent. Actors diminish blameworthiness by defining their acts as occurring without real awareness or intent; that is, they attempt to absolve themselves of responsibility by denying having knowingly intended to cause the untoward consequences. Had they known otherwise, they would have acted differently.

Scapegoating

Scapegoating involves an attempt to blame others for one's untoward behavior. Scapegoating is available as a form of excuse in the professional-client relationship because of the contextual opportunity for the therapist to shift personal responsibility to the client. The therapist contends that his or her actions were the product of the negative attributes or will of the client, e.g., deceit, seduction, or manipulation. The therapist in the following example recognizes the wrongfulness of his behavior but deflects responsibility by holding the client culpable for her actions:

> I am not denying that this sexual activity took place, nor am I trying to excuse or justify it. It was wrong. However, the woman who complained about me is a psychologist. She was counseling me as well, on some vocational issues. So if anyone had cause for complaint under the regulations, it seems it would be me.

Another example of an account where the therapist attempts to "blame the victim" for the improper sexual activity reveals the focus on his diminished personal control of the relationship:

> That I became involved in a sexual relationship with her is true. While my actions were reprehensible, both morally and professionally, I did not mislead or seduce her or intend to take advantage of her. My fault, instead, was failing to adequately safeguard myself from her seductiveness, covert and overt.

Here we have a therapist recognizing the impropriety of his actions yet denying personal responsibility because of the client's overpowering charms. The message is that the therapist may be held accountable for an inadequate "self-defense" which left him vulnerable to the client's seductive nature, but that he should not be culpable for the deviant sexual behavior since it was really he who was taken in and thus "victimized." The therapist's account for his predicament presumes a "reasonable person" theory of behavior; that is, given the same set of circumstances, any reasonable person would be expected to succumb to this persuasive client.

Justifications

Sad Tale

The sad tale presents an array of dismal experiences or conditions that are regarded—both collectively and cumulatively—as an explanation and justification for the actor's present untoward behavior. The therapists who presented sad tales invariably focused on their own history of family problems and personal tribulations that brought them to their present state of sexual affairs with clients:

> Ironically, her termination from therapy came at one of the darkest periods of my life. My father had died that year. I had met him for the first time when I was in my twenties. He was an alcoholic. Over the years we had worked hard on our relationship. At the time of his dying, we were at peace with one another. Yet, I still had my grief. At the time I had entered into individual therapy to focus on issues pertaining to my father's alcoholism and co-dependency issues. I then asked my wife to join me for marriage counseling. We were

having substantial problems surrounding my powerlessness in our relationship. Therapy failed to address the balance of power. I was in the worst depression I had ever experienced in my entire life when we began our sexual involvement.

Therapists who employ sad tales admit to having sexual relations with their clients, admit that their actions were improper, and admit that ordinarily what they did would be an instance of the general category of the prohibited behavior. They claim, however, that their behavior is a special case because the power of circumstance voids the defining deviant quality of their actions. This type of account is similar to Lofland's (1969, p. 88) "special justification," where the actor views his current act as representative of some category of deviance but does not believe it to be entirely blameworthy because of extenuating circumstances. One therapist outlines the particular contextual factors that help explain his misbehavior:

The following situations are not represented as an excuse for my actions. There is no excuse for them. They are simply some of what I feel are circumstances that formed the context for what I believe is an incident that will never be repeated.
(1) **Life losses:** My mother-in-law who lived with us died. My oldest son and, the next fall, my daughter had left home for college.
(2) **Overscheduling:** I dealt with these losses and other concerns in my life by massive overscheduling.

Other therapists offer similar sad tales of tragic events that are seen to diminish their capacity, either physically or mentally, to cope with present circumstances. Two cases illustrate this accounting strategy:

In the summer of 1988, my wife and I separated with her taking our children to live out-of-state. This was a difficult loss for me. A divorce followed. Soon after I had a bout with phlebitis which hospitalized me for ten days.

My daughter, who lived far away with my former wife, was diagnosed with leukemia; and my mother had just died. Additional stress was caused by my ex-wife and present wife's embittered interactions.

Sad tales often incorporate a commitment to conventionality whereby one's typical behavior is depicted as conforming to generally approved rules or practices—the instant deviant act being the exception. The imputation is that "the exception proves the rule"; that is, one's normally conventional behavior is confirmed or proven by the rare untoward act.[5] The transgression may thus be viewed as an exception to the deviant classification to which it would justifiably belong if the special circumstances surrounding the enactment of the behavior in question did not exist. Given such circumstances, individuals depict themselves as more acted upon than acting. In the next case the therapist outlines the special circumstances that account for his behavior:

I had "topped out" at my job, was being given additional responsibilities to deal with, had very little skilled staff to work with, and received virtually no support from my supervisor. I was unconsciously looking for a challenging case to renew my interest in my work, and she fit that role. My finally giving into her seduction was an impulsive act based on my own hopelessness and depression.

Sad tales depict individuals acting abnormally in abnormal situations. In short, their instant deviance is neither typical nor characteristic of the type of person they really are, that is, how they would act under normal conditions. They are victims of circumstance, for if it were not for these dismal life events, their sexual improprieties would never have occurred.

Denial of Injury

Denial of injury is premised on a moral assessment of consequences; that is, the individual claims that his or her actions should be judged as wrong on the basis of the harm resulting from those acts. Again, the actor acknowledges that in general the behavior in which he or she has engaged is inappropriate but asserts that in this particular instance no real harm was done. This type of account was prevalent among the therapists who had engaged in sexual relations with clients following the termination of therapy.

A good therapy termination establishes person-to-person equality between participants. Blanket condemnations of post-therapy relationships also are founded on a belief that such relationships invariably cause harm to the former patient. I defy anyone to meet Gerry, interview her, and then maintain that any harm was done to her by me.

The issue of sexual involvement with former clients represents an unresolved ethical controversy among therapists. On the one hand, the American Psychiatric Association has no official policy which categorically bans sexual relations between a psychiatrist and a former patient; instead, there is a case-by-case analysis of such relationships conducted by an ethics committee to determine their propriety. On the other hand, some states have enacted statutes that expressly prohibit any sexual relations between psychotherapists and former clients during a specified posttherapy time period. The statutory period in Colorado is six months following the termination of therapy.

Despite this explicit restriction, some therapists in the present study still insist that their sexual relationships with former clients are neither in violation of professional standards of conduct nor in conflict with state law.

Her psychotherapy with me was successfully concluded two months prior to her seeking a social relationship with me. She herself was unequivocal in her desire for a social relationship with me, which was entirely free from any therapeutic need or motivation. . . . I expressly clarified to her that in becoming socially involved I no longer could every again function as her therapist. With the dual relationship problem laid aside, strictly speaking, such relationships are not unethical since no ethical rule of conduct has ever been formulated against them. . . . I hope that I have convincingly demonstrated that there is no generally accepted standard of psychological practice in . . . post-termination relationships, and so I cannot have violated the statute.

In denial of injury one seeks to neutralize the untoward behavior by redefining the activity in such a way as to reduce or negate its negative quality, such as injury, harm, or wrong. To some extent this involves structuring one's accounts to alter the dominant conceptions of what the activity is. Accounts thus sometimes go beyond the "linguistic forms" that Scott and Lyman have emphasized. Deviance reduction often involves manipulation of various symbols as a basis for one's behavioral account. As seen in the preceding denials of injury, therapists sought to have their sexual relations with former clients redefined according to a professional code of conduct that is subject to individual interpretation. This ethical code may be seen as symbolically governing the therapist-client relationship, establishing the grounds on which the therapist may make autonomous moral judgments of his or her own behavior.

Apology

Scott and Lyman (1968, p. 59) assert that "every account is a manifestation of the underlying negotiation of identities." In a sense, it is probably more accurate to conceive of accounts as referring to desired outcomes rather than as negotiating techniques. They indicate a sought-after definition of the situation, one in which the focus on the deviant act and the shame attached to the individual are lessened. For example, Goffman (1971) argues that the apology, as an account, combines an acknowledgment that one's prior actions were morally reprehensible with a repudiation of both the behavior and the former self that engaged in such activity.

Two consequences of an accused wrongdoer's action are guilt and shame. If wrongful behavior is based on internal standards, the transgressor feels guilty; if the behavior is judged on external normative comparisons, the person experiences shame. Shame results from being viewed as one who has behaved in a discrediting manner. In the following three cases, each therapist expresses his remorse and laments his moral failure:

I find myself in the shameful position that I never would have thought possible for me as I violated my own standards of personal and professional conduct.

I feel very badly for what I have done, ashamed and unprofessional. I feel unworthy of working in the noble profession of counselling.

I entered into therapy and from the first session disclosed what I had done. I talked about my shame and the devastation I had created for my family and others.

Schlenker and Darby (1981) observe that the apology incorporates not only an expression of regret but also a claim of redemption. An apology permits a transgressor the opportunity to admit guilt while simultaneously seeking forgiveness in order that the offending behavior not be thought of as a representation of what the actor is really like. One therapist expresses concern for his actions and proposes a way to avoid such conduct in the future:

I continue to feel worry and guilt about the damage that I caused. I have taken steps I felt necessary which has been to decide not to work with any client who could be very emotionally demanding, such as occurs with people who are borderline or dependent in their functioning.

This account seems to imply that one's remorse and affirmative effort to prevent future transgressions are sufficient remedies in themselves, preempting the need for others to impose additional sanctions. Self-abasement serves a dual purpose in the apology. First, it devalues the untoward behavior, thus reaffirming the moral superiority of conventional conduct. Second, it represents a form of punishment, reprimanding oneself consistent with the moral judgments of others. The message is that the actor shares the views of others, including their assessment of him or her, and both desires and deserves their acceptance. As Jones and Pittman (1982, pp. 255–56) contend,

To the extent that the threatened actor sustains his counteractive behavior or to the extent that the counteractive behavior involves effort and costly commitments, social confirmation will have the restorative power sought.

Several elements of self-management combine when an apology is offered. While confessing guilt and expressing shame, the individual directs anger at himself or herself— denouncing the act and the actor. The actor then attempts to insulate his or her identity from the stigma of the deviant act, reconfirming an allegiance to consensual values and standards of conduct. As Goffman (1971, p. 113) observes, the deviant

. . . splits himself into two parts, the part that is guilty of an offense and the part that dissociates itself from the delect and affirms a belief in the offered rule.

In the following account, the therapist accepts responsibility for his behavior but attempts to make amends by demonstrating a desire to learn from his mistakes:

I am firmly aware that my judgment at the time was both poor and impaired. I am also aware that my thinking was grandiose and immature. One cannot hold a position of public trust and violate community standards. I have incorporated that knowledge into my thoughts and acts.

The demonstration of shared understandings may also be seen as consistent with a desire to preserve self-respect; moreover, self-initiated or proactive response to one's own deviance may serve as a mechanism to lessen the actor's feelings of shame and embarrassment, to militate against negative affect, and to foster a more favorable image of self. Goffman (1971) calls this ritual attempt to repair a disturbed situation "remedial work." In the next account the therapist reveals his effort to repair his spoiled identity:

I have been grieving for Betty and the pain I have caused her. I am deeply distressed by my actions and am doing everything within my power for personal and professional discipline and restoration. I have tried through reading, therapy, and talking with other men who had experienced similar situations to understand why I allowed this to happen.

Such impression-management strategies involving remedial work convey to others that the actor is "solicitous for the feelings of and sensibilities of others and . . . willing to acknowledge fault and accept or even execute judgment for the untoward act" (Lyman and Scott 1989, p. 143). Hewitt and Stokes (1975, p. 1) further note that actors "gear their words and deeds to the restoration and maintenance of situated and cherished identities." The vast majority of apologies were offered by therapists who sought restoration of self by immediately entering therapy themselves. In the following case, the therapist's realization of the emotional damage resulting from her homosexual affair with a client led to her self-commitment to a mental hospital:

> I truly had no prior awareness of my vulnerability to a homosexual relationship before she became a client. In fact, it was such an ego dystonic experience for me that I soon ended up in the hospital myself and had two years of psychotherapy. From this therapy, as well as some follow-up therapy, I have come to understand the needs which led to such behavior. I regret the negative impact it has had on both of our lives.

Efforts to gain insight into their sexual transgressions appear critical to the therapists' transformation of self. By entering therapy, the individual becomes the object of the therapeutic process, whereby the "act" and the "actor" can be clinically separated. The very therapeutic context in which the initial deviance arose is now seen as the means by which the therapist can be redeemed through successful treatment. Through therapy individuals gain awareness of the causal processes involved in their deviant activity and are thus empowered to prevent such transgressions in the future. Introspective accounts convey a commitment both to understand and to change oneself. In this way, the therapist disavows his or her former discredited self and displays the new enlightened self.

DISCUSSION

The consequences of deviant activity are problematic, often depending on a "definition of the situation." When a particular definition of a specific situation emerges, even though its dominance may be only temporary, individuals must adjust their behavior and views to it. Alternative definitions of problematic situations routinely arise and are usually subject to negotiation. Thus, it is incumbent upon the accused therapist to have his or her situation defined in ways most favorable to maintaining or advancing his or her own interests. When "transformations of identity" are at stake, such efforts become especially consequential (Strauss 1962). The imputation of a deviant identity implies ramifications that can vitally affect the individual's personal and professional life. As noted earlier, the negotiation of accounts is a negotiation of identities. The account serves as an impression-management technique, or a "front," that minimizes the threat to identity (Goffman 1959). If the therapist can provide an acceptable account for his or her sexual impropriety—whether an excuse, justification, or apology—he or she increases the likelihood of restoring a cherished identity brought into question by the deviant behavior.

There is a close link between successfully conveying desired images to others and being able to incorporate them in one's own self-conceptions. When individuals offer accounts for their problematic actions, they are trying to ease their situation in two ways: by convincing others and by convincing themselves. An important function of accounts is to make one's transgressions not only intelligible to others but intelligible to oneself. Therapists sought to dispel the view that their deviation was a defining characteristic of who they really were; or, to put it another way, they attempted to negate the centrality or primacy of a deviant role imputation. The goal was to maintain or restore their own sense of personal and professional worth notwithstanding their sexual deviancy. In a way, laying claim to a favorable image in spite of aberrant behavior means voiding the apparent moral reality, that is, the deviance-laden definition of the situation that has been called to the attention of significant others (Grievance Board) by a victim-accuser (former client).

Goffman (1959, p. 251) maintains that individuals are not concerned with the issue of morality of their behavior as much as they are with the amoral issue of presenting a moral self:

> Our activity, then is largely concerned with moral matters, but as performers we do not have a moral concern with them. As performers we are merchants of morality.

The presentation of a moral self following deviance may be interpreted as an attempt by the individual to reaffirm his commitment to consensual values and goals in order to win the acceptance of others (Tedeschi and Riorden 1981). The demonstration of shared standards of conduct may also be seen as consistent with the wish to redeem oneself in the eyes of others and to preserve self-respect. The desire for self-validating approval becomes more important when circumstances threaten an individual's identity. In these instances an actor will often make self-presentations for purposes of eliciting desired responses that will restore the perception of self by others that he or she desires. If discredited actors can offer a normal presentation of self in an abnormal situation, they may be successful in having their instant deviant behavior perceived by others as atypical, thus neutralizing a deviant characterization.

Individuals seek a "common ground" in accounts of their deviant behavior, explaining their actions in conventional terms that are acceptable to a particular audience. These accounts should not be viewed as mere rationalizations. They may genuinely be believed in. While accounts do not themselves cause one's behavior, they do provide situationally specific answers about the act in question and manifest a certain style of looking at the world.

Finally, it should be noted that, as retrospective interpretations, accounts may have little to do with the motives that existed at the time the deviance occurred. In this case accounting for one's deviant behavior requires one to dissimulate, that is, to pretend to be what one is not or not to be what one is. As Goffman (1959) asserts, social behavior involves a great deal of deliberate deception in that impressions of selves must be constantly created and managed for various others. Thus, it is not logically necessary that one agree with others' moral judgments in order to employ accounts. Even where no guilt or shame is consciously felt, one may offer accounts in the hope of lessening what could be, nonetheless, attributions of a deviant identity. When used convincingly, accounts blur the distinctions between "appearance and reality, truth and falsity, triviality and importance, accident and essence, coincidence and cause" (Garfinkel 1956, p. 420). Accounts embody a mixture of fact and fantasy. As shown in the accounts provided by therapists, what is most problematic is determining the mixture best suited for a particular situational context.

NOTES

1. Clients who suffer sexual exploitation often experience self-blame and are reluctant to disclose their victimization to others. Some are unsure what to do or to whom to report, and others simply believe it would do no good to report (Brown 1988).

2. Twenty of the 33 complaints involved unlicensed therapists. There were no discernible differences between the licensed and unlicensed therapists in type of account presented. Moreover, there was no association between type of account and disposition of the case by the Grievance Board.

3. The written statements submitted by therapists to the State Board have been obtained under provisions of Colorado's Public Records Act, which provides "any person the right of inspection of such records or any portion thereof" unless such inspection would violate any state statute or federal law or regulation or is expressly prohibited by judicial rules or court order. The first author serves as a public member on the Grievance Board. The third author is Program Administrator of the Mental Health Licensing Section in the Department of Regulatory Agencies and is directly responsible for the administration of the Grievance Board.

4. Some written explanations contained elements of more than one type of account, and we classified the response according to what we judged to be the controlling theme of the account taken in its entirety. A breakdown of the 24 cases according to their thematic account is as follows: defeasibility, $n = 8$; scapegoating, $n = 2$; sad tale, $n = 5$; denial of injury, $n = 2$; and apology, $n = 7$. It should be pointed out that our classification scheme is exhaustive of all accounts offered by therapists. The other types of accounts in Scott and Lyman's typology were not exhibited in the present data. This is not unexpected because therapists are attempting to present specific accounts that are considered appropriate to a specific type of situation. Thus, other excuses (e.g., appeal to accidents and appeal to biological drives) or justifications (e.g., denial of the victim, condemnation of condemners, and appeal to higher loyalties) were deemed either inappropriate or improper by therapists in explaining their sexual improprieties to the Grievance Board. As Scott and Lyman (1968, p. 53) argue,

accounts may be regarded as illegitimate or unreasonable; consequently, "The incapacity to invoke situationally appropriate accounts, i.e., accounts that are anchored to the background expectancies of the situation," may actually exacerbate the deviant's predicament. It is thus conceivable that some therapists have simply attempted to avoid employing those types of accounts that could only make matters worse for themselves (or at least are perceived that way).

5. "The exception that proves the rule" is a form of illogical reasoning that individuals use to interpret observations that contradict their preconceived views. In the present case, however, this rationalization process may be seen as an accounting scheme used by actors to explain the apparent contradiction between their typical "normal behavior" and the "deviant exception."

References

Akamatsu, J. T. 1987. "Intimate Relationships with Former Clients: National Survey of Attitudes and Behavior Among Practitioners." *Professional Psychology: Research and Practice* 18: 454–58.

Benetin, J., and M. Wilder. 1989. "Sexual Exploitation and Psychotherapy." *Women's Rights Law Reporter* 11:121–35.

Brown, L. S. 1988. "Harmful Effects of Post-Termination Sexual and Romantic Relationships Between Therapists and Their Former Clients." *Psychotherapy* 25:249–55.

Finkelhor, D. 1984. *Child Sexual Abuse: New Theory and Research*. New York: Free Press.

Garfinkel, H. 1956. "Conditions of Successful Degradation Ceremonies." *American Journal of Sociology* 61:420–24.

Gartell, N., J. Herman, S. Olarte, M. Feldstein, and R. Localio. 1986. "Psychiatrist-Patient Sexual Contact: Results of a National Survey. I: Prevalence." *American Journal of Psychiatry* 143:1126–31.

——. 1987. "Reporting Practices of Psychiatrists Who Knew of Sexual Misconduct by Colleagues." *American Journal of Orthopsychiatry* 57:287–95.

Goffman, E. 1959. *The Presentation of Self in Everyday Life*. Garden City, NY: Doubleday.

——. 1963. *Stigma: Notes on the Management of Spoiled Identity*. Englewood Cliffs, NJ: Prentice-Hall.

——. 1971. *Relations in Public: Microstudies of the Public Order*. New York: Basic Books.

Hewitt, J. P., and R. Stokes. 1975. "Disclaimers." *American Sociological Review* 40:1–11.

Holsti, O. R. 1969. *Content Analysis for the Social Sciences and Humanities*. Reading, MA: Addison-Wesley.

Jones, E. E., and T. S. Pittman. 1982. "Toward a Theory of Strategic Self-Presentation." Pp. 231–62 in *Psychological Perspectives on the Self*, edited by J. M. Suls. Hillsdale, NJ: Erlbaum.

Lofland, J. 1969. *Deviance and Identity*. Englewood Cliffs, NJ: Prentice-Hall.

Lyman, S. M., & M. B. Scott. 1989. *A Sociology of the Absurd* (2nd ed.). Dix Hills, NY: General Hall.

Mills, C. W. 1940. "Situated Actions and Vocabularies of Motive." *American Sociological Review* 5:904–13.

Pope, K. S., and J. Bouhoutsos. 1986. *Sexual Intimacy Between Therapists and Patients*. New York: Praeger.

Pope, K. S., P. Keith-Spiegel, and B. G. Tabachnick. 1986. "Sexual Attraction to Clients: The Human Therapist and the (Sometimes) Inhuman Training System." *American Psychologist* 41:147–58.

Ray, M. C., and R. L. Simons. 1987. "Convicted Murderers' Accounts of Their Crimes: A Study of Homicide in Small Communities." *Symbolic Interaction* 10:57–70.

Rothman, M. L., and R. P. Gandossy. 1982. "Sad Tales: The Accounts of White-Collar Defendants and the Decision to Sanction." *Pacific Sociological Review* 25:449–73.

Schlenker, B. R., and B. W. Darby. 1981. "The Use of Apologies in Social Predicaments." *Social Psychology Quarterly* 44:271–78.

Scott, M. B., and S. M. Lyman. 1968. "Accounts." *American Sociological Review* 33:46–62.

Scully, D., and J. Marolla. 1984. "Convicted Rapists' Vocabulary of Motive: Excuses and Justifications." *Social Problems* 31:530–44.

Spencer, J. W. 1983. "Accounts, Attitudes, and Solutions: Probation Officer-Defendant Negotiations of Subjective Orientations." *Social Problems* 30:570–81.

Starosta, W. J. 1984. "Qualitative Content Analysis: A Burkean Perspective." Pp. 185–94 in *Methods for Intercultural Communication Research,* edited by W. Gudykunst and Y. Y. Kim. Beverly Hills, CA: Sage.

Strauss, A. 1962. "Transformations of Identity." Pp. 63–85 in *Human Behavior and Social Processes: An Interactional Approach*, edited by A. M. Rose. Boston: Houghton Mifflin.

Tedeschi, J. T., and C. Riorden. 1981. "Impression Management and Prosocial Behavior Following Transgression." Pp. 223–44 in *Impression Management Theory and Social Psychological Research*, edited by J. T. Tedeschi. New York: Academic Press.

11

Sex Offender Community Notification: Managing High Risk Criminals or Exacting Further Vengeance?

Richard G. Zevitz

Mary Ann Farkas

Studying the subject of sex offenders' return to the community has become quite controversial in recent years. Zevitz and Farkas focused their research on the impact that stringent release conditions have on these ex-inmates. With the current trend of making sex offenders more visible to the public, community notification laws have been enacted, which put the public on notice that a convicted sex offender has moved into their neighborhood. The authors examined the social and psychological effects of such a notification policy on sex offenders' adjustment in communities where public notification was instituted. After interviewing thirty released offenders in Wisconsin, they concluded that public notification often has a negative impact on sex offender reintegration within the community. There are many detrimental effects that these offenders experience, and the issue of rehabilitation and rights of privacy as opposed to the public's right to know needs to be reexamined.

In response to widespread public concern about convicted sex offenders being returned from prison, federal and state laws have been passed authorizing or requiring the notification of local communities where sex offenders will be living. These laws are collectively referred to as "community notification statutes" and are currently in place in all 50 of the United States and the district of Columbia. The common purpose behind these community notification laws is to prevent sexual victimization by notifying potential victims that a convicted sex offender lives nearby.

The danger posed to society by lack of disclosure was raised by the rape and murder of seven year old Megan Kanka by a twice convicted pedophile living anonymously across the street from her New Jersey home. The case set off a national debate on the issue of sex offenders in the community and led to calls for legislative reform. Community notification was heralded as a common sense solution to this perceived threat of undisclosed sex offenders, but that solution itself presents a problem. The basic dilemma associated with community notification is balancing the public's right to be informed with the need to successfully reintegrate offenders within the community.

On the one hand, public notification empowers community members to protect themselves and their children from a sex offender living next door. On the other hand, public notification invades the privacy of the offender who has "served his sentence" and paid his debt to society. Notification may have anti-therapeutic consequences for the social and psychological adjustment of sex offenders. It may have an adverse impact on treatment for those who might otherwise respond favorably. Notification also disrupts the stability of residence and employment as well as the support network necessary for successful reintegration. Family and other personal relationships are strained and irreparably damaged in many cases. Furthermore, negative reactions to the notification process and excessive media coverage for the release of a sex offender can result in further stigmatizing, ostracizing, and even harrassment (Myers, 1996). From 1993 to 1996, the State of Washington experienced 30 incidents of harassment against sex offenders who had undergone community notification (Matson & Lieb, 1996a). Thus, notification may "exact further vengeance" on sex offenders who have already been punished for their crimes.

Wisconsin is one of the states that have tried to strike a balance between the competing interests of public safety and offender reintegration through enactment of its sex offender community notification statute. Under the Wisconsin Sex Offender Registration and Community Notification law (1996), sex offenders convicted of felony sex offenses since 25 December 1993 are subject to a complex set of informational procedures, including long term registration and collection of a DNA sample. In addition, the Wisconsin Department of Corrections (DOC), as a matter of policy, requires all sex offenders under active field supervision to report to and register with local law enforcement officials within 10 days of their release from confinement.

LITERATURE REVIEW

Notwithstanding the extensive media attention and widespread public support the enactment of sex offender disclosure requirements have generated, research on the impact of community notification laws is notably lacking. Little is known about how law enforcement agencies tasked with implementing such notifications have adapted to the change in the way sex offenders are dealt with in the community or how probation/parole officers have responded to increased demand for enhanced supervision and greater control over sex offenders. There is also scant information reported in the social science literature about the effect of the notification process on the people being informed and the consequence of notifications on sex offenders and the communities where they reside.

To date, almost none of the empirical studies on community notification has examined the effects of notification on sex offenders, their experience in the community, and their reaction to the law. One study, however, focused on notification and its relationship to subsequent sex offending. Schram and Milloy (1995) studied recidivism rates for sex offenders who had experienced community notification in Washington State over a 4½ year period. They found no statistically significant difference in the arrest rates for sex offenses between sex offenders subject to notification and a comparison group of sex offenders who were not.

The present study is an in-depth assessment of a single state's experience from the vantage point of those most affected by the community notification process. They are convicted sex offenders who are at high risk to reoffend and, as such, have been designated level III offenders, making them eligible for news media releases, flyers, and community notification meetings. Their experiences are the subject of this paper. Thirty face-to-face, individual interviews with recently released level III sex offenders were conducted in 13 counties within Wisconsin. To the extent that there has been no research to date on the experiences and reactions of sex offenders who have undergone community notification, this aspect of the study is exploratory in nature.

METHODOLOGY

Interview participants were selected based on their status as level III sex offenders and/or their notification exposure in the community at large. The study sample consisted of 30 sex offenders who consented to be interviewed from the population of 44 offenders who had been the subject of community notification meetings and/or news media exposure. This level III population was generated largely through DOC sources. Notification meetings and newspapers throughout the state were also researched. Two incarcerated sex offender interviewees were in revocation status due to violation of parole conditions. The others were under community supervision. Interview data were collected beginning in August 1998 and ending in December 1998.

The sample of 30 sex offenders consisted of all adult males. The median age was 37 years and the mean age was 40. Twenty-one or 70% were identified as European-Americans. The remaining interview subjects were African-American (five or 16.7%), Hispanic (three or 10%) or Native American (one or 3.3%). At the time interview data were collected, all but three interviewees lived in cities and towns. Thirteen interviewees had been convicted of sexually assaulting children, ten for sexual assaults where the victims were minors (including one incest victim), and seven for sexual assaults involving adults (including elderly females or other vulnerable persons other than children). Six interviewees targeted victims who were strangers. The others were acquainted with their victims before their offenses. All interviewees had been sentenced to prison, with terms ranging from 18 months to 20 years. The average term imposed was 7.3 years. Twenty-one had at least one prior sexual assault conviction. Nine

had no prior convictions for registerable sex offenses. Characteristics of the sex offender sample are representative of the age, gender, race, and offender profile of level III sex offenders taken as a whole. When compared to the total population of registered sex offenders in the state, the sample tends to be slightly older with a somewhat higher percentage of whites (70% versus 65%).

The sex offender subjects were first questioned about their participation in face-to-face registration with local law enforcement (as required by Department of Corrections (DOC) policy) and whether this process affected them in any way. They were then asked a series of questions about their experiences with community notification and the impact that it had on their lives and the lives of their families and friends. They were also asked to assess how notification affected their supervision and treatment experiences.

FINDINGS

The first few questions dealt with their experience in the sex offender registration process with local law enforcement. Only a few of those interviewed felt the enhanced registration requirements under the new law or the face-to-face meeting with local police and/or county sheriffs were major impediments in their lives. Most of the respondents reported that they were merely "somewhat embarrassed" or "uncomfortable." The majority of the respondents had registered with local law enforcement only once, but some had registered several times as they were moved to other locations. Similarly, the federally mandated requirement that they provide a sample of their DNA was viewed by most offenders as "minimally invasive" and as a protection against being wrongly accused for other unsolved sexual assault crimes. According to one respondent: "I see it as a safeguard, primarily because if something does happen, a victim does claim that I was the victimizer and there's a DNA sample, I can prove it wasn't me."

Sex offenders in the sample were also queried about their experience in the community after notification and their overall reaction to the law. They identified several consequences that affected their post-prison adjustment. All but one of the interviewed subjects stated that the community notification process had adversely affected their transition from prison to the outside world.

Effects on Employment and Living Arrangements

Loss of employment and exclusion of residence were frequently mentioned as consequences of expanded notification actions and ensuing detrimental publicity. Most of these sex offenders attributed their loss of employment directly to their high profile status. Statements by two of those interviewed were typical:

> I've found that there have been people that have gone to the library and read SBN information on me through the local paper and have, as a result, not given me work opportunities.
>
> I was working temporary at a place in Whitewater called _____ and I busted my butt. I always do when I work and I thought I was going to be hired on. Well, someone asked me about it there. You know, asked if I was the guy in the paper, and the next day, the boss called and canceled my job. Fired me basically.

This loss or inability to obtain employment had critical consequences for the offenders. One respondent explained:

> All I want to do is get a job and save money. I have no transportation. They talk about buying me a bus ticket, but I got to pay half. I ain't got no money, no money to pay half... I don't have enough money for food or clothes or anything else... I don't understand how they can expect somebody to make it.

Finding a place to live was also an enormous obstacle according to most sex offenders. As one interviewee noted:

> They [community members] picketed against the landlord and all of that. They made up signs telling they don't want the "sexual predator" in the neighborhood with pictures and stuff like that.

Another respondent also pointed to the difficulty of finding a residence under the restrictions imposed by the Department of Corrections along with his fear of the recurring notification process.

> I was evicted from my apartment. I found another apartment that I could afford. The DOC said, no, you can't live here because it was fairly close to a school. We found another place, but it was kind of close to a park. So then we came out here only because my girlfriend's mother owns the place. So, yeah it's had a big impact. It's like I'm stuck here because I'm afraid to move. As soon as I move, they're going to renotify and it's going to be the whole shebang again. So I'm stuck paying $750.00.

Yet another sex offender told of his seven or so relocations within a 5 month period, describing one in particular:

> On _____ Street, I was there for 22 hours and the police chief personally came with my PO and the supervisor, and handed me a piece of paper for the neighborhood saying I was removed from the neighborhood.

Some sex offenders were angered that they were left with nowhere to live except in minimum-security prisons/correctional centers because of the absence of housing for them in the community. One sex offender discussed his difficulty with placement and how he had to accept placement in correctional facilities and drug halfway houses when there was no other alternative:

> The contract for the _____ Hotel was canceled. I had no place to go, so they sent me back to Dodge Correctional Facility for two weeks. I was placed in a drug halfway house in Oshkosh. I was then sent to the Sanger B. Powers Correctional Facility.

Effects on Family, Friends and Victims

Twenty-three out of the 30 told of being humiliated in their daily lives, of being ostracized by neighbors and lifetime acquaintances, and of being harassed or threatened by nearby residents or unknown strangers. All expressed varying degrees of concern for their safety. This fear was described in the following separate statements by three respondents in the study:

> I feared for the longest time going out on the street, that there might be some type of vigilante attitude. . . . There were people in the building trying to get a petition together to have the sex offenders that were in the paper ousted from the building, which made it very hard for me to behave in a neighborly fashion with other people. It caused me to be more confined and I felt ostracized from everyone there.
>
> Just wondering . . . they know? It kind of induces paranoia, you get all worried every time you see someone looking at you like they read it. You think—they know. You wonder, if someone confronts me, what am I going to say?
>
> It's just a very scary, a frightful situation—to have your life put in the hands of inexperienced people. And the community has no experience as far as I'm concerned with dealing with sex offenders. And all they see is a sex offender.

Only one interviewee was on the receiving end of what might be described as vigilante actions. The following is his description of the incidents:

> The neighbor across the street, he called me on the phone. He said he's going to kill me and this and that. Well, lo and behold, he was out in my driveway and he was getting ready to start a fire in the back of my truck . . . , he had a rag with some gas on it and his lighter and some wood in my truck. . . . In another incident, somebody spray painted some stuff on my truck. They spray painted the mirrors and stuff on the doors.

Twenty of the 30 interviewed sex offenders also spoke of how community notification unfavorably affected the lives of family members, such as parents, siblings, and offspring. Several cited emotionally painful examples. One interviewee talked of his mother's "broken heart," her anguish and depression following newspaper accounts stemming from notification. Another spoke of his son's decision to quit his high school freshman football team because of ridicule from teammates and a third related how

his sister was shunned by her former friends. Still another offender stated that his wife threatened suicide because she could not handle the stress of constant media exposure.

Community notification was also blamed for irreparably harming personal relationships. As one interviewee related:

> It really hurts. I lost a lot of good friends. I even lost my in-laws and all that. They don't want nothing to do with me because of family respect. I was divorced from my wife because of this.

Another remarked:

> My ex-girlfriend left because of it. Because of being in the newspaper, she was afraid she's gonna be attacked. It got to the point where she was scared to even go out to the store for milk. So she went her way, I went mine.

A third interviewee recounted the difficulty of being a level III sex offender and trying to maintain rapport with a woman:

> I have a relationship with a woman . . . other people in the building keep harping at her about—read the information, read the information (notification flyer). So it constantly keeps her under pressure from others. It's like, is it okay to love this man because the community hates him?

Five of those interviewed mentioned their concern for what expanded notification and renewed public attention might do to their crime victims in view of the fact that these offenders and their victims lived in the same community. The mention of the word "incest" in news articles was criticized. One incest perpetrator related the devastating effect on his daughters of his notification:

> My daughters went to school and had a situation where there was a newspaper that was on the table and some of the kids came back up to my oldest daughter and basically started teasing her, saying, "You know, I heard that your daddy played sex with you." The impact of that goes beyond measure.

These experiences illustrated how people other than sex offenders may be hurt by the public disclosure process. Of particular concern are the victims of sex offenses. Nor should the antitherapeutic effects of notification on the social adjustment of returning sex offenders be overlooked.

Effects on Supervision in the Community

As to what effect community notification had on the manner in which they were supervised in the community, the opinion of sex offenders in the study was mixed. Nineteen interviewees characterized their relationship with their probation/parole agent as "supportive" and "fair," while the other eleven described their dealings with their agents in less favorable terms. The latter felt that their agents tried to place unnecessary constraints on them, to vilify them to law enforcement and the public, and to generally "make their lives hell." Many deeply resented being subjected to certain special conditions of supervision and to conditions that did not apply to them. As one interviewee observed:

I don't drink, I never drank in prison. Never had a dirty urinalysis. There was no reason to put me on alcohol monitoring. All that time I was on alcohol monitor and finally they took me off. . . . The conditions should be individualized with each parolee as to the terms of this parole.

The respondents talked of how these parole conditions for sex offenders impacted their lives and the lives of their family members. In the words of one interviewee:

I can't be in the house alone with my own kids, man, that's bad. I mean I don't have a child molester charge. I can't be in the house with my kids unless there's someone there with me—like they need to watch me or something.

Another remarked:

The other day, I got called. A female is over at my house, my cousin's girlfriend. This woman is 36 years old. They tell me that she has to leave or I'll go up town. My PO told me that. So I had to make her leave—that was embarrassing. She's a grown woman—a friend of the family. I could see if she was under 18. I can't be alone with anyone unless

I contact them (Department of Corrections) and let them know and they got to be interviewed by them.

Others protested the legality or constitutionality of their parole conditions. As one man noted:

Well, I mean, in a court of law, the plethysmography and the lie detector test can never be used. But under probation and parole rules, it can be used and it can be used to revocate somebody. The basic principle should still stand that you're innocent until proven guilty, but the Constitution has been turned around to the point that you're guilty until proven innocent.

A second interviewee commented:

Another rule says that you shall consume any medication as ordered by a psychiatrist. I flatly see that as a direct violation, period. To anybody else walking on the street, there's absolutely no way that they can force you to take any type of medication at all. And just because I'm on probation or parole, they can basically dictate that.

Conditions of parole for sex offenders in general were viewed as an impediment to "moving on" with their lives. One interviewee's remarks typified this sentiment, when he elaborated:

The rules are humiliating—a constant reminder. It's hard to, in a manner of speaking, to move on and try to put things behind when you're constantly reminded by the rules that you are a sex offender and the rules more or less make you feel like it just happened yesterday. . . . The rules don't allow you to have a normal life and the rules are a constant reminder that you're not a normal person. . . . The only thing is when, *when* does there come a time to move on?

Some sex offenders felt that their agents were reacting in a punitive manner to pressure brought about by the high profile nature of their cases. An interviewee noted:

Everything she [parole officer] does—she has to make damn sure because she's being watched.

And I got to be damn sure I'm doing the right thing because I'm being double watched.

Another expressed this commonly held opinion:

There is no leeway—barely any discretion—for transfer, for getting off the monitor, for going out of the county. This is all directly related to my being an SBN. My agent is required to treat me like I'm a more outstanding risk because I'm posted that way. So he's held more accountable to the public regardless of whether I really am more of a risk or not.

The majority of the sex offenders in the sample met with their agents on a weekly basis that included periodic home visits. Most did not feel that their supervision contacts with their agents had increased since their notification occurred. The job demands on their parole officers were acknowledged by several of the sex offenders. They expected their agents to be continually "checking on them," sometimes more than they actually were. In one interviewee's words:

It seemed like in a lot of ways she [parole officer] was unavailable and it seemed to me that the workload was way too great. I mean if you're going to put somebody on that level of restriction— of knowing specifically where they are every minute of the day and you make the decision to do that. Consequently, you need to accept the responsibility that you are completely controlling that person's life, that person's movements and you need to make yourself available, you know, for that.

Two of the sex offenders in the sample were revoked upon recommendation of their parole agents. Both revocations were for rule violations of parole. The first had fallen short of treatment requirements and the second had violated drinking and curfew conditions. None of the 30 was revoked for a new sexual offense.

Effects on Sex Offender Treatment

Most of the interview subjects in the study reported that they had received treatment at one time or another for deviant sexual behavior.

Those currently undergoing such treatment indicated that, for the most part, community notification was not antitherapeutic. The public reaction to their release in the community, aside from drawing initial comments from others in their therapy group, was discounted as a negative influence on their ability to "open up" in treatment. As one man stated:

There was one group where we sat after I appeared in the paper. You know, we talked about it, just to see how I was dealing with it, if it bothered me. But it, you know, it's a closed environment and all the people know exactly where I'm coming from and I don't have to worry about discrimination. I would have to worry about it otherwise but they're all in the same boat.

A few interviewees struggled with the psychological effects of notification. One such expression was the following:

I have my days. Depends on what frame of mind I'm in that day. I think when the community notification first came out I felt really depressed. I got to feeling well now everybody knows. I wanted to go back down the old road. I can't change, so why even try? I know that's a dangerous road for me. And I had to pull myself back out of it and say, "wait—I'm not going to throw away three years of programming, because my name is going to appear in the paper or has appeared in the paper."

Another's words reflected a similar sentiment:

I'm worthless, you know, why am I even alive? Your thoughts run that way, but you have to change them. And it doesn't help when you start to get on your feet and you get knocked on your ass again. Two days later I have to move out. After you think you're doing okay, you're asked to leave again.

However, one interviewee felt that community notification actually furthered his progress in treatment by helping him to fully understand and take responsibility for his crime: "Without sex offender treatment there would be no acceptance of responsibility. I believe that it is necessary to have treatment." Another interviewee also stressed the need to be completely "open and honest" about his offenses and his responsibility for his actions.

The subjects who were interviewed were also asked if they would consider appearing at a community notification meeting. The majority indicated that they would welcome the opportunity, with the most common reason being the chance to clear up negative perceptions about who they were and what they were like. The following comment was illustrative:

> I wanted to go to the meeting because it could be looked at as though I'm showing up and accepting responsibility and saying, yes, I'm the person that did this and you know, I'm showing up of my own free will.

Another interviewee stated:

> I would want to go, so I can speak my mind. So they would know what's going on and they [community members] can judge for themselves. You can't judge a book by its cover. You can't judge people by something you hear, what's written, that type of thing—because it ain't like that.

The three respondents who did attend community notification meetings, however, did not find the experience to be such a productive one. They complained of the meetings "getting out of control" and attendees "shouting insults" at them and of fear for their safety.

Attitudes Toward Notification Law

Sex offender interviewees were also asked about their specific attitudes toward the requirements of the new notification law and the overall impact of this law on their lives. Several complained of being arbitrarily singled out from among the hundreds of sex offenders in the state for community notification. For many, the news media was to blame for treating all sex offenders as though they were sexual predators and inaccurately reporting and sensationalizing their crimes. One such expression was the following:

> There's two types of sex offender, you can say. Ones in the media and ones that aren't. Ones in the media are all level IIIs. Anybody makes it in the media. You can be a level I, it's up to the

media. Originally when I came home, I wasn't going to be that high of a risk level—the Department of Corrections originally rated me only a "one." The Madison Police Department picked it up right away and did it [notification] as a "three."

Only a few of the interviewed offenders thought that the new law on community notification would prevent reoffending by making sex offenders' actions more visible to the public. Most believed that the law would have no deterrent effect on future law violations. The following response was typical:

> If you're going to reoffend, it doesn't matter if you're on TV, in the newspaper, whatever, you're going to reoffend. And there's nothing to stop you. It's a choice you make. . . . The only person that can stop it is the sex offender himself. And that's one of the choices he makes. If he chooses not to offend anymore and he chooses to take part in treatment and deal with the situation like a real human being and to have empathy in his life, then he won't reoffend.

Some interview subjects were of the opinion that the law would even have the opposite effect. As one sex offender explained:

> If you have any familiarity with links and patterns of the cycle of sexual offense, much of it revolves around an individual being under pressure and his behavior under pressure. Well, there is no more pressure than being exploited by the media, the people you work with, the people you live with, relatives, and so the pressure is constantly there. And because they're [sex offenders] miserable, then that would put them in that cycle to recommit an offense.

Another put it even more bluntly:

> If these people know that you're a sex offender and they keep saying—keep pointing at you and everything else, everything breaks under pressure, everything. No matter what. No matter how strong he thinks he is. You taunt a dog long enough, no matter how calm and cool—calm and collected that dog might have been the whole time, it might have been the most loving dog with children and

everything else, but you taunt that dog long enough, it's going to bite. And that's exactly what this law does. It makes John Q. Public taunt the sex offenders. And sooner or later something is going to snap.

Others drew from their own embittered experience with community notification in suggesting that the tremendous pressure placed on sex offenders by the public and the media would drive many back to prison. For most offenders in the sample, their overall reaction to the law was negative. The law was regarded as a "public humiliation," or "another obstacle to overcome." For others, it was viewed as an "insurmountable obstacle" preventing their chance to ever succeed in society.

DISCUSSION AND CONCLUSIONS

There is nothing in this data that would indicate different results in different locations of the United States. While the notification law's primary goals of community awareness and protection are no doubt being served, the findings point to a high personal cost for the law's main focus, namely those sex offenders being identified. Any given law may have unanticipated consequences of either a therapeutic or antitherapeutic nature for offenders that may be express or more subtle (see Wexler, 1995), and it is critical to consider these effects in any analysis of the law's impact.

The social and psychological adjustment of sex offenders is one such consideration. According to respondents in the sample, housing and employment have become nearly impossible for sex offenders. Stable residence, productive work activity, and effective treatment are essential prerequisites for managing the behavior of this group of offenders in the community (Cumming & Buell, 1997).The notoriety created by the notification process has resulted in the inability or loss of residence and employment. Sex offenders continually worry about harassment, over having to move again, and about the possibility of placement in a correctional facility in lieu of residence in the community. They are deeply concerned also about the stress on their families and the loss of

relationships resulting from community notification. This network of supportive relationships is critical to successful reintegration.

The impact of the notification law on sex offender treatment is another matter of importance. A handful of sex offenders identified therapeutic effects of notification for their treatment. Acceptance of responsibility and minimization of denial were related to public disclosure of their crimes. The respondents mentioned how notification had forced them to be honest and accept responsibility for their past behavior. Still these very same sex offenders and other respondents as well struggled with feelings of worthlessness, hopelessness, and other emotional consequences of community notification. Their attitude about attending treatment groups and how they respond to that treatment may be affected by this stress.

The effects of community notification on sex offenders who have experienced this disclosure process were most profound in those cases where the news media were involved. More than anything else, sex offenders were disturbed with media coverage of their post-release circumstances. Publicity about the details of their crimes, including those situations where family members had been victims, greatly disturbed respondents. In such cases, public disclosure of the crime may undermine the therapy of offender and victim alike. While most newspaper and news broadcasts have acted responsibly, many have not. Working to avert future misunderstandings and problems, such as sensationalizing or misclassifying a sex offender or inadvertently identifying a victim, would seem to be in the best interests of all concerned.

What is proposed is a reintegrative approach which suggests that stable housing and employment would mitigate the disruptive and antitherapeutic effects of community notification. This approach also emphasizes maintaining the offender's ties to family and community (Clear & Cole, 2000). It entails that offenders be linked with needed resources that will enable them to transition from an incarcerative setting to society. As McCarthy and McCarthy (1997) maintain, providing the ex-offender with the minimum essentials needed to reintegration will undoubtedly prove more effective in the long run than returning him to the community with no regard for his ability to locate a job or place to live. The majority of sex offenders eventually are released back into the community. For those who have spent a significant amount of time in a prison cell, transition to the community can be apprehensive enough (Cumming & Buell, 1997, p. 56), but notifying the public of a sex offender's conviction and residence may be overwhelming. Critical to an effective transition is an appropriate residence and a job while on community supervision. Absent these essentials, transition to the community is fertile ground for high risk behavior.

REFERENCES

Clear TR, Cole GF. 2000. *American Corrections*, 5th ed. West-Wadsworth: Belmont, CA.

Cumming G, Buell M. 1997. *Supervision of the Sex Offender*. Safer Society: Brandon, VT.

Matson S, Lieb R. 1996a. *Sex Offender Registration: A Review of State Laws*. Washington State Institute for Public Policy: Olympia, WA.

McCarthy BR, McCarthy BJ. 1997. *Community-Based Corrections*, 3rd edn. Wadsworth: Belmont, CA.

Myers J. 1996. Societal self-defense: News laws to protect children from sexual abuse. *Child Abuse and Neglect* 20(4): 255–258.

Schram D, Milloy C. 1995. *Community Notification: A Study of Offender Characteristics and Recidivism*. Urban Policy Research: Seattle, WA.

Wexler DB. 1995. Reflections on the scope of therapeutic jurisprudence. *Psychology, Public Policy, and Law* 1(1): 222–236.

Wisconsin Sex Offender Registration and Community Notification Law of 1996. Wis. Stat. §§ 301.45.

12

AURAL ASSAULT: OBSCENE TELEPHONE CALLS

PRISCILLA KIEHNLE WARNER

In this article Warner looks at the illegal act of obscene calls, a subject that has rarely received much attention by researchers, but that affects many women in this country. Abused women's conceptions of their assailants and their reactions to the perceptions of their handling of such offensive phone calls affects the meaning that women victims attach to these frightening experiences. Warner uses interview data on 83 obscene call incidents and analyzes how callers manipulate normal conventions of telephone conversation etiquette to assault their victims. An important finding reveals the victims' conceptions of callers as threatening, unpredictable strangers, and socially distant from themselves and others who they know. Often women recipients of such assaults blame themselves for their victimization. They begin to question their moral character and are sensitive to the moral judgments made by family members, police, and friends that often challenge a victim's sense of worthiness.

The experiences of people who have received obscene telephone calls provide a chance to examine how actors make sense of an event and how their perceptions of a situation affect their feelings and self-concepts (Cooley, 1981; Thomas, 1923). Interactors *interpret* each other's behaviors. As Blumer states, "Their 'response' is not made directly to the actions of one another but instead is based on the *meaning which they attach* to such actions" (1969:79, emphasis added).

The anonymity that telephone technology affords callers and the conventions of interaction which have developed as a response furnish men with an opportunity for a special type of harassment of women. Phone conventions supply a ready sense-making framework within which speakers operate. Analysis of call structure and content is necessary for an understanding of how an obscene call is accomplished. Callers and recipients use their familiarity with phone conventions and categories of social relationships to make a call happen as it does. In addition, women's responses during calls and their experiences in discussing calls with others give insight into the meaning obscene calls have for them.

EDITOR'S NOTE: From Warner, P., "Aural assault: Obscene phone calls," in *Qualitative Sociology, 11,* pp. 302-318. Reprinted with permission from Kluwer Academic/Plenum Publishers.

Using interview data on 83 call incidents, the following analysis reveals how obscene callers manipulate conventions of telephone interaction and take advantage of categories of social relationships to assault their victims. The second section describes recipients' conceptions of their callers and shows how these characterizations locate callers in a socially distant category e.g. "stranger." The analysis closes by considering the assessments that family, friends, and personnel of control agencies make about recipients' responses to calls. A woman assaulted by an obscene caller suffers victim-blaming experiences similar to those of other victims of sexualized aggression. Judgments of others cause a woman to feel defensive about the moral acceptability of her behavior, challenging her sense of credibility and degrading the value of her judgment.

TELEPHONE INTERACTION CONVENTIONS: CAPTURING AN AUDIENCE

Manipulation of Conventions by Obscene Callers

In the initial demand for identification, it is typically the *called* party who attempts to take control of the conversation. One approach by obscene callers disrupts this effort and defies the answerer's control. After hearing "hello," the caller immediately fulfills his purpose for the call. For example, the answerer says, "Hello?" and the caller declares, "I want to suck your pussy."

This approach is a direct violation of conversational form and makes the call doubly deviant. Although the caller observes the distribution rule, he introduces the purpose of the call too abruptly, subverts the identification demand in the answerer's first "hello," and commonly inspires a recipient to hang up.

The obscene caller, like conventional callers, thus uses the organization of turn taking sequences to capture a receptive audience. The caller anticipates, because of experience with conversational conventions, that after an answerer says "hello" he or she will be listening for a response (which the caller summarily provides).

A second approach is taken when an obscene caller begins to use conventional conversational forms to make the call more difficult to recognize. The caller manipulates the answerer through the use of deceptive "hooks." An obscene caller who uses a standard response to a greeting increases a call's ambiguity and the likelihood that the answerer will stay on the line.

Obscene callers who use an intimate greeting cause recipients to sift through the identifies of callers who ordinarily address them in this manner. A respondent explained, "But other people don't identify who they are right away either so I thought it could be someone [known] . . . I was running through them . . . Is this Colette, or who?"

Obscene callers play upon speakers' preferences to accomplish recognition parsimoniously. A caller who continues in the "friend" style reinforces a recipient's hesitancy to solicit more information. The caller's "hello" is linked with a phrase that deepens an answerer's conversational involvement. A frustrated respondent reported, "He said, 'Hey,—how ya doin'?' You know, like he knew me, like he was a friend. I would have felt like an asshole if I said I didn't know who it was."

Before resorting to the "Who is this?" query, an answerer may give herself a second chance to recognize the caller's voice by engaging in a deceptive ploy herself (Schegloff, 1979:42–45). She pretends she has already made an identification by continuing to talk to her caller in a familiar manner. Ordinarily this lets her figure out who is calling so that she does not have to reveal that she has not recognized a person whom she should. However, in the case of an obscene call, the recipient's deception works against her in that it keeps her engaged in conversation with the obscene caller for several conversational turns, giving the caller multiple chances to say something offensive.

When the ploy has not allowed the call recipient to recognize her caller, she questions him directly. Three respondents reported that when they gave up guessing and asked who was calling, the caller replied with a name. Puzzled, the women paused to think once again. The caller then seized his opportunity to strike.

The following quotation illustrates the caller's use of a conversational hook, the respondent's use of the deception ploy, her reluctance to ask who is calling, the caller's use of a name, and the caller's abuse of his audience.

Respondent: Hello.

Caller: How're you doin', R—?

Respondent: Oh . . . I'm fine.

Caller: Whatcha doin'?

Respondent: Oh . . . (hesitantly) . . . just hanging around . . . (long pause) . . . Who is this?

Caller: It's Kevin . . . I want to come in your mouth. [Respondent put receiver down]

Content

Recognizable categories of social relationships have appropriate phone styles associated with them. Callers and call recipients draw upon this knowledge since it provides a contextual frame in which to operate during a call. An obscene caller has a range of possible relationships he may exploit to sustain a conversation.

As a third approach to capturing an audience, an obscene caller may either play upon answerers' attempts to place a call in an understandable context or supply a more elaborate deception himself. A third of the respondents had callers who attempted to influence them to remain on the line by elaborating content. Callers constructed accounts that possessed varying degrees of legitimacy to help them retain an audience until they had delivered their messages. These callers used conventional forms although call recipients ultimately defined the content as obscene.

Illustrations drawn from respondents' experiences show how respondents and callers interact with reference to various social relationships. Content of calls ranged from virtually non-existent (breathing) to quite complex (mock social science research). Among the more complicated gambits were:

"Suicide"

This respondent said that while continuing to reassure her apparently "suicidal" caller, she had been "frantically flipping through the phone book" to find the number for a help line. "Little did I realize," she said, "what the 'crimp' in his voice was—until he said, 'I'm coming' . . . I felt so used."

"Kidnapper"

Caller: "I have E—here with me, so you'd better do what I say so she doesn't get hurt." This caller had the respondent standing near the kitchen window stripped down to her underclothes before she was able to determine he didn't have her friend captive.

"Relative"

Respondent: I said, "Hello," and a man said, "Hello." He asked me how I was. I didn't recognize him so I said, "Who's this?" He said his name was D—. I wasn't sure if I *didn't* know him. There are so many grandchildren now . . . But then he said, "Would you just talk to me for a few minutes?" I said, "Well, I don't know . . ." *Then* he said, "I'm lying on the couch here and I'm rubbing my cock for you." I *banged* the phone down!

"Social Scientist"

Respondent: "He said he was calling from the University of —. They were doing a survey of sexual practices in the Northeast corridor. He gave his name . . . it wasn't sexy. It was like a real survey. He asked things like age, the number of people in the residence, but he asked if I was alone . . . The first weird question was, "Did I ever engage in sex with multiple sex partners at the same time?" And then he went into a part about emotional responses—like a survey is organized into related questions—but they were real strange mixed up with normal questions, like, he asked did I live with someone, how frequently during the week I had intercourse with my husband . . . but he asked, when I had sex, did my nipples harden? I was thinking, "this is a *real* strange call" . . . The questions were, he asked me to answer, like a survey, a question with categories. [One was] How long was my husband's penis: one to six inches, six to twelve, more than twelve? I asked him to repeat it again. He did. I said that I didn't carry a tape measure, but I put him in the right category."

Call recipients made sense of and attached meaning to a telephone encounter through their understandings of conventions of telephone interaction. Through their recognition of violations of these conventions, victims came to define a call as obscene and question whether or not a caller threatened their safety.

CONCEPTIONS OF CALLERS

Respondents' imagined conceptions of their callers are part of the process of interpreting and handling calls. Respondents constructed images of and ascribed motivations to callers that reveal how they thought of callers in relation to themselves. Because they had minimal information about their callers, women added the impressions they gathered during calls to folklore laced with official versions of reality. Respondents' characterizations located callers in a socially distant category ("not like 'us' . . . people we 'know'").

Categories of Callers

Everyone with a telephone must search for callers' identities since contact begins without the clues available during other social encounters. An answerer's search separates callers into categories depending upon the style and content of the response they make to recipients' initial greetings. The following analysis extends Schegloff's (1979) discussion of familiar callers whom he names "recognizables." His work chiefly concerns speakers' recognition of specific people, not the recognition of and responses to social categories here described.

Call recipients' initial assessments divide all possible callers into two groups, people who are either "known" or "unknown." A caller whose response indicates a previous relationship inspires a memory search for callers who (1) are socially close enough to the recipient to use an intimate greeting style and (2) satisfy other miscellaneous particulars (e.g., sex, time of day). A respondent's realization that a caller is *unknown* to her or him provokes a renewed search for information about identity and intent.

The conceptions of obscene callers that respondents develop over the course of a call and during its aftermath concern the "unknown" category. To "know" people is to have awareness of them, to assess them as having understandable motivations, and to consider them predictable. ("Predictable" meaning that a person could be recognized and believed likely to behave in ways that can be anticipated.) Conversely, to be "unknown" places a person in the status of stranger, with social distance, unpredictability, and less understandable, less desirable, motivations.

Characteristically, call recipients perceive obscene callers to be anonymous and possessed of abusive intent directed at the victim because of her gender. Women interpret callers' use of sexually explicit language as the vehicle by which they express this intent. Recipients consider the language obscene because of the context in which it is used (over the telephone) and because it is spoken by a person who lacks the "right" to do so—a right that, alternatively, could be granted by a different social relationship, usually one characterized by intimacy (lover, peer).

Inferred Conceptions of Callers

By describing callers as incompetent and deficient, recipients minimized the degree of threat they felt. In *all* the interviews, respondents expressed negative conceptions of callers. Invariably, recipients assessed caller behavior as socially unacceptable.

Mental Illness: Respondents assumed that callers were "mentally ill" to some degree. Their behaviors were taken as evidence of sexual inadequacy or pathology. Call recipients regularly described callers as "sick" or "disturbed" and "in need of help." When asked how they thought callers should be dealt with if caught, few respondents thought that imprisonment would be an appropriate remedy. It would not "treat" the behavior.

> Respondent 7: Mostly I think they need psychiatric treatment. They have to be sick in a way to do things like that, to upset others. Just to lock them up won't do the trick. They're just paroled in five months—and [laughing] they'll go to the first phone booth! Maybe they need their voice boxes cut out!

Socially Marginal: Respondents seldom described callers in the context of a social network, i.e. having families or participating in close relationships. They assessed callers as unable to "relate" and lonely—characteristics which respondents believed could motivate call behavior.

Call recipients also imagined their callers to be physically unattractive: "strange looking, greasy hair, pale, overweight," or "doesn't keep himself up." Respondents regularly described callers in detail, including specific hair colors or the presence of a moustache. The socially disvalued physical descriptions that respondents created suggest a negative halo effect—bad character embodied in repugnant visual form.

Threatening: This characteristic pertains both to a caller's demeanor and a respondent's interpretation of how she felt about her experience. A caller's manner could be evaluated as manipulative, hostile, or aggressive. But even without making overtly threatening remarks, callers were perceived as possible assailants. Several respondents stated that they believed callers had "sinister" or "criminal" motivations and that these "presumptuous" men were "scuzzballs" and "assholes." Women worried about how callers had discovered their telephone numbers and whether, as a result, callers knew where they lived.

Curiously, callers were seen as threatening even when they were characterized as "wimps," "weak," "pathetic," or of "lesser mentality"— images of powerless people. However, their threatening quality can relate to the *unpredictability* ascribed to emotionally disturbed individuals. Stereotyping is a way to minimize unpredictability, a difficult thing to accomplish with sparse information.

Gender and Moral Issues

The obscene phone call is similar to activities that indicate lack of deference for women's "personal space." Women are subject routinely to nonverbal and verbal communications that indicate their relative powerlessness (Henley, 1975). Women are touched by others more often than are men. They experience street badgering and "complimentary" pats and pinches—"little rapes" according to Medea and Thompson (1974). Research findings indicate that women are encroached upon more frequently than men through less formal modes of address and through interruptions in conversation, examples of verbal behavior that emphasize status differentials (Zimmerman & West, 1975; Kollack et al., 1985).

One way to understand obscene calls is to see the interaction between the male caller and the female recipient as an extension or distortion of conventional gender roles. When women are abiding by the rules that shield them from undesirable male attention, men are often motivated to attempt to make contact.

> It follows, then, that females are somewhat vulnerable in a chronic way to being "hassled;" for what a male can improperly press upon them by way of drawing them into talk or by way of improperly extending talk already initiated stands to gain him (and indeed her) a lot, namely, a relationship, and if not this, then at least confirmation of gender identity. (Goffman, 1977:329)

Obscene calls do not appear to be attempts at initiating relationships beyond masturbatory accompaniment. They *do* provide clearcut confirmation of gender identity. Men, in fact, use calls to inform both heterosexual and lesbian women about their status as objects for sexualized violation.

Women consider whether their awkward handling of a call reflects on their own moral character. They assess their experiences from the perspective that the handling of an obscene call is a private problem. This creates opportunities for women to blame themselves for their victimization (they didn't handle it "right" or they attract this kind of abuse). Depending upon the feedback a woman receives from people she tells about her experience, a woman may find herself defensive about the moral acceptability of her behavior and, by extension, her character. Her definition of the event as possibly having significant consequences for her safety may not be taken seriously. She may be left to internalize a sense of blame for her own conduct.

Moral Assessments

Responding to the telephone's ring would not appear to have a potential for negative moral implications. For women, however, this commonplace behavior can have such consequences.

Goffman recognized that a woman does not gain if she returns an insult in response to an attack. "She is faced with the dilemma that any remonstrance becomes in itself a form of self-exposure, ratifying a connection that theretofore had merely been improperly attempted" (1977:328). A woman risks, merely by speaking to her caller, subsequent vulnerability to a moral assessment of *her* behavior. Family and police or telephone company personnel repeatedly told call recipients not to "encourage" their callers. However, it was rather difficult to follow this directive in that agency personnel, some family members, and friends categorized almost *any* response, negative or positive—banging down the phone, seeking information, being defensively polite to callers—as "encouragement." One respondent reported (not atypically) that her husband " . . . got a little angry. He said, 'Why didn't you just hang up?'"

Others viewed even an angry or abusive retort as encouragement on a woman's part. Recognition of the caller's presence is not part of the gender code that includes the value-weighted instruction "nice women don't talk to strangers." (Note that one meaning of "to recognize" someone is to acknowledge their *right* to speak.)

An implication of the rule is that if a woman does speak to males with whom she is unacquainted, she opens herself to unpredictable, but no longer unjustifiable improprieties. She has *allowed* herself to become morally implicated through contact with the other. Respondents became vulnerable to evaluations by others and by themselves that they provoked calls and/or that they didn't handle them properly.

> Respondent 1: The police asked if we'd made anyone angry with us. [An officer] accused us of doing it ourselves [making calls] and of writing the letters. [Threatening letters accompanied the calls.]
>
> Respondent 39: I had a friend who was really offended that I could handle those calls. . . . She really thought that I should have been more upset. . . . She was angry at me.

My findings parallel those in Davis's (1978) account of the feelings of women who encountered exposers and Janoff-Bulman's analysis (1979) of self-blaming reactions among rape victims. Janoff-Bulman distinguishes between *behavioral* and *characterological* self-blame: the first concerns a woman's judgment of her behavior in a specific situation, the second involves her negative evaluation of herself as a person. The different types predicted a victim's sense of having the control to prevent a rape in the future. Obscene call recipients could not change their behaviors to prevent obscene calls, but planning a new response strategy (e.g. blowing an ear-splitting whistle or changing to an unlisted phone number) could produce a moderate sense of control. A victim who felt that she had an unalterable characterological flaw was hindered in her ability to cope emotionally and burdened by a sense of powerless vulnerability.

> Respondent 22: I felt like—"Here it is again." I felt like I might be a freak, that there was something about *me* that attracted this.

Defenses

Aware that conversation with callers could be construed as morally suspect, several respondents attempted to deflect blame and potential condemnation. One professed a state of naiveté, a commonplace excuse available to women.

> Respondent 11: I didn't know. I didn't live in the city . . . What do I know about what goes on in the Northeast corridor?

Scott and Lyman call this defense an "appeal to defensibility." Actors disavow responsibility through its use by claiming they lack information that would have affected their behavior.

Respondents also presented their behaviors as excusable. Women who adopted this approach claimed to have been "fooling around" while talking to callers. One declared,

> Respondent 6: I went along for the joke. . . . I was just teasing him. I was only playing around . . . not serious.

Women also defended themselves by asserting that they had stayed on the phone believing it important to their safety not to offend the caller—a double-bind situation. Women are socialized thoroughly not to be rude to others for fear of offending them and possibly provoking an aggressive response. The maintenance of polite behavior, however, makes women vulnerable to harassment and can facilitate the escalation of a potentially dangerous situation. A woman could seldom assess reliably whether or not a caller knew her location, whether she was alone, or her identity. One worried,

> Respondent 38: He knew what I looked like. He had all the advantages; he could find me. I mean, he could be walking down the street right next to me and I wouldn't know it—or he could be watching when I left work.

Validation

A few respondents found it difficult to give credence to evidence heard with their own ears. It was not that they could not distinguish the words callers used, but respondents questioned their own perceptions. Obscene calls are a form of interaction characteristically carried on between two individuals. A call recipient cannot turn to others for immediate consensual validation of her experience in the same way that, for example, Davis's (1978) respondents could when they doubted that they had "really" seen an exhibitionist.

> Respondent 17 [female]: Is she [female caller] saying what I think she is? But she repeats it. I thought, "Oh my god, is this really happening?"

Other respondents anticipated and received the feedback from family, friends, or personnel in social control agencies that they had *overreacted* to the experience. This process further undermined the view the respondent had of (1) herself as a credible interpretive witness and (2) the value of her judgment.

> Respondent 33: A few [other women] told me about their experiences. But it was so long ago they just laughed about them. I was surprised, how could they laugh? . . . When I talked with the ones who had had calls they hadn't had their numbers changed. I felt a little foolish that I had my number changed so fast.

Respondents least often told their mothers or social control agencies about the calls and most often told their male partners and friends. Boyfriends and friends frequently sympathized, but equally often they discounted respondents' fears or anger. Respondents anticipated that their mothers would take the calls even more seriously than they themselves did. As a result, after only 13 of the 83 incidents did women discuss their experiences with their mothers. Also, only 16 incidents provoked respondents to call the telephone company about number changes and/or to call the police to find out about line tracing.

Many respondents who did not report their calls realized that a nonrepeat caller was untraceable. However, the women who received multiple calls did not necessarily report them. For some, the anticipation of testifying in court was too intimidating—if they believed callers would be prosecuted at all. Or, like other victims of gender-related crimes (indecent exposure, voyeurism, rape), respondents thought that the police might not take their complaints seriously—fearing that they might themselves be implicated or blamed, as women subject to sexual harassment often are. (See Martin, 1976:90–118; Medea & Thompson, 1974; Rafter & Stanko, 1982:8–10, 63–82; Weis & Borges, 1973.)

> Respondent 4: I didn't think [the police] would believe me, especially since I didn't call them after the first one. . . . They might think he kept calling because of something I said.

Women also saw family, agency personnel, and friends as providing alternative interpretations of their experiences. One male interpretation classified obscene calls as within the range of normal gender relationships. Women tended to see such teasing as malicious rather than amusing.

> Respondent 27: My brother makes jokes, stuff like, "Does he say anything good?" or "Are you gonna go out with him?"

Some of the women also interpreted men's comments as discounting their experiences as consequential *because they were women*.

> Respondent 16: At first the man who was asking questions was skeptical, like we were silly women or girls . . .

Women reported that "these things happen" is a routine "cooling-out" phrase that men offered to counter their concerns. Women understood this to mean that "these things happen" *to women* and because they are a part of the experience of being women this means they are *not important.* Women's beliefs that others discount their interpretations of their experiences has repercussions for views of self (which include gender identity, a woman's sense of her credibility, and the value of her judgment), all reportage rates, and further, whether obscene calls are an activity "worth" trying to suppress—from the point of view of social control agents and women who have internalized similar values.

It is difficult to get many men to share women's definitions and treat gender-related crimes as illegal. The treatment of call activities by agencies charged with controlling it as well as the responses of people whom women tell about their experiences create conflict over the definition that these acts are unambiguously wrong. The fact that it is women who are offended by gender-related harassment while the perpetrators, ordinarily, are socially more powerful men is a condition of this conflict.

Summary and Conclusion

The nonvisual aspect of telephone communication affects interaction by forcing speakers to begin a conversation before participants' identities have been established. Speakers rely on familiar patterns of conversational structure to accomplish a conversation. These patterns and the anonymity allowed by telephone technology place a call recipient in a position where she or he is vulnerable to caller abuse. Obscene callers victimize women by manipulating or violating conversational and gender interaction conventions and drawing upon recognized categories of social relationships that have particular telephone styles associated with them.

Abused women's conceptions of their assailants and their sensitivities to others' reactions to their handling of calls affect the meaning that women attach to their experiences. Recipients created images of physically unattractive, psychologically unstable men. These images reveal that women consider their callers fearsome, unpredictable outsiders, socially distant from the people whom they "know." In addition to threatening conceptualizations of callers, moral judgments by family members,

control agency personnel, or friends challenge a woman's sense of self. The assessments of others often put her on the defensive about her own decisions and behavior, an ordeal familiar to the social scientist from the experiences of other victims of sexualized assault.

References

Blumer, H.
   1969   Symbolic Interactionism, Perspective and Method. Englewood Cliffs, NJ: Prentice Hall, Inc.

Cooley, C.H.
   1981   "Self as Social Object." In Gregory P. Stone and Harvey A. Farberman (Eds.). Social Psychology Through Symbolic Interaction. Pp. 169–173. New York: Wiley and Sons.

Davis, S.K.
   1978   "The Influence of an Untoward Public Act on Conceptions of Self." Symbolic Interaction 1:106–123.

Goffman, E.
   1977   "The Arrangement Between the Sexes." Theory and Society 4:301–331.

Henley, N.
   1975   "Power, Sex, and Nonverbal Communication." In Barrie Thorne and Nancy Henley (Eds.). Language and Sex: Difference and Dominance. Pp. 184–203. Rowley, MA: Newbury House Publishers.

Janoff-Bulman, R.
   1979   "Characterological Versus Behavioral Self-blame: Inquiries into Depression and Rape." Journal of Personality and Social Psychology 37:1798–1809.

Hopper, C.
   1987   "Obscene Calls: Impersonal Sex in Private Places." Unpublished paper. Presented at the annual meeting of The Popular Culture Association, Montreal, March 26.

Kollock, P., Blumstein, P. & Schwartz, P.
   1985   "Sex and Power in Interaction: Conversational Privileges and Duties." American Sociological Review 50: 34–46.

Leidig, M.W.
1981   "Violence Against Women: A Feminist-Psychological Analysis." In Sue Cox (Ed.). Female Psychology: the Emerging Self. Pp. 190–205. New York: St. Martin's Press.

Martin, D.
1976   Battered Wives. San Francisco: Glide Publications.

Medea, A. & Thompson, K.
1974   Against Rape. New York: Farrar, Straus and Giroux.

Murray, F.S.
1967   "A Preliminary Investigation of Anonymous Nuisance Telephone Calls to Females." Psychological Record 17: 395–400.

Murray, F.S. & Beran, L.C.
1968   "A Survey of Nuisance Calls Received by Males and Females." Psychological Record 18:107–109.

Nadler, R.P.
1968   "Approach to Psychodynamics of Obscene Telephone Calls." New York State Journal of Medicine 68:521–526.

Rafter, N.H. & Stanko, E.A.
1982   Judge, Lawyer, Victim, Thief: Women, Gender Roles, and Criminal Justice. Boston: Northeastern University Press.

Ratliff, D.H.
1976   Minor Sexual Deviance: Diagnosis and Pastoral Treatment. Dubuque: Kendall/Hunt.

Roper Poll Organization
1978   Roper Reports 78–5. New York: The Roper Organization.

Russell, D.H.
1971   "Obscene Telephone Callers and Their Victims." Sexual Behavior 1:80–86.

Sadoff, R.
1972   "Anonymous Sexual Offenders." Medical Aspects of Human Sexuality 6:118–123.

Schegloff, E.A.
1979   "Identification and Recognition in Telephone Conversation Openings." In G. Psathas (Ed.). Everyday Language, Studies in Ethnomethodology. Pp. 23–78. New York: Wiley and Sons.

Scott, M.B. & Lyman, S.M.
1981   "Accounts." In Gregory P. Stone and Harvey A. Farberman (Eds.). Social Psychology Through Symbolic Interaction. Pp. 343–361. New York: Wiley and Sons.

Southern New England Telephone Co.
1984   Personal communication with the author, February 17.

Stanley, L.
1976   "On the Receiving End." OUT 1:6–7.

Stanley, L. & Wise, S.
1979   "Feminist Research, Feminist Consciousness and Experiences of Sexism." Women's Studies International Quarterly 2:359–374.

Thomas, W.I.
1923   The Unadjusted Girl. Boston: Little Brown.

United States Congress
1966   Senate. Committee on Commerce. Subcommittee on Communications. Abusing and Harassing Telephone Calls, Hearings . . . May 11 and June 14, 1966. Washington, D.C.: U.S. Government Printing Office.

Weis, K. & Borges, S.S.
1973   "Victimology and Rape: The Case of the Legitimate Victim." Issues in Criminology 8:71–115.

Zimmerman, D.H. & West, C.
1975   "Sex Roles, Interruptions, and Silences in Conversation." In Barrie Thorne and Nancy Henley (Eds.). Language and Sex: Difference and Dominance. Pp. 105–109. Rowley, MA: Newbury House Publishers.

Part IV

WHITE COLLAR–OCCUPATIONAL CRIME

White-collar crimes are committed by middle-class individuals as well as the upper economic strata of our society in the course of their professional occupations or business life. These offenders often are in positions of trust and experience less risk of being caught than do street criminals. Most of their illegal opportunities occur within their occupational environment as a result of little internal oversight and even less external control. In short, it is often difficult at best to even know that these persons are involved in illegal activity because their crimes are an extension of their everyday work responsibilities.

The readings in this area of criminal behavior deal with persons who are considered to be white-collar occupational offenders. As one can tell by the first article, not all occupational crime is committed for the sole objective of monetary profit. In their study of pharmacists who became addicted to prescription drugs due to the access they have to them, Dabney and Hollinger interviewed two distinct types of pharmacist drug abusers: recreational users and therapeutic self-medicators who started their drug abuse by using prescription narcotics for medicinal purposes. Recreational and therapeutic type abusers had different paths of entry to drug use, but eventually both types merged into one common criminal trajectory with common causal themes that are found among all criminals that attempt to explain their illegal behavior. Last, the article provides answers to the underlying causes for the illegal abuse of prescription drugs among practicing pharmacists.

The criminal behavior by medical doctors in a government health insurance program (Medicaid) was examined by Jesilow, Pontell, and Geis. They interviewed 42 medical physicians who were sanctioned for violating Medicaid policies. Two-thirds of their study population of doctors was suspended from practicing medicine. A comparative sample of sanctioned doctors' responses was compared with that of nonsanctioned doctors. The authors found that even when the sanctioned physicians related the basic illegal and unethical facts pertaining to their untoward and often illegal behavior, they attempted to neutralize their behavior through the use of excuses and justifications in a manner similar to the one described in Pogrebin, Poole, and Martinez's study of psychotherapists. By doing this, the sanctioned doctors were attempting to lessen the wrongfulness of their unprofessional actions.

Although a number of states have attempted to add various types of gambling, including sports betting, to their menu of legalized games, Nevada remains the single state where wagering on sports events is legal. However, there are many gambling operations throughout the United States run illegally by bookmakers. One of the most popular forms of illegal gambling is wagering on all forms of sporting events. Coontz, in her study of bookmakers, examined the social, organizational, and occupational features of sports bookmaking. Through in-depth interviews with 47 operating bookmakers, she found that the social organization of this illegal enterprise insulated bookies from outside law enforcement agencies as well as from having to associate with other criminal types. Coontz concludes that sports bookmakers resemble legitimate businessmen more than criminals and operate with honesty and integrity in their interactions with their customers. Bookies are knowledgeable of the fact that the relationship between them and their clientele is based on trust; and further, if they cheat their customers by not paying off a winning bet, they will lose them.

Moving from white-collar occupational crime to corporate crime, Benson examines how these offenders deny their criminal activity once they are found out by the authorities. The author analyzes the offender's reactions to being caught for white-collar crimes that involved antitrust violations, tax violations, fraud, and making false statements. Benson discusses his interview findings by showing how those convicted persons attempt to lessen their guilt by deflecting blame for their offending activity in order to avoid the process of a status degradation ceremony caused by the public declarations of their misdeeds during the criminal process of arrest and adjudication. By their attempts at denial of wrongdoing, these offenders are minimizing their deviant identity that now defines them as white-collar criminals.

13

DRUGGED DRUGGISTS: THE CONVERGENCE OF TWO CRIMINAL CAREER TRAJECTORIES

DEAN A. DABNEY

RICHARD C. HOLLINGER

Dabney and Hollinger studied 50 drug-addicted pharmacists who were in recovery. The authors examined the life histories of these professionals and focused on various aspects of their career, particularly use of drugs, and the intertwined nature of their legitimate and illegitimate behavior within the profession. Two paths of entry into drug use were found. One group consisted of recreational abusers who enjoyed the euphoric effects of prescription narcotics, and the other group the authors termed "therapeutic self-medicators," who began their use of self-prescribed drugs for medicinal purposes. Eventually, these two differing paths of entry to the use of narcotics converged into a single, illegal career trajectory. Explanations pharmacists offered for their self-perceived drug addiction were similar to those most criminals use to account for their offenses.

This is the first empirical inquiry designed to identify systematically the reasons why pharmacists begin and subsequently continue to use unauthorized prescription medicines. We begin by exploring the ways in which criminal-deviant roles and associations develop and evolve within the personal and professional worlds of these addicts. Specifically, we draw on the personal experiences of 50 recovering drug-impaired pharmacists using lengthy, face-to-face interviews to map the criminal career trajectory of pharmacists who became addicted to prescription medicines. We then apply the lessons learned from these data to contemporary theory on criminal careers to expand the conceptual understanding of this phenomenon.

The principal objective of this research was to ascertain which of these two competing

EDITOR'S NOTE: From Dabney, D. & Hollinger, R., "Drugged druggists: The convergence of two criminal career trajectories," in *Justice Quarterly, 19,* pp. 182-213. Copyright © 2002. Reprinted with permission of the Academy of Criminal Justice Sciences.

models—recreational or therapeutic—is the most accurate in explaining the illicit prescription drug use careers of practicing pharmacists. Despite the pervasiveness of both explanations in the literature, we quickly recognized that perhaps neither was entirely correct. In fact, we ultimately concluded that both models were partially accurate. As will be seen, one can readily identify two different paths of entry to drug use. However, as the theft and use of drugs gives way to drug abuse and eventually uncontrollable addiction, these two deviant career trajectories converge, so that the motivational and behavioral patterns of mature deviants appear very much the same.

METHODOLOGY

Procedure

We used structured personal interviews to examine the individual life histories of a group of pharmacists, all of whom had previously been in treatment and were now in recovery for the past illicit abuse of prescription drugs. Individuals were accessed through a "snowball" sampling technique. Each interview was conducted using a loosely structured guide (Berg, 1998). The guide was divided into 13 "topical areas" that allowed the interviewer to probe various aspects of the individual's pharmacy career, deviant drug use, and the intertwined nature of these two conflicting worlds. Interviews generally lasted 90–120 minutes.

After talking at some length about their personal and professional backgrounds, the respondents were asked to revisit the initial onset of their deviant drug use—to identify their earliest motivations and to locate their drug abuse behavior within the context of their professional situation and personal relationships at that time. Next, the respondents were prompted to talk about subsequent changes in the nature and extent of their pharmaceutical drug use and to chart the significant events and transitions on a written time line. By detailing the ways in which these motivational and behavioral variations corresponded with changes in their professional and personal lives, we were able to document the various stages in their "deviant careers." Moreover, the graphic representation of these events provided the interviewer with a reference tool that was routinely used to revisit or expand on temporal holes or additional substantive themes.

The Respondents

The pharmacists who were interviewed represented a variety of social and demographic backgrounds. For example, 78 percent were men and 22 percent were women. Of the 50, 48 were white, 1 was Hispanic, and 1 was African American. Furthermore, the respondents varied widely in age: 8 percent were 30 or younger, 38 percent were in their 30s, 36 percent were in their 40s, 12 percent were in their 50s, and 6 percent were over age 60.

With regard to professional status, 86 percent had bachelor's degrees in pharmacy and 16 percent had some advanced pharmacy degree. Although many of the pharmacists moved from job to job and crossed over different practice settings, they categorized their "primary practice setting" as follows: hospital pharmacy (36 percent), independent retail pharmacy (28 percent), chain retail pharmacy (26 percent), home infusion pharmacy (4 percent), and nursing home (4 percent); the remaining 2 percent were temporary-contract pharmacists. These demographic characteristics closely resemble the descriptive elements that were revealed in a 1992 nationwide study of 179,445 of the 194,570 pharmacists who were licensed to practice at the time (Martin, 1993).

TWO PATHS OF ENTRY INTO A DEVIANT CAREER

Given our specific interest in the various career aspects of deviant behavior, a significant portion of each interview was focused on the pharmacists' entry into illicit drug use. An examination of the transcripts of the interviews quickly revealed that their initial deviant drug use took two distinct forms. One group (23 pharmacists)—classified as recreational abusers—began using prescription drugs recreationally to "get high." These individuals usually had a history of "street" drug use before they began the formal pharmacy education process. The other group (27 pharmacists)—classified as therapeutic self-medicators—described how they began using prescription drugs much later for therapeutic purposes when they were confronted with some physical malady while on the job.

Recreational Abusers

One of the defining characteristics of recreational abusers is that they all began experimenting with street drugs, such as marijuana, cocaine, alcohol, and various psychedelics, while in high school and during their early college years. The motivation behind this use was simple: they were adventurous and wanted to experience the euphoric, mind-altering effects that the drugs offered. Because of procurement problems, these individuals reported that they engaged in little, if any, prescription drug use before entering pharmacy school.

Initial use of prescription drugs. For the recreational abusers, the onset of their careers in the illicit use of prescription drugs usually began shortly after they entered pharmacy training. These respondents were quick to point to the recreational motivations behind their early prescription drug use. As one 42-year-old male pharmacist stated, "I just wanted the effect; I really just wanted the effect. I know what alcohol is. But what if you take a Quaalude and drink with it? What happens then?" Similarly, a 36-year-old male pharmacist said:

> It was very recreational at first, yeah. It was more curiosity . . . experimental. I had read about all these drugs. Then I discovered I had a lot of things going on with me at that time and that these [drugs] solved the problem for me instantly. I had a lot of self-exploration issues going on at that time.

Trends in the data indicate that pharmacy school provided these individuals with the requisite access to prescription drugs. The respondents recalled how they exploited their newly found access to prescription drugs in an effort to expand or surpass the euphoric effects that they received from weaker street drugs. For example, a 27-year-old male pharmacist said:

> It was a blast. It was fun. . . . It was experimentation. We smoked a little pot. And then in the "model pharmacy" [a training facility in college], there was stuff [prescription drugs] all over the place. "Hey this is nice . . . that is pretty nice." If it was a controlled substance, then I tried it. I had my favorites, but when that supply was exhausted, I'd move on to something else. I was a "garbage head!" It was the euphoria. . . . I used to watch Cheech and Chong [movies]. That's what it was like. I wasn't enslaved by them, [or so I thought]. They made the world go round.

Pharmacy as a drug-access career choice. It is important to note that the majority of the recreational abusers claimed that they specifically chose a career in pharmacy because it would offer them an opportunity to expand their drug-use behaviors. For example, a 37-year-old male pharmacist said: "That's one of main reasons I went to pharmacy school because I'd have access to medications if I needed them." Further evidence of this reasoning can be seen in the comments of a 41-year-old male pharmacist:

> A lot of my friends after high school said, "Oh great, you're going into pharmacy school. You can wake up on uppers and go to bed on downers," all that stuff. At first, [I said] no. The first time I ever [used prescription drugs] I thought, "No, that's not why I'm doing it [enrolling in pharmacy school]. No, I'm doing it for the noble reasons." But then after a while I thought, "well, maybe they had a point there after all." I [had to] change my major. So I [based my choice] on nothing more than, "Well, it looks like fun and gee, all the pharmacy majors had drugs." The [pharmacy students] that I knew . . . every weekend when they came back from home, they would unpack their bags and bags of pills would roll out. I thought, "Whoa, I got to figure out how to do this." [I would ask:] "How much did you pay for this?" [They would respond:] "I haven't paid a thing, I just stole them. Stealing is OK. I get shit wages, so I got to make it up somehow. So we just steal the shit." Well, I thought, "This is it; I want to be a pharmacist." So I went into pharmacy school.

This trend was observed over and over again among the recreational abusers. Namely, access to prescription drugs was a critical factor in their career choice.

Learning by experimentation. Once in pharmacy school, the recreational abusers consistently described how they adopted an applied approach to their studies. For example, if they read about a particularly interesting type of drug in pharmacy school, they often would indicate that they wanted to try it. Or if they were clerking or interning in a pharmacy that offered access to prescription medicines, they would describe how they stole drugs just to try them. If a teacher or employer told them about the unusual effects of a new drug, they would state that this piqued their interest. This pattern of application-oriented learning is exemplified in the comments of a 49-year-old male pharmacist:

I began using [prescription drugs] to give myself the whole realm of healing experience . . . to control my body, to control the ups and the downs. . . . I thought I could chemically feel, do, and think whatever I wanted to if I learned enough about these drugs and used them. Actually, I sat in classes with a couple of classmates where they would be going through a group of drugs, like, say, a certain class of muscle relaxants, skeletal muscle relaxants, and they would talk about the mechanism of pharmacology and then they would start mentioning different side effects, like drowsiness, sedation, and some patients report euphoria, and at a high enough dose hallucinations and everything. Well, hell, that got highlighted in yellow. And then that night, one of us would take some [from the pharmacy], and then we would meet in a bar at 10:00 or in somebody's house and we would do it together.

This quote is an example of how recreational abusers often superimposed an educational motive onto their progressive experimentation with prescription drugs in pharmacy school. The respondents explained that they wanted to experience the effects of the drugs that they read about in their pharmacy textbooks. They adeptly incorporated their scientific training and professional socialization in such a way that allowed them to excuse and redefine their recreational drug use. Many went as far as to convince themselves that their experimental drug use was actually beneficial to their future patients. This adaptation strategy is illustrated in the comments of a 59-year-old male recreational abuser:

In a lot of ways, [college drug use] was pretty scientific. [I was] seeing how these things affected me in certain situations . . . [just] testing the waters. I thought that I'll be able to counsel my patients better the more I know about the side effects of these drugs. "I'll be my own rat. I'll be my own lab rat. I can tell [patients] about the shakes and chills and the scratchy groin and your skin sloughing off. I can tell you all about that stuff."

Socially acceptable use of recreational drugs in pharmacy school. The recreational abusers unanimously agreed that there was no shortage of socially acceptable experimental drug use while in pharmacy school (both alcohol and street drugs). Moreover, all 23 claimed that it was not uncommon for students to use amphetamines to get through all-night study sessions once or twice each semester. Many of the recreational abusers recalled that they were not satisfied with this type of controlled drug use. They were more interested in expanding their usage. One 48-year-old male pharmacist described the makeup of his pharmacy school cohort as follows:

There were a third of the pharmacy students in school because Mom and Dad or Grandfather or Uncle Bill were pharmacists. They looked up to them and wanted to be one [too]. A [second] third had been in the [Vietnam] war. They were a pharmacy tech in the war or had worked in a pharmacy. They had the experiential effect of what pharmacy is and found a love for it or a desire to want it. Then you had the other third . . . and we were just drug addicts. We didn't know what the practice was all about, but we did know that we got letters after our names, guaranteed income if we didn't lose our letters. And we had access to anything [prescription drugs] we needed.

Many of the recreational abusers claimed that they specifically sought out fellow pharmacy students who were willing to use prescription drugs. The most common locus o these peer associations was pharmacy-specific fraternities. The respondents said that there was usually ample drug use going on in these organizations to allow them the opportunity to search cautiously for and identify other drug users. Once they were connected with other drug users, the prescription drug use of all involved parties increased. This type of small-group drug use gave them access to an expanded variety of drugs, a broader pharmacological knowledge base, and even larger quantities of drugs. However, numerous respondents clearly stated that these drug-based associations were tenuous and temporary in nature. Over time, as the intensity of their drug use increased, the recreational abusers described how they became more reclusive and guarded and selective in their relationships, fearing that their heightened use of prescription drugs would come to be defined as a problem by their fellow pharmacy students. One 43-year-old male pharmacist said:

> You get the sense pretty quickly that you are operating [using] on a different level. Those of us [who] were busily stealing [prescription drugs] from our internship sites began to tighten our social circle. We might party a little bit with the others, but when it came to heavy use, we kept it hush, hush.

Unlike other pharmacy students who were genuinely experimenting with drugs on a short-term basis, these recreational abusers observed that there was an added intensity associated with their own use of prescription drugs. Although most of these recreational abusers entered pharmacy school with some prior experiences in the recreational use of street drugs, these experiences were generally not extensive. It was not until they got into pharmacy school that they began to develop more pronounced street and prescription drug use habits. A 38-year-old female pharmacist had this to say about her transition to increased usage:

> I went off to pharmacy school. That was a three-year program. I had tried a few things [before that], but I would back off because it was shaming for me not to get straight As. The descent to hell started when I got to pharmacy school. There were just so many things [prescription drugs] available and so many things that I thought I just had to try.

It might be a different high; it might be a different feeling—anything to alter the way that I just felt. I was pretty much using on a daily basis by the time I got to my last year.

Pharmacy practice yields even more access and use of drugs. Pharmacy school was just the beginning of the steep career trajectory for the recreational abusers. School was followed by pharmacy practice, which offered even greater access to prescription medicines. Daily work experiences meant exposure to more new drugs. Introduction to a newly developed compound was followed by some quick research on the effects of the drug and then almost immediate experimentation, as is illustrated by a 37-year-old male pharmacist's description of his early work experience:

> I remember I came down here and applied for a job. . . . in May of '82. I remember even then, I went out to the satellite [pharmacy facility] and I heard about this one drug, Placidil. As soon as I got to the interview, they were showing me around. A friend took me around . . . and I saw Placidil on the shelf there. . . . I took a chance, kind of wandering around and I went back and took some off the shelf. So even then, I was [stealing and misusing]. You know, why would you do that in the middle of interviewing for a job? I took it even then. I just jumped at the chance.

Once the recreational abusers got into their permanent practice setting, most described how they quickly realized that they had free rein over the pharmacy stock. At first, they relied on other, more experienced pharmacists for guidance in gaining access to (or using) newly available prescription drugs. Later, their nearly unrestricted access meant that they could try any drugs that they pleased. And most did. More important, increased access allowed the pharmacists-in-training to secretly use the drugs that they most liked. No longer did they have to worry about others looking over their shoulders. Thus, it is not surprising that the level of their drug use usually skyrocketed shortly after they entered pharmacy practice. This trend is demonstrated in the comments of a 41-year-old male pharmacist:

By the time I got to pharmacy school in 1971, I was smoking dope [marijuana] probably every day or every other day, and drinking with the same frequency, but not to the point of passing out. . . . Then in 1971, that was also the year that I discovered barbs [barbiturates]. I had never had barbs up until I got to pharmacy school. So it was like '75 or '76 [when I got out of pharmacy school], I was using heavy Seconals and Quaaludes and Ambutols [all barbiturates]. I withdrew, and it [the heavy abuse] just took off.

At the start, the recreational abusers' drug use was openly displayed and took on an air of excitement, much like others' experimentation with street or prescription drugs. However, as it intensified over time, the majority described how they slowly shielded their use from others. They thought it important to appear as though they still had the situation under control. As physical tolerance and psychological dependence progressed, these individuals began to lose control. Virtually all the recreational abusers eventually developed severe prescription drug-use habits, using large quantities and sometimes even multiple types of drugs, and their prescription drug use careers were usually marked by a steep downward spiral. This trend was clearly evidenced in the time line that each respondent drew. What started out as manageable social experimentation with drugs persistently progressed to increasingly more secretive drug abuse. In almost all the cases, it took several years for the drug use to reach its peak addictive state. The intense physical and psychological effects of the drug use meant that the recreational abusers' criminal-deviant careers were punctuated by a "low bottom." Commonly identified signs of "bottoming out" included life-threatening health problems, repeated dismissal from work, having actions taken against their pharmacy licenses, habitual lying, extensive cover-ups, divorce, and suicide attempts. By all accounts, these recreational abusers' personal and professional lives suffered heavily from the drug abuse. In the end, most were reclusive and paranoid— what started out as collective experimentation ended in a painful existence of solitary addiction.

Therapeutic Self-Medicators

The criminal-deviant career paths of the 27 (54 percent) therapeutic self-medicators fit a different substantive theme. One of the defining characteristics of this group was that they had little or no experience with street or prescription drugs before they entered pharmacy school. In fact, many of these individuals did not even use alcohol. What little drug involvement they did report was usually occasional experimentation with marijuana. If they had ever used prescription drugs, they had done so legitimately under the supervision of a physician. Members of this group did not begin their illicit use of prescription drugs until they were well into their formal pharmacy careers.

The onset of the therapeutic self-medicators' drug use was invariably attributed to a difficult life situation, accident, medical condition, or occupationally related pain. When faced with such problems, these pharmacists turned to familiar prescription medicines for immediate relief. Rather than a recreational, hedonistic, or pleasure motivation, they had simply decided to use readily available prescription drugs to treat their own medical maladies.

Therapeutic motives for using prescription drugs. The therapeutic self-medicators unanimously insisted that their drug use was never recreational—that they never used drugs just for the euphoric effects. Instead, their drug use was focused on specific therapeutic goals. This trend is illustrated in the comments of a 33-year-old male pharmacist:

There was no recreation involved. I just wanted to press a button and be able to sleep during the day. I was really having a tough time with this sleeping during the day. I would say by the end of that week I was already on the road [to dependence]. The race had started.

Other pharmacists stated that they began using drugs as a way of treating insomnia, physical trauma (e.g., injury from a car accident, a sports injury, or a broken bone), or some chronic occupationally induced health problem (e.g., arthritis, migraine headaches, leg cramps, or back pain).

It is important to point out that during the earliest stages of their drug use, these individuals appeared to be "model pharmacists." Most claimed to have excelled in pharmacy school and continued to be successful after they entered full-time pharmacy practice. Personal appraisals, as well as annual supervisory evaluations, routinely described these individuals as hardworking and knowledgeable professionals.

Since they were usually treating the physical pain that resulted from the rigors of pharmacy work, all the therapeutic self-medicators described how their early use of prescription drugs began under seemingly innocent, even honorable, circumstances. Instead of taking time off from work to see a physician, they chose simply to self-medicate their own ailments. Many felt that they could not afford to take the time off to get a prescription from physicians who often knew less about the medications than they did. A 50-year-old male pharmacist described this situation as follows:

> When I got to [a job at a major pharmacy chain], the pace there was stressful. We were filling 300 to 400 scripts a day with minimal support staff and working 12- to 13-hour days. The physical part bothered me a lot. My feet and my back hurt. So, I just kept medicating myself until it got to the point where I was up to six to eight capsules of Fiorinol-3 [a narcotic analgesic] a day.

Peer introductions. Without exception, the therapeutic self-medicators described how there was always a solitary, secretive dimension to their drug use. Although they usually kept their drug use to themselves, many claimed that their initial drug use was shaped by their interactions with coworkers. That is, they got the idea to begin self-medicating from watching a coworker do or merely followed the suggestion of a concerned senior pharmacist who was helping them remedy a physical malady, such as a hangover, anxiety, or physical pain. For example, a 38-year-old male pharmacist described an incident that occurred soon after he was introduced to his hospital pharmacy supervisor:

> I remember saying one time that I had a headache. [He said] "Go take some Tylenol-with-Codeine elixir [narcotic analgesic]." I would never have done that on my own. He was my supervisor at the time, and I said to myself, "If you think I should?" He said, "That's what I should do." I guess that started the ball rolling a little bit mentally.

Members of the therapeutic self-medicator group took notice of the drug-related behaviors and suggestions of their peers but never accepted upon them in the company of others. Instead, they maintained a public front of condemning the illicit use of prescription drugs but quietly following through on the suggestive behaviors in private.

Perceived benefits of self-medication. Whereas the recreational abusers used drugs to get high, the therapeutic self-medicators saw the drug use as a means to a different end. Even as their drug use intensified, they were able to convince themselves that the drugs were actually having a positive effect on their work performance. This belief was not altogether inaccurate, since they began using the drugs to remedy health problems that were detracting from their work efficiency.

Some therapeutic self-medicators looked to their notion of professional obligation to justify their illegal drug use. For example, in describing his daily use of Talwin, a Schedule II narcotic analgesic, a 43-year-old male pharmacist maintained: "I thought I could work better. I thought I could talk better with the nurses and patients. I thought I could socialize better with it."

A slippery slope. At first, these pharmacists reported that their secretive and occasional therapeutic self-medication seemed to work well. The drugs remedied their problems (e.g., pain or insomnia) and thus allowed them to return to normal functioning. However, over time, they invariably began to develop a tolerance for the drugs and thus had to take larger quantities to achieve the same level of relief. In the end, each had to face the fact that the regular use of a seemingly harmless therapeutic medicine had resulted in a serious and addictive drug habit. The following comments of a 50-year-old male pharmacist offer a good overview of the life history of a therapeutic self-medicator:

Well, I didn't have a big problem with that [early occasional self-medication behavior]. I wasn't taking that much. It was very much medicinal use. It was not an everyday thing. It really was used at that point for physical pain. But that's when I started tampering with other things and started trying other things. I would have trouble sleeping, so I would think, "You know, let's see what the Dalmane [benzodiazepine] is like?" When I was having weight problems.... "Let's give this Tenuate [amphetamine] a try." And I just started going down the line treating the things that I wanted to treat. And none of it got out of hand. It wasn't until I came down here [to Texas] . . . that things really started to go wild.

It generally took between 5 and 10 years for these pharmacists to progress to the later stages of drug abuse. Such a time frame suggests that the therapeutic self-medicators were able to prevent their drug abuse from interfering with their personal or professional lives for a considerable time. For example, consider the exchange that occurred between the interviewer (I) and a 42-year-old male pharmacist (P):

P: Every time I [drank] even two martinis I [would] throw up. I [would] get diarrhea and [would be] . . . sick. So I took some Zantac [antacid]. I tried to cure my hangovers a little bit.

I: These were just for medicinal purposes?

P: Yeah, medicinal. Zantac [antacid], I mean how can that hurt? And I go to work, but I'm sick and I don't want to go in smelling like alcohol. Now I am deeply trying to just make it by. So now I begin to take pills to cure being sick so I can go to work. First, I'm taking things strictly to cure hangovers, which began happening with practically drinking nothing and it's scaring me to death. . . . So I start working and I start to take a few pills. I feel a little better. Now the [mood-altering] meds start to happen. I take a couple Vs [Valium, a benzodiazepine] now and then. I'm taking a few Xanax [benzodiazepine]. Next, I'm taking some Vicodin [narcotic analgesic]. It took years [for the usage pattern] to go anywhere. Then somebody comes in with drugs and says, "These are my mother-in-law's prescriptions, she passed away, she had cancer." I look at it, and it's all morphine. She says, "I don't know

what to do. Will you please take it for me?" [P replies, laughing], "We'll destroy the drugs, don't worry."

With the exception of their unauthorized self-medication, most of these individuals continued to be "model pharmacists." Despite their progressive drug use, they usually continued to garner the respect and admiration of their peers and employers alike. It was not uncommon for them to be promoted to senior management positions even after they began using prescription drugs daily. The bulk of the self-medicator group experienced a slow, progressive transition from the occasional use of therapeutic drugs to a schedule of repeated daily doses. In retrospect, they attributed their increased usage to the body's tendency to develop a chemical tolerance to the medications. This situation necessitated larger and more frequent dosage units to achieve the desired therapeutic effects.

A handful of therapeutic self-medicators were not so lucky; they had less time between the onset of their use and manifestation of drug addiction. The faster progression of their abuse is illustrated by the comments of a 49-year-old male pharmacist:

About two or three years after I had my store, I was working long, long hours. Like 8:00 to 8:00 Monday through Saturday and some hours on Sunday. And my back hurt one day. It was really killing me. . . . I started out with two Empirin-3 [narcotic analgesic], just for the back pain. I mean I hurt, my back hurt, my head hurt. I don't know why, but I just reached for that bottle. I knew it was against the law to do that, but I did it anyway. Man, I felt good. I was off and running. This was eureka. This was it. It progressed. I started taking more and more.

The key to a self-medicator's fast-paced progressive drug use seemed to lie in the person's perceived need to treat a wider and growing array of physical ailments. In fact, it got to the point that many "drug-thirsty" pharmacists now recognized that they were actively inventing ailments to treat. As a 40-year-old female pharmacist put it, "I had a symptom for everything I took."

These 27 therapeutic self-medicators had entered their pharmacy careers admittedly as extremely naive about drug abuse. They were either counseled or had convinced themselves that there was no harm in the occasional therapeutic use of prescription medicines. In short, the normative and behavioral advances in their criminal and deviant behavior were largely the result of a well-intentioned exploitation of their professional position and knowledge. The justifications for their drug use were firmly entrenched in their desire to excel in their jobs and to care for their patients efficiently. The therapeutic self-medicators always used their drugs in private, carefully disguising their addiction from others. Over time, their false confidence and self-denial allowed their drug use to progress significantly into addiction. Once their facade was broken, these pharmacists awoke to the stark reality that they were now chemically dependent on one or more of the drugs that they so confidently had been "prescribing" for themselves.

COMMON COGNITIVE AND BEHAVIORAL THEMES

Although there were clearly two different modes of entry into drug abuse for the recreational abusers and therapeutic self-medicators, these two groups of offenders were *not* mutually exclusive categories; that is, these two categories were not completely dichotomous. Rather, we identified a number of cognitive and behavioral themes that were common to almost all the respondents, regardless of how they initially began their illicit drug abuse careers. The existence of these common themes suggests that pharmacy-specific occupational contingencies play a central role in the onset and progression of the illicit use of prescription medicines. The three most common of these cognitive and behavioral themes are discussed next.

"I'm a Pharmacist, So I Know What I Am Doing"

Intuitively, it should not be surprising that pharmacists would steal prescription medicines to treat their own physical ailments. After all, they have been exposed to years of pharmacy training that emphasized the beneficial, therapeutic potential of prescription medicines. Each pharmacist has dispensed medicines to hundreds of patients and then watched the drugs usually produce the predicted beneficial results. They have all read the literature detailing the chemical composition of drugs and studied the often-dramatic beneficial, curative effects of these chemical substances. Pharmacists, more so than any other members of the society, are keenly aware of how and why drugs work. There was strong evidence to suggest that both the therapeutic self-medicators and the recreational abusers actively used the years of pharmacological knowledge they had acquired. So, when they developed health or emotional problems, it made perfect sense to them that they should put their knowledge to work on themselves. This personal application of pharmaceutical information can be seen clearly in the comments of a 40-year-old female self-medicator:

> In 1986 I was sent to the psychologist. That was when I was forced to recognize that I had an alcohol problem. And I recognized that I had to do something. And in my brilliant analysis, I made a decision that since alcohol was a central nervous system depressant, the solution for me was to use a central nervous system stimulant. That would solve my alcohol problem. So I chose the best stimulant that I had access to, and that was [pharmaceutical grade] cocaine. I started using cocaine in 1986. I never thought that it would progress. I never thought it was going to get worse. I thought, "I'm just going to use it occasionally."

Similar trends were observed among the recreational abusers, but their applied use of drugs was based on more recreational motives.

Virtually all the therapeutic self-medicators and the recreational abusers described how they became masters of quickly diagnosing their own ailments or emotional needs and then identifying the appropriate pharmacological agent that would remedy the problem. Moreover, as professionals, they were confident that they would be able to limit or self-regulate their drug intake so they would never become addicted. All the respondents drew on their social status as pharmacists to convince themselves that their drug use would not progress into dependence. As a

40-year-old female self-medicator put it, "I'm a pharmacist; I know what I am doing." The respondents all agreed that a well-trained, professional pharmacist could not possibly fall prey to drug addiction. They recalled being even more adamant in their view that they were immune from such problems, believing that only stupid, naive people became addicted to drugs. This distinct form of denial is illustrated in the remarks of a 35-year-old male recreational abuser:

Yeah, I thought, "It [addiction] can't happen to me because I know too much." We somehow think that knowledge is going to prevent it from happening to us when [we know that] knowledge has nothing to do with it. It's like heart disease or anything else. It's like "Well, I know about this, so it can't happen to me." . . . Now I teach pharmacy. I developed a chemical dependency curriculum at our pharmacy school. I do a clerkship in it, and I don't think there is one in the country, except for mine, that deals with some of that. Maybe they [the students] can personalize [pharmacists' drug abuse] a little bit.

A 39-year-old male self-medicator went so far as to say: "I mean, we know more [about the effect of drugs] than doctors. We have all the package inserts. We have the knowledge. We know a lot about the drugs, so what's the big deal?" The respondents did not understand why they should use sick leave to go to a physician and pay money to acquire a written prescription to dispense a drug that was on the shelves right behind them—especially since they firmly believed that pharmacists know more about dispensing medicines than most physicians. Elsewhere (Dabney & Hollinger, 1999), we referred to this denial mechanism as a "paradox of familiarity," arguing that familiarity can breed consent, not contempt, toward the use of prescription drugs.

No Cautionary Tales or Warnings

Remarkably, the vast majority of both the recreational abusers and therapeutic self-medicators claimed that they had never been warned about the dangers of drug addiction. Rather, they insisted that their formal training had stressed only the beneficial side of prescription medicines. For example, a 48-year-old male recreational abuser stated:

I never had anybody come right out and tell me that [prescription drug abuse] was probably unethical and illegal because they assumed that we knew that. But nobody ever said this is something that is not done.

Left without precise ethical guidance on the issue, some pharmacists assumed that their drug use was acceptable behavior. In explaining this point of view, a 39-year-old female self-medicator stated:

It's [self-medication] . . . just part of it [the pharmacy job]. It's just accepted because we know so much. I'm sure it's the same way when the doctors do it. It wasn't a big stretch to start [thinking], "You know, I got a headache here; maybe I should try one of these Percocets [narcotic analgesic]?"

In fact, many pharmacists spoke about their theft of prescription drugs as if the drugs were a fringe benefit that went along with the job. Much like a butcher who eats the best cuts of meat or a car dealer who drives a brand-new automobile, pharmacists always have access to free prescription drugs. This theme is illustrated in the comments of a 45-year-old male pharmacist:

Why take plain Aspirin or plain Tylenol when you've got this [Percocet—a narcotic analgesic]? It works better . . . [so] you don't even have to struggle with it. I really believed that I had license to do that . . . as a pharmacist. I mean with all that stuff sitting there, you know. Oh, my back was just killing me during that period . . . and this narcotic pain reliever [was] sitting right there. I thought, "Why should [I] suffer through back pain when I have this bottle of narcotics sitting here?"

Out-of-Control Addiction

The aforementioned themes involve cognitive dimensions of the pharmacists' drug abuse in that they speak to common motivational and justification themes that were present in all the interviews. Perhaps more important is the fact that there was a common behavioral characteristic that all 50 pharmacists shared. In every case, the occasional abuse of prescription drugs eventually gave way to an advanced addictive state that was marked by an enormous intake of drugs, unmistakable habituation, and the constant threat of physical withdrawal. Both the recreational abusers and therapeutic self-medicators routinely reported daily use levels exceeding 50–100 times the recommended daily dosage. One pharmacist noted that his drug-use regimen progressed to 150 Percocets [a strong narcotic analgesic] per day, another reported injecting up to 200 mg of morphine each day, and still another described a daily use pattern that, among other things, included 5 grams of cocaine.

Invariably, these advanced levels of drug use led to clear signs of habituation and the constant threat of physical withdrawal. At this point, the respondents recalled, they grew increasingly desperate. Consider the following quote from a 44-year-old male pharmacist who was in charge of ordering the narcotics at the independent retail pharmacy where he worked:

I was ordering excessive quantities and chasing down drug trucks. That's what I used to do. I was really reaching my bottom. I would chase these delivery trucks down in the morning, because I didn't come to my store until midafternoon. I was in withdrawal in the morning, and I was without drugs, so I had to have it. I was just going nuts. Many mornings I had gone to work sweating. It would be 30 degrees, it would be January, and the clerk would say, "You look sick," and I would say, "It's the flu." So I would pay the delivery guys extra money to deliver my drugs first, or I would chase the delivery trucks down in the morning. I knew the trucks delivered at 6 in the morning, they came by my area, and I would get up early and chase the trucks down the highway. I would go in excess of 100 miles an hour trying to catch up with this truck and flag it down.

The advanced stages of addiction almost always produced traumatic physical and psychological events, as in the following comments by a 39-year-old male pharmacist:

I was out of control for four years. I was just lucky that I never got caught. I don't know how I didn't

get caught. I fell asleep twice coming home on Interstate 95. I fell asleep at the wheel doing 70 once, and then I scraped up the side of the car and blew out the tires. I also tried to kill myself with a shotgun. She [my wife] was going to leave me. My world was falling apart, but I couldn't do anything about it. I didn't know what to do.

These out-of-control drug-use patterns, along with the realization of their chemical dependence, left the pharmacists in a problematic mental state. It was at this point that all the respondents recalled coming to grips with their addiction. This realization was accompanied by a shift in the way they thought about their drug use. They no longer denied the situation by drawing on recreational or therapeutic explanations. Instead, they finally admitted the dire nature of their situation and became more and more reclusive. In short, all the respondents grew to realize that they had a drug problem, turning to fear and ignorance to foster the final weeks or months of their addiction.

CONCLUSIONS

The pharmacists who we classified as therapeutic self-medicators resembled loners with defensive motives who engage in acts of individual offending. However, the criminal-deviant careers of the recreational abusers were not so static. These pharmacists followed an evolving career path, starting out as "peers," growing as "colleagues," and then spending their mature years as "loners." Corresponding changes were also observed in the nature of their criminal-deviant transactions and motives. Although ideological issues appeared to remain constant throughout, the longer the habitual drug use continued, the more similar the traits of the recreational abusers and therapeutic self-medicators became.

Drug treatment professionals have long been aware of the propensity for pharmacists to abuse the drugs that they are entrusted to dispense (Hankes & Bissell, 1992). At present, members of the drug treatment community and state licensure agencies alike generally operate under the assumption that there are two separate, mutually exclusive types of drug-using pharmacists: those who use for therapeutic reasons and those who use for recreational reasons. Although we were able to confirm this general trend, our data suggest that treatment professionals can benefit from taking a fluid approach to prescription drug users. It appears that considerable time, effort, and money can be saved by evaluation and treatment modalities that are sensitive to the evolving and overlapping qualities of prescription drug use among pharmacists. Most important, our research, which targeted individuals with severe histories of drug abuse, provides a glimmer of hope that focused prevention efforts may well deter early drug-use behavior from maturing into full-blown addiction.

REFERENCES

Dabney, D. A., & Hollinger, R. C. (1999). Illicit prescription drug use among pharmacists: Evidence of a paradox of familiarity. *Work & Occupations, 26,* 77–106.

Hankes, L., & Bissell, L. (1992). Health professionals. In J. H. Lowinson, P. Ruiz, R. Millman, & J. G. Langrod (Eds.). *Substance abuse: A comprehensive textbook* (2nd ed., pp. 897–908). Baltimore: Williams & Wilkins.

Martin, S. (1993). Pharmacists number more than 190,000 in United States. *American Pharmacy,* NS*33*(7), 22–23.

14

Doctors Tell Their Stories of Medicaid Fraud

Paul Jesilow, Henry Pontell, and Gilbert Geiss

The establishment of the Medicaid program provided many opportunities for physicians to violate the administrative regulations and commit criminal acts for monetary gain. The authors interviewed 42 medical doctors who were arrested for scams against the Medicaid insurance program. In questioning the study participants, the research team focused on the participants' interactions with Medicaid enforcement officers and found that doctors who committed Medicaid fraud offered explanations in the way of justifications and excuses to rationalize their white-collar crimes. These physicians had a vested interest in explaining their side of events in order to present a normal self and to convince others that their unprofessional and illegal behavior was done in the best interests of their patients. These rationalizations are quite similar to other offenders' accounts presented in other chapters.

In the course of our research we interviewed forty-two physicians apprehended for Medicaid scams. To hear them tell it, they were innocent sacrificial lambs led to the slaughter because of perfidy, stupid laws, bureaucratic nonsense, and incompetent bookkeepers. At worst, they had been a bit careless in their record keeping; but mostly they had been more interested in the welfare of their patients than in deciphering the arcane requirements of benefit programs. Certainly, the Medicaid laws are complex and, by many reasonable standards, unreasonable. But we were surprised by the number of rationalizations that these doctors offered, by the intensity of their defenses of their misconduct, and by their consummate skill in identifying the villains who, out of malevolence or ineptitude, had caused their downfall. In these doctors' system of moral accounting, their humanitarian deeds far outweighed their petty trespasses against Medicaid.

Rationales and Rationalizations

The sanctioned doctors generally appeared open and candid, at ease and involved with the subject. Most were perfectly accurate in

EDITOR'S NOTE: From Jesilow, P., Pontell, H., & Geiss, G., *Prescription for Profit: How Doctors Defraud Medicaid.* Copyright ©1993. Reprinted with permission from the University of California Press.

response to our opening question—which asked them to provide the factual details of their cases—though these recitals were interladen with a plethora of self-excusatory observations. Throughout the interview, we gave the respondents a great deal of leeway in responding, and we sought to avoid putting words in their mouths or guiding them in any particular direction. They could be rude or polite to us (most were very polite), satisfied or disgusted with the government, and optimistic or pessimistic about the futures of their careers. At times, some doctors told us more than they realized. It is difficult in a long, sometimes emotional interview to camouflage strongly held convictions.

All the doctors in the sanctioned group had been suspended from billing the Medicaid program, and about two-thirds had been convicted of a criminal offense. Nonetheless, a doctor often would ask us, "What did I do that was so bad?" Clearly, their interpretations of the ethical and legal character of their actions were quite unlike those made by the law enforcement authorities.

Because we carried out our interviews several years after the offenses had taken place, we could not determine whether the doctors had fashioned their explanations before or after they committed the abuses—an analytical issue that has bedeviled all researchers attempting to verify the importance of neutralization techniques in lawbreaking. Most likely, we heard explanations of both types. Our data do tend to support the hypothesis that neutralization often constitutes an important element of what has been called the "drift" into illegal behavior, a period during which the perpetrator's episodic lawbreaking often goes unattended and thus begins to lose whatever unsavory moral flavor it might have possessed.[1]

Denial of Responsibility

Few physicians took full personal blame for their violations in the sense of describing them as volitional, deliberate acts of wrongdoing. They were apt to call their activities "mistakes," and some blamed themselves for not having been more careful. This neutralization practice corresponds to what Sykes and Matza call denial of responsibility: "Denial of responsibility . . . extends much further than the claim that deviant acts are an 'accident' or some similar negation of personal accountability. . . . By learning to view himself as more acted upon than acting, the delinquent prepares the way for deviance from the normative system without the necessity of a frontal assault on the norms themselves."[2]

The wrongdoing physicians did engage in a frontal assault on Medicaid norms, but they typically laid the blame on a wide variety of persons other than themselves. Several blamed patients' demands, portraying their own behavior as altruistic. One insisted she was doing no more than trying to see to the essential health of needy people: "Some of the kids didn't have any Medicaid, and you get a mother saying, 'Look, my child is sick. I don't have any Medicaid. Could you put it on the other kid's Medicaid?' It probably wasn't their child to begin with. It was like a sister's child." The physician admitted that she complied with the mother's request, and her "goodheartedness" got her into trouble with the government. She says that during the investigation, "the mother who brought in her sister's child forgot that this kid was treated because it was like a year ago." The mother's lapse of memory or fabrication, the doctor suspected, occurred because the mother hoped to avoid implicating herself in the fraud.

The same physician also insisted she had been victimized by thieves who stole Medicaid cards from beneficiaries and then presented themselves for treatment. Her practice was in "a bad area," and such thefts were common, she pointed out. When the itemization of treatment services came to the legitimate cardholders, they would complain to the authorities. Even if this was true, however, her explanation sidestepped the issue of her responsibility to match the Medicaid card with the person presenting it.

A psychiatrist also argued that his problems resulted from the irresponsibility of a patient, a young woman with whom he had had sexual relations. He had charged Medi-Cal for the time he spent with his mistress only so that his wife, who handled some of his billing, would not become suspicious:

People knew I was seeing this individual after-hours, when the staff went home. My wife knew I was down at the office. So the only way to cover that visit and not let there be any suspicion was to bill Medi-Cal. The purpose was not to defraud Medi-Cal in the pure sense of making money. It was to simply protect the relationship, so my wife wouldn't be suspicious. Why am I seeing a person and not charging them anything? She [the mistress] was totally in agreement. I mean, it was the only basis on which we would be allowed to continue. It was her suggestion. "Why don't you just continue to put in the charges under Medi-Cal? I am not going to blow the whistle on you." So she was in the scheme as well as I was.

By blaming the patient for initiating the sexual relationship and the Medi-Cal fraud, the psychiatrist was able to neutralize his guilt about violating ethical and legal standards. He further buttressed his self-image as an altruistic human being by suggesting that he was doing his mistress-patient a favor, in view of how demanding his practice was.

Another physician got into trouble for accepting kickbacks from a laboratory. (Physicians can bill Medicaid for laboratory tests if they own the testing facility; otherwise the laboratory bills Medicaid.) This physician told us he had decided to do his own lab work in order to increase his income. A former employee, whom the physician held responsible for his misfortune, offered the doctor a deal:

My lab technician quit to start his own laboratory. He said: "Why don't you give me all the lab work." I said: "Fine, you bill Medicaid, and I'll bill my private patients."

They were doing the tests for so much, and I charged them the going rate, and he was giving me a good deal, and that was a private deal. The Medi-Cal, he was doing it all and billing it himself.

Then I told him: "I'm going to get a technician so I can have the benefits of the laboratory, of Medi-Cal too." [But] eventually he says: "I'll give you some benefits on your private patients. For instance, your bill is $300. I'll cut it down to $200 or something so we'll make it up somehow." I said: "All right."

Another physician who took kickbacks portrayed himself as an unwary, passive participant, motivated only by amiability and generosity, in a plan hatched by a hospital. The initiative came from the hospital, and the direct beneficiaries were his employees, so, as far as he was concerned, he had done nothing wrong in allowing the hospital to underwrite his payroll:

Hospital, privately owned, was in the habit of giving kickbacks to physicians using the hospital. I had arranged for three of the girls that worked for me to receive part-time pay to the tune of about $250 each, per month.

This lasted about two years before it was stopped, and the amount of work that they did for the money they received was negligible. So they were able to show in court that this was an indirect type of kickback. Even though the money was not paid to me, by the girls receiving this money, it obviously made them happier or better employees or whatever you want to call it.

I felt that if I didn't accept it, that if I let the girls take it, and then made sure that they got their full salaries and their Christmas bonuses, that I wasn't actually getting any benefit out of it. I thought that therefore I was immune.

And I really wasn't getting any benefit out of it. I had three employees—two of them were getting divorces, and one of them had a third child. The hospital wanted to give kickbacks, let them have it, you know.

An obstetrician, perhaps truthfully, cast responsibility on the welfare department, which had told him how to circumvent an inconvenient regulation:

Even the [welfare] department told me to change dates . . . to be within the letter of the law. For example, some girl delivers the baby, she decides to have her tubes tied. Now, according to the state, there has to be an application thirty days ahead of tubular ligation, and it has to be submitted, approved, and thirty days given for the patient to make up her mind, to decide.

If she hadn't given us any indication to the ligation, and she has to have one, what do we do now? I can't say, "You have to go home and come back in thirty days." And so they [the welfare caseworkers] told me, as well as the other doctors for Medicaid, they'd just say: "Backdate the request thirty days."

Commonly, denials of responsibility were blended with other self-justifications. A psychiatrist who illegally submitted bills under his name (and took a cut) for work done by psychologists not qualified for payment under Medicaid blamed the therapists but also added that he did it for the benefit of his patients:

> There were times I wanted to quit, but the therapist would say that these people are in need of therapy, and it is going along well. It seemed to make sense at the time. They were qualified people. I couldn't do it myself; I wasn't there all the time. It was partly a moral thing. I was persuaded to keep doing it, and a good percentage of patients were getting something out of it. I should have been more responsible, but it also had to do with my trust in people, and that trust was misplaced.

This physician, perhaps as a reflexive bow to a major postulate of his vacation, commented, "I don't want to think of myself as a victim, so I want to take responsibility for what I did." Yet he found irresistible the idea that it was his essential goodness—his trust in people and his sympathy for patients—that had led him astray.

Denial of Injury

Justifying lawbreaking by citing the superordinate benefits of the act—such as the psychiatrist's comment on the value of the therapy for his patients—is called denial of injury in the roster of Sykes and Matza's techniques of neutralization. As they point out, "wrongfulness may turn on the question of whether or not anyone has clearly been hurt by [the] deviance, and this matter is open to a variety of interpretations.[3] Physicians often depicted Medicaid regulations, which assuredly can be both onerous and mercilessly nitpicking, as bureaucratic obstacles, erected by laypeople, that threatened patient care. By breaking or bending the Medicaid rules, the sanctioned physicians argued, they were responding to their higher calling and helping—not hurting—patients.

Taking this tack, a physician who treated obesity by performing surgery emphasized that he was motivated only by "medical reasons" and "didn't give a goddamn what Medi-Cal said":

> Consider patients on welfare. There's a huge group of people who are on welfare because they cannot work. Nobody will give them a job. Their obesity serves as an excuse for remaining in the welfare system. To themselves they just say: "Well, I'm fat; I cannot get a job; therefore, I have to be on welfare." And they're satisfied with it. To alter that situation is hazardous, both from the emotional standpoint, but particularly in terms of physical aspects of it because if they eat enough, no matter how you loused up their gut, they're going to manage to remain obese. Earlier on, fifteen years ago, I did a few welfare patients, and I soon recognized the problem that particular group has especially. The bottom line is that there is a subconscious need for obesity.
>
> Well, I came up with the idea many years ago of saying: "OK, if you want this done, I've got no handle on what your subconscious state is, how much you need your obesity; there's no test for it. But the pocketbook is pretty close to the subconscious mind. If you are willing to pay for something, chances are you want it." Doesn't work that great; but at least it's a way. So what I started doing was charging them in advance: You want surgery, you come up with the money. Your insurance company happens to pay it back, you know, pay the full fee; well, I'll give it right back to them.
>
> I must admit that I did some Medi-Cal patients before I came up with this gimmick. And one or two of them worked very well. They became employable and very successful. But on the other hand, for every successful one, I would find two or three that became a disaster and had to be taken down and reoperated, all kinds of problems.
>
> So, anyway, I started this business. So I go: "OK, you want this surgery; you pay half of it." Now, at that particular point in time, I didn't give a goddamn what Medi-Cal said. I mean, if the patient wants this done, whether it's legal or illegal. I said I'm doing this for my reasons— medical reasons. What they'd have to come up with was maybe $100 or $500, whatever. The fee that Medi-Cal was paying at that time, plus the $500, was still less than what it would be for a private patient. Medi-Cal was paying maybe $500 or $600.

I didn't give a damn about the money. It wasn't as if I got a patient in the emergency room and Medi-Cal only paid me $150, but my fee was $600, and so I tried to collect the balance. That wasn't my intent at all. Welfare patients do not expect to pay you. But this was a volitional thing. They knew about it in advance. It was not an emergency. It was purely elective, cosmetic. I knew I couldn't bill. That's the only part of the regulations I knew and recognized.

It wasn't as though there was fraud involved. I wasn't defrauding anybody. If you want this surgery, you pay me before surgery, not afterwards. If I were billing them afterwards, I recognize that was bad. But this isn't the same thing. I still don't think so.

I really thought I had some really nice results on a few of them. I did have some disasters. That is why I started this. If I had just had the disasters and said "the hell with it, there's no answer to this, get out," I would have avoided it, because I didn't need it. It didn't amount to that much money. I could have been doing a private patient and come out way ahead.

Of course, the physician could have employed tactics other than reaching into a beneficiary's wallet to determine the patient's motivation—for example, adherence to a diet and exercise regimen prior to the operation. That the preoperative payment was intended to preclude reneging on the fee, rather than to measure motivation, is implicit in the physician's subsequent statement:

Now, plastic surgeons, for example, have done this for years. If you want to have your nose fixed, you pay them right up front. And for the same reason. Because they know that if you have to try to collect afterwards, the patient's going to find five hundred reasons why their nose isn't the way they thought it was going to be. If you've already paid for it, they'll be happy and satisfied.

Such comparisons with other specialties were common among doctors seeking to explain away their violations of the law. The Medicaid rules were capricious, and their own interpretation of fairness was far more sensible than that of the bureaucrats. Consider this psychiatrist's self-righteous indignation:

My wife is an eminently qualified psychiatric nurse. She had a medical teaching appointment on a medical school staff, supervisor of their inpatients, very qualified individual. So, anyhow, the basic problem was, she worked for me. They accused me—I don't know what the hell they accused me of—charging [Medicaid] for her services.

She was my nurse employee, just like I've had nurse employees everywhere I've been. I have always billed, like when the nurse gave a shot, you didn't bill it through her name. I don't even understand this concept, you know. I still don't understand what basis they can say arbitrarily that she is any different than a nurse that works for an obstetrician.

The way people are supervised in psychiatry is different than for general medicine. They never understood the difference, and still don't, and don't want to know. You can't talk with other physicians about it because they don't know anything about psychiatry, and don't want to know. You are in an esoteric field, that you have to be in to understand.

In a similar vein, another psychiatrist found the regulations senseless and the work he was doing eminently valuable for his patients. Besides, he had been able to bill in another setting for therapy provided by nonphysicians, so he could not comprehend why such an action was not permitted under Medi-Cal:

I worked as a convalescent lead psychiatrist at one time. Now that means working in one of these county clinics where you are seeing patients and the social worker is seeing patients. These people are being charged the full rate, and it is only the psychiatrist who has a Medi-Cal number. If it is all right in the agency, why isn't it all right in private practice?

I was aware of putting down my name and not any other therapist, you know, who wouldn't be honored. But nevertheless, they were still my patients, and everything that went on was under my signature and my supervision. Some social workers are better than some psychiatrists in the analysis of a problem. Some psychiatrists are not that good in understanding human behavior.

I will stand by unequivocally that the patients that were seen by me, in conjunction with others, were getting far more for their dollar, whether it is paid for by them, their company, or by Medi-Cal. They were getting far more from my clinic than they would get anywhere from one practitioner, and it was because there were certain areas in my training and intuition, my skills, where someone else could do better, and vice versa. As far as I am concerned, they are quality people—something I insisted upon.

My idea was with Medi-Cal, or with whatever I was doing, do what was right for the patients and for the patients' good. I consider four eyes better than two eyes, four arms better than two arms, and four ears better than two ears, and these patients were getting more in their hourly fee.

Another physician in his seventies got into trouble for failing to keep adequate records. In his patients' files he would note a prescription, but little else—a serious violation. The doctor explained his disagreement with the government this way: "They felt that I had not documented sufficiently the patients' records. I don't think so. I recorded notes each time the patients visited. Then I would record the medications prescribed. There is no need to write volumes because records are solely for my own purposes." To his way of thinking, the records were his private papers, not documents that might be needed by other doctors to ensure continuity of care.

Denial of a Victim

A third neutralization technique, denial of a victim, occurs when an offender grants that his or her behavior caused injury but insists "the injury is not wrong in light of the circumstances."[6] In our interviews, the sanctioned physicians claimed that although the law had been broken, the excess reimbursement they had received represented only what they deserved for their work. They saw overcharging for services and ordering excessive tests as ways to "make back" what they *should* have been paid.

One physician, for example, granted that his excessive billings were wrong, especially because Medicaid participation was voluntary; but he maintained that the regulations and payment schedules encouraged—even necessitated—cheating, so that doctors in the program could earn fees equivalent to those paid by private insurance:

If you voluntarily choose to accept Medicaid patients, you have to put up with their baloney, and if you're not willing to put up with their baloney, then maybe you shouldn't take Medicaid patients. So in that sense, it's difficult to say something is not fair and you shouldn't do it. . . .

But the system has got many flaws in it and loopholes and irregularities which necessitate abuses to take place. Otherwise, you can't see patients because of the reimbursement attitude.

Let's take an example. A patient comes in for the first time and is examined. For that it would be, let's say, $60. Now, if the patient goes to a general practitioner for the first time with a cold, he [the doctor] will get that amount. If the patient goes to an internist, who has to evaluate the patient for a complicated situation, such as diabetes, heart disease, god knows what, and spends a lot of time with that patient, he will get compensated the same amount. So it's all the same because it is a new patient visit.

Now for that reason, it is virtually impossible to [receive treatment] at this time in this area. There are virtually no internists in this area that I know who accept Medicaid, because if a patient comes to an internist, they expect a thorough going-over, which they devote anywhere from half an hour to forty-five minutes, and yet the reimbursement rate is exactly the same as if the patient went to a general practitioner with a cold and spent five minutes with him.

Now the system, of course, does ask, when you bill for this visit, whether you spent a lot of time with the patient or a little time. You are supposed to voluntarily say that it was a brief visit and, if you state that, they will pay you a less amount. However, very few people I know do that. They will always bill the maximum amount because that maximum amount is actually less than we charge our private patients. This is one door of abuse that virtually everyone I know who takes Medicaid is using. If they billed the patient with a cold for a very brief visit, then they get paid as little as $12. There's no one I know who can function in this area, with an office and a staff and insurance and all these things, and accept a patient for $12. I would say that form of abuse exists in 90 to 100 percent of doctors that I know who take

Medicaid. They're all using the maximum [reimbursement] levels.

This physician's conviction that virtually all his colleagues engaged in billing scams provided fuel for self-justification. As Sykes and Matza note, such a belief allows the perpetrator to transform the violation from "a gesture of complete opposition" to one that represents no more than "an extension of common practice."[4]

An anesthesiologist, caught billing the government for excess time, argued that he was reasonably charging for the patients' recovery time—a charge he knew was against Medicaid regulations:

> We saw no reason why we should do abortions on the garbage of the ghettos and barrios and be responsible for their recovery time and not be paid for it. I really think that it was gray, not black. I'm not defending it. In the context of the time, it wasn't really that bad. But it was stupid to try and do it considering how little money was involved and the horrible consequences. I should have known better. If you are going to steal from the system, it was a very stupid act. I don't think it was really a basically crooked act. I guess you could say it was, but it depends on how you look at it.
>
> When I did it, it was being done by at least 50 percent or 70 percent of the anesthesiologists in Southern California—at least 50 percent. People were fudging time, particularly on Medicaid.

Another physician also blamed the government for creating intolerable conditions that pressed practitioners toward fraud in order to meet patients' needs. The doctor illustrated what he saw as his dilemma by telling of a fourteen-year-old girl who was having her third abortion in less than six months. He had given the girl birth control pills, but she obviously hadn't taken them. When she came back for the third abortion, he coaxed her into allowing him to insert an IUD: "It's very simple, very easy. We'll put it in right now, immediately after the abortion."

The doctor then billed Medicaid for the IUD and its insertion, but the program would not pay because the insertion was done at the same time as the abortion for a fee the agency decreed reasonably covered both procedures. The doctor was irritated.

So they don't give a damn. It has to be a separate visit. So, then there's the question of getting this patient back. She's already got a local anesthetic in for the abortion. She says, "OK, do it. I won't feel it." You try to convince that same girl a week later or two weeks later? "Oh, no, I don't want those shots again" or "No, it's going to hurt. I don't want an IUD. I'll take the pill." It's another way of driving up costs. It drives up costs because they're going to have the girl pregnant again, and they're going to be perfectly willing to pay for more abortions rather than violate their rule. I paid $7, $8, $9 for an IUD. If I've been foolish enough to insert it immediately after an abortion—I can just forget getting paid. That's too bad. That's my problem. I should have made her come back in two weeks.

Now that's a medical decision that they have no business getting involved in. One of the ways I can handle this problem with this fourteen-year-old, if I've decided that the most important thing is her welfare, is I'm going to put in the goddamn IUD and put on the chart that I put it in tomorrow. Right? Because your health really should come before some bureaucrat, and there's no reason why I should be asked to throw away my money.

Another physician had essentially the same lament, one that also included an assessment of the kinds of people who receive Medicaid:

> They say you cannot bill for more than one service rendered on one day. They forget what kind of patients the Medicaid patients are. They don't keep appointments. They don't keep regular checkups. I know that. And being conscientious, when a patient came with a problem, or I saw that it was three years that they had not yet come in, and they were coming in because they had a vaginal itch, I would treat for the vaginitis which prompted that visit, but also I performed the yearly checkup—Pap smear, breast examination, all that—urinalysis, et cetera—complete physical.
>
> I knew that if I wanted to be within the strict letter of the law, I had to tell them to come tomorrow for the extra things—knowing that they would never come. They were poor. They didn't have transportation. So I used my better judgment and did everything they needed then. And to be within the program, I used next day's date, and that's where the problem was.

Various complaints about the nature of Medicaid recipients were offered by physicians in support of their view that Medicaid practice was more demanding than "normal" medicine and, by implication, should pay more rather than less. One doctor flaunted his disgust to us: "The Medicaid patients are filthy. They keep the place in turmoil. They are the toughest type to treat. The Medicaid patient is more demanding as a rule, and is not as cooperative in their treatment programs. I found this to be very troublesome at times."

Even if one were to accept the doctors' premise that the regulations were too inflexible given the difficulties of working with Medicaid patients, one has to wonder about the structural conflict between the physicians' interest in their patients' well-being and their own financial self-interest. For example, the physician who inserted IUDs after performing abortions argued that he was offering important care to patients. Yet he was unwilling to assume the small cost of the IUD and the minimal extra time to insert it; instead, he chose to cheat Medicaid and reap illegitimate profits. Before the advent of Medicaid, of course, many physicians performed services for indigent patients without charge. They spread the cost of care for the indigent among their fee-paying and insured patients. Government benefit programs now offer—if one is willing to cheat—the opportunity both to proclaim a humanitarian interest in the welfare of poor patients and to get paid at or above the going rate for that interest.

One physician we spoke with harkened to the theme of pro bono service, but quickly added that he was always ready to circumvent the law in order to obtain his fee from Medicaid:

> I would say personally I am disappointed with my colleagues in medicine. All of them are interested in their business. They are not concerned about the health of their patients. They want to make money. And, of course, they will make money. But the primary objective should be the care of patients. In medicine, the fee you get is a side effect.
>
> I was satisfied with the Medicaid reimbursement because I always knew that if somebody didn't pay me, my conscience would make me treat them anyway. I figured it was better to get

Medicaid than to treat them free. Some of the regulations are annoying, but you could always get around them. I would first treat the patient, then deal with Medicaid.

This blend of decency, self-righteousness, and a thoroughly high-handed attitude about "annoying" regulations that "you could always get around" nicely satisfied this doctor's conscience and cash flow.

Getting around the rules, playing the Medicaid game, working the system for maximum profit—many of the doctors we spoke with defined their illegal activities in such terms. As one doctor observed:

> The ones that use the system play it, just like a musician plays an organ, you know. You play it to get the maximum response out of it. You know which buttons to push in order to get the kind of response that you want. . . . I've always looked upon it as a game; you're dealing with this big nebulous monstrosity called the health care administration. The game was trying to extract the maximal amount of money out of them to see if you could raise it up to the level that you get from others. . . . I would get absolutely no satisfaction in following the regulations to the letter, which give you very low reimbursement.

Condemning the Condemners

Sykes and Matza describe condemning the condemners as a fourth neutralization technique: "The delinquent shifts the focus of attention from his own deviant acts to the motives and behavior of those who disapprove of his violations."[5] This shift enables violators to minimize responsibility for their behavior by construing it as trivial compared to the misdeeds of the rule makers or as rational compared to the irrational expectations of the rule makers.

In a typical condemnation of the Medicaid program, one of our respondents insisted that Medicaid not only invited but demanded cheating:

> It's not related to reality, you know. It's done by people who are not medical people, who know nothing about the services being provided. One of the peculiarities that they do is that they make

arbitrary decisions about things totally unrelated to the services you provide. They're constantly irritating and aggravating the doctors and their staff.

They say, "What we used to do, we're not going to do anymore." And you're already three months into your new billing system. I could keep you here the rest of the day giving you examples of their kind of idiocy, that they somehow manage to make sense out of in their little peculiar world that's unrelated to ours. They've built in systems that either ask for somebody to cheat, you know, or to cheat the patient on the type of care that's provided. You put somebody in the position where lying is the most reasonable course, and they will lie. The patients will lie; the doctor may even lie on what they say about what happened.

This physician illustrated his point by citing the often-criticized Medicaid rule that three months must elapse between compensable abortions. Suppose, he argued, a young woman had undergone an abortion in his office one week short of three months ago. According to the rules, he should tell her to come back the following week. But suppose she insists that she has to visit her sick mother in another state. This puts "everybody in a position because somebody has made some rule that doesn't make sense." A "naive" doctor would do the young woman "a favor" and postdate the reimbursement form. But, he concluded, "I don't know whether there's any naive doctors around anymore; they've been so hassled and harassed by this system."

That the Medicaid regulations might represent an attempt, however flawed, to control abuse, rather than an effort to harass or second-guess doctors, did not enter into the thinking of doctors who condemned the system as arbitrary, capricious, and unreasonable. One doctor could only fall back on the word "ridiculous":

I think they are ridiculous. We ask permission; we send them proof; we send them everything. One of them even asked me for a picture of the patient. I told him: "Who do you think I am, a crook or what? I am telling you this big long hernia there is hanging out of the testicles." I said: "Well, what do I send a picture of? Oh man, you are crazy." And that's what I do; I sent a picture, but it was absolutely ridiculous.

They are spending so much money. So many secretaries they have. They check all the cases in the hospitals, the Medicaid cases that go in, all the welfare cases they check.

A psychiatrist expressed a similar sense of frustration with what he regarded as petty interference with his professional autonomy: "There is a problem with the number of sessions. You had to make it sound like the person was really in bad shape. And then, they always sent it back, saying, 'Please give us more information,' which is more harassment. So you write it again, and send it in, and often that would be OK'd the second time."

Appeal to Higher Loyalties

The fifth and final neutralization technique discussed by Sykes and Matza is the appeal to higher loyalties. Delinquents engage in law-breaking, they say, to benefit smaller and more intimate groups to which they belong, such as their gangs or their friendship networks. Laws are broken "because other norms, held to be more pressing or involving a higher loyalty, are accorded precedence."[6] For the sanctioned physicians, such higher loyalties included service to patients and adherence to professional standards. In an unusual case, one physician insisted that being diagnosed with cancer prompted his cheating. He was worried about whether his infant son would have an adequate inheritance. In addition, he said he was despondent, angry, and bitter and wanted to get caught because of a wish to destroy himself or "to get back at the world" for his illness.

A Subculture of Delinquency?

Every sanctioned doctor we interviewed relied on one or more neutralization techniques to explain what had happened, and only rarely did we hear even the most elemental acknowledgment of self-serving motives. The structure of Medicaid, as we have noted, offers more than ample opportunities to harvest rationalizations that locate blame on factors other than the offender's lack of restraint. At times, doctors agreed that they might have been more careful and diligent about supervising others or challenging unusual goings-on, but such admissions were most often accompanied by claims to have been concerned with more important, socially valuable matters. On occasion, we heard physicians suggest that their own stupidity led to their apprehension—but it was the method of cheating, not the cheating itself, that they regretted.

The tenor of the interviews indicated that the cavalier attitudes these doctors had adopted toward the government benefit programs had been at least partially absorbed from others in the profession, and that professional values may effectively neutralize conflicts of conscience. Here we took our cue from Matza's discussion of a "subculture of juvenile delinquency"—"a setting in which the commission of delinquency is common knowledge among a group" and which provides norms and beliefs that "function as the extenuating conditions under which delinquency is permissible."[7] A subculture of medical delinquency, we concluded, arises, thrives, and grows in large part because of the tension between bureaucratic regulation and professional norms of autonomy.

Physicians who cheated government programs were not committed to a life of crime and undoubtedly did not cheat on all their billings. Nor did they always steal from Medicaid or from private insurance programs; they probably were honest in much of their work. But when these doctors did defy Medicaid's legal requirements, they typically offered professional justifications in lieu of defining their activities as deviant, illegal, or criminal.

CONCLUSION

The establishment of the Medicaid program provided new opportunities for doctors and other medical practitioners and organizations to commit criminal acts and to violate administrative regulations. Our research did not yield a composite portrait of physicians who typically get into trouble with Medicaid, though there are some recurring traits in the roster of physicians dealt with by the authorities. The stereotypical image of the violator as an inner-city doctor associated with a Medicaid mill is misleading: Offenders include some of the most respectable members of the profession and physicians of all ages, specialties, and attitudes toward patients and government medical benefit programs.

NOTES

1. David Matza, *Delinquency and Drift* (New York: Wiley, 1964).

2. Gresham Sykes and David Matza, "Techniques of Neutralization: A Theory of Delinquency," *American Sociological Review 22* (1957): 664–70.

3. Ibid.

4. Ibid., 668.

5. Ibid.

6. Sykes and Matza, "Techniques of Neutralization," 669.

7. Matza, *Delinquency and Drift*, 33, 59.

15

MANAGING THE ACTION: SPORTS BOOKMAKERS AS ENTREPRENEURS

PHYLLIS COONTZ

In her exploration of the social, organizational, and occupational features of the booking enterprise, Coontz interviewed 47 bookies about their careers and experiences. Her findings dispel many stereotypical beliefs about this illegal profession having ties to organized crime or that bookmakers are unscrupulous in their dealings with customers. To the contrary, the author claims, bookmakers were found to be hardworking and dedicated to their business and are skilled and function with a high degree of integrity. Successful sports bookies are honest with their clientele and operate in a businesslike manner. Their relationship with customers is based on mutual trust, and bookies are well aware that cheating behavior on their part will result in losing the trust of clients, which they must maintain to survive in this criminal operation.

INTRODUCTION

As a cultural phenomenon, gambling in the United States has been an enduring contradiction—while it is widespread it is also heavily regulated. Despite the growth of commercialized gambling since the 1960s, some forms of gambling remain off limits. One of these is sports betting. Betting on sports events is illegal in every state except Nevada. This is somewhat puzzling given the popularity of both gambling and sports.

One of the ironies of this is that while Nevada is the only state that allows sports betting, one of the accoutrements for gambling on sports, i.e., the "line" or "point spread," can be found everywhere. It appears daily in the sports section of most major newspapers; it is a link on numerous sports news websites; it is central to sports commentary about winners, losers, and outcomes on network and cable television; it is packaged as advice that can be purchased through various sports services through 900 telephone numbers; and it is woven into conversations that take place at work, in restaurants, classrooms, dorms, country clubs, bars, and parties every day around the country. It is even available through one of the more than 700 gambling casinos operating on the Internet.

Some have suggested that sports betting might be the number one form of gambling (legal or illegal) in the United States (Scarne 1974; Reuter 1983; Congressional Hearings 1995). Industry analysts (e.g., Moody's and Standard and Poor) report that sports betting is one of the biggest growth areas in commercialized gambling (Moody's Industry Review 1995:L46). For example, by the end of fiscal year 1996, book-makers in Las Vegas handled $2.5 billion in wagers (O'Brien 1998). And according to the Gaming Control Board of Nevada, the amount wagered on sports events has been increasing every year and more sports events are being added to the menu. For example, sports books in Las Vegas added last year's Women's World Cup soccer final to their array of events. Focusing on only a single event, the Super Bowl, the amount wagered almost doubled from $40 million in 1991 to $70 million in 1997 (personal communication, Research Division, June 1998). While illicit wagers are far more difficult to track, it is estimated that in 1997 alone, over $100 billion was wagered illegally on sports (U.S. News & World Report 1997).

Given the magnitude of legal or illegal sports betting and the popularity of sports, it is difficult to understand why it continues to be outlawed. Of course, sociologists have long recognized that a behavior's legal status has little to do with its popularity, intrinsic harm, or even desirability. Far more important in determining legal status is the way the behavior is socially constructed. In the case of sports betting, we see that one of the dominant social constructions involves bookmakers as unsavory characters with ties to the underworld. For example, supporters of the Professional and Amateur Sports Protection Act of 1992 argued that "[I]nstead of standing for healthy competition through team work and honest preparation, professional sport contests would come to represent the fast buck, the quick fix, and the desire to get something for nothing. Betting would undermine the integrity of team sports and public confidence in them . . . Fans could not help but wonder if a missed free throw, a dropped fly ball, or a missed extra point was part of a player's scheme to fix the game" (S-474, pg S-7302 Temp. Record). Other advocates of criminalization allege a link between bookmaking and organized crime and argue that it is a major source of its total gambling revenue (Cressey 1969; The President's Commission on Law Enforcement and Administration of Justice 1967; Clark 1971). The criminal construction is also common in our popular culture, particularly in films such as *Wiseguy 1986, Goodfellas 1990,* and *Casino 1995* and novels such Gay Talese's *Honor Thy Father,* Mario Puzo's *The Godfather,* and most recently Bill Bonanno's *Bound by Honor.*

How closely do the social constructions presented in the public debate and the popular culture mirror reality? The research on book-making shows that what is known about sports bookies, their methods of operation, career trajectories, life styles, and relationships with each other is limited, fragmented, and largely inferred. Some research on bookmaking generally suggests that instead of being members of a larger criminal syndicate, bookmakers are more likely to operate as a loose confederation of independents that may at times cooperate with one another (Chambliss 1978; Anderson 1979; Reuter and Rubinstein 1982; Reuter 1983; Rosencrance 1987; Sasuly 1982). Despite the lack of empirical evidence about the nature of sports bookmaking, the image of the sports bookie as criminal is pervasive.

This study set out to learn about sports book-making from those who do it. The assumption was that if sports bookies are a part of an organized crime syndicate, then there should be some evidence of it in the day-to-day operations of bookmaking. The study examines the social organization of bookmaking and the occupational features of being a sports bookie. This approach is useful because it allows us to move beyond the criminal aspects of sports bookmaking and consider what is actually involved in operating a sports book. That is, what sorts of skills, knowledge, processes, and contexts are needed to get into and manage a sports book.

The analysis draws upon face-to-face interviews with 47 different bookmakers operating in Western Pennsylvania. This researcher examines how one becomes a sports bookmaker; how sports bookmakers manage and grow their business; their relationships with bettors, other sports bookmakers, law enforcement officials, and family; and finally, how they exit the business. If sports bookmaking is indeed a part of a larger criminal organization then it is reasonable to expect that entry into and departure from the business would be determined by the larger organization; that the day-to-day operations would follow norms and practices dictated and monitored by an outside organization; and that the conditions under which bookmakers transact business with bettors would involve criminal elements such as coercion, dishonesty, and bribery.

BACKGROUND FOR THE STUDY

Although I have been interested in deviant behavior for more than 20 years, I became interested in sports betting only recently through a friend. It happened during dinner with this friend about six years ago when she excused herself so that she could call her bookie to place a bet on a football game. I didn't know she gambled on sports, and she certainly did not fit my image of someone who would know a bookie. She was a professional and civic minded and did not resemble my image of someone who associated with bookies. This contradiction between my friend and my beliefs piqued my curiosity and is what led to this research.

When I thought about the source of my beliefs, I realized they came mostly from the literature on organized crime and popular culture. I did a thorough literature review and discovered that there was little empirical knowledge about sports bookmaking. In order to learn what sports bookmaking was all about I would need to do some field work and find bookies who were willing to talk about bookmaking. Like most field work on illicit activities, one needs a contact to gain access. I explained my growing interest to my friend and asked whether she would vouch for me with her bookie. My plan was to place bets with her bookie, develop a relationship with him, and then explore whether he would talk to me about the business. For the next several months I placed bets on different football games for an entire season. Eventually, I shared my research interest with my bookie and he agreed to be interviewed. The first interview took place in my home. Simultaneously, I began talking more openly and frequently with friends,

acquaintances, and colleagues about betting on sports events and discovered that many of the people I came into contact with—both socially and professionally—bet on sporting events. These interactions led to contacts with other bookies, many of whom agreed to be interviewed. Over the course of a two-year period I interviewed a total of 47 bookies at least once and had follow-up interviews with 11 of them. All of the interviews were tape recorded, transcribed, and analyzed using NUD*IST (a computer based text analysis program). The findings presented here reflect major aspects of the organizational and occupations dimensions of the bookmaking business.

With the exception of the interviews with my bookie, all other interviews took place in locations selected by the bookmakers. Typically we met in restaurants, private clubs, and bars. The interview format focused on the process by which sports bookmakers got involved in the business; their strategies for building and maintaining their business; the duration of their involvement in the business; their relationships with other bookmakers; experiences with outside pressure (from law enforcement, other bookmakers, or professional criminals); the size of their business; the volume of the action; their families and friends; and their perceptions of the business, themselves, and its legal status. I begin the analysis with a general profile of the bookies that participated in this study.

DEMOGRAPHIC AND GENERAL INFORMATION ABOUT SPORTS BOOKIES

All bookmakers were male and ranged in age from 22–72 years old. The mean age for this group was 46. When asked about female bookmakers, I was told that none operated in this region, although some had heard about female sports bookies working elsewhere. All of my informants were originally from the region. All but two of my informants were currently married or living with someone, and all but five had children. Five had gotten divorced as a result of their bookmaking activities, and four remarried. Ninety-one percent ($n = 43$) were currently employed or self-employed in a legitimate job and five reported relying solely on earnings from bookmaking at different times. Four of my informants did not finish high school, 34 completed high school, and 7 completed college. Of those who completed college, one had a graduate degree and another was studying for an MBA.

On average these men had been involved in bookmaking for 17 years, ranging from a low of 2 years to a high of 45 years. None would reveal the amount of income earned from their bookmaking activities explaining their actual income was contingent on a number of factors such as the amount of a bet, the kind of bet made, the number of winners and losers, and whether losers paid their losses. However, all of the bookies in this study reported making money from bookmaking. Most experienced occasional "off" periods meaning they paid out more than they brought in. And one informant revealed he lost $96,000 the first year he went into business.

Like other illegal markets, sports betting is a cash business—all transactions are conducted with cash. Payouts were usually made midweek (typically Wednesdays) while collections were made the first part of the week (typically Mondays) and took place at designated locations, which were typically public. If a customer was unable to pay their losses in full, terms were established without an interest penalty. Although there are numerous kinds of wagers that can be made, the four most common were: straight bets, over and under bets, teasers, and parlays. The amount wagered varied by customer, but the average wager was $50.00 per game. All of these bookies established a minimum and maximum amount they would accept with the majority not willing to accept wagers over $5,000. Five, however, accepted bets for $10,000 and occasionally more. The majority accepted wagers only on the day of the game (and usually approximately 30 minutes before the start of a game), but some would accept bets when the line first came out.

The Mechanics of Sports Bookmaking

The "line" which is the basis for taking bets serves two general purposes: 1) to encourage the gambling public to gamble on both opponents playing a game and 2) to balance the amount of money wagered equally between both teams so that amount is equally divided between both teams. According to the "Roxy" Roxborough, the guru of the Las Vegas Line, "the fundamental concept of bookmaking is to force people to gamble . . . at odds favorable to the house. The line is used to encourage gambling by equalizing the contest" (Cook 1992). Sports bookies could care less about who wins or loses a game; they are interested in the final score and whether they have more losers than winners for any given game. The point spread then is the bookie's management tool for balancing the betting action on individual games.

The following illustrates how the point spread works. The Super Bowl is the final championship game between the two Conferences in the National Football League (NFL). Although each Conference has a champion, one is usually perceived to be stronger than the other (this is based on a combination of record, performance, and hype). In Super Bowl XXX (for the 1995–96 season) the Dallas Cowboys (the National Football Conference [NFC] champions) and the Pittsburgh Steelers (the American Football Conference [AFC] champions) played for the championship. At the time, the NFC and Dallas Cowboys were perceived to be superior to the AFC and Pittsburgh Steelers. Based on this alone, the outcome of the game was seen as a foregone conclusion—Dallas would win.

If you know the outcome of a game before it starts, it will not be very interesting. Because so many matchups are between unequal teams, the line equalizes the contest by giving the underdog an advantage. When the line came out for Super Bowl XXX it was set at 14 points, favoring Dallas. Translated, a 14-point spread meant that the Cowboys would need to win by more than two touchdowns. The question raised by such a high point spread for the bettor, and the question the bookie hopes the bettor will ask, is whether the Cowboys can beat the Steelers by more than 14 points. The line alters what is otherwise a foregone conclusion and entices bettors to put money on the underdog.

Getting Into the Business

As argued earlier, if sports bookmaking is controlled by outside interests then getting into the business should be determined by those associated with these interests. At a minimum, if the external control allegation has merit, there should be accounts of being recruited, coerced, pressured, or getting permission to get started from bookies. None of the bookies reported external pressure—all entered the business voluntarily and most viewed their entry as an opportunity to make money. The oldest bookie in the study, Artie, started out as a numbers runner for another bookmaker when he was only 8-years-old. In fact, he told me that during his early years (which he characterized as "hard times") he was the sole support for his family (including his parents and siblings). Artie was the only bookie I interviewed who progressed from numbers to sports bookmaking. He worked in the numbers business for over 30 years and then moved into sports bookmaking when he was in his early 40s.

> I ran numbers and made a pretty nice income for a kid. If you hustled $400–$500 worth, you could make a nice piece of change. I took care of my family—I literally paid the bills. I learned from one of the smartest guys in the area. He's dead now, but he was smart, very wise—he was a cousin of my mother's. He liked me, and he took me under his wing. I used to hang around him and one day he asked me if I wanted to do something for him. He sent me out to pick up the day's business—in a paper bag. I brought it back and he put me to work. He trusted me even though I was just a kid, he had confidence in me. You can't imagine how good I felt cause I looked up to him, he had something and everybody, they admired him too. He trusted me—this skinny kid. My mother didn't like it, but at the time, I was the only one working and bringing any money in. I made enough to pay the rent, food, and utilities. Times were hard back then.

Before setting up their own bookmaking operations, betting on sports had been an integral part of every bookie's everyday experiences. Joe, who is in his 50s and operated a legitimate sports book in the Caribbean, grew up around bookies and bettors. He entered the business as a young man while working in the steel mills and began his involvement as a favor to a coworker.

> I was helping a guy I worked with out. A lota guys bet with him and he couldn't handle it one weekend cause he was sick. We were buddies, he trusted me and asked me to help him out. I took all the action that weekend even though I didn't know what I was doin'. After spending that weekend on the phone I told my buddy I wanted to work with him and I did. The rest is history as they say. After a while I went on my own, I worked every day which is tough when you're workin too, especially during basketball and baseball season. You're takin action all day and all night long. You gotta time everything. You're on the phone all day and sometimes all night. How many games are played in a day during March Madness? You could have a couple dozen games in one day. My first wife really hated it cause I couldn't leave, we didn't go out—I worked all week and I was on the phone all weekend till the games were over. And if I had a bad day, if I got upset—I was an animal. One time I got so upset, I threw a chair through the TV. That was a long time ago, I'm still embarrassed by it. My kids hated it, too. Like now, they'd never get involved with it cause they saw what it did to me—my wife left me 'cause of it—it's your whole life. Things are different now though. The business in the islands, it's legitimate, it's totally different. I have time to enjoy all my hard work—I'm making so much money down there it'd make you crazy. I'm paying taxes, everything's legitimate. My second wife and I can travel, we can go wherever we want and do whatever we want. I could never make this kind of money at anything else. And the way I look at it, I've earned it cause I worked hard for over 30 years.

Steve, a student studying for his MBA at an area university at the time I interviewed him, said he was motivated by what he saw was a way to make money. Steve explains:

> There was a guy in the dorm and everyone bet with him. I started watching him and saw the potential for making money and thought to myself, 'hey I can do that.' So, one Saturday I started, just like that, with just a few guys I knew. At first I used the local line in the newspaper and I'd talk to other guys to see what they were gettin. I only deal with students around here, so it's not too complicated. I mean, at first I kept track of

everything in my head, but as I took on more customers I had to write things down—I couldn't keep them straight. I was doin pretty well and after a couple of months I had over 70 regulars. Then I got a couple of friends of mine involved, 'cause I needed the help—I give 'em a few bucks to help me out. You know, I'm not doin this forever, but for right now it pays the bills—I'm paying for school. I really don't want to have any debt when I graduate.

There was no evidence of coercion, pressure, or an apprenticeship to get into the business. The two main routes through which bookies got into the business were through their personal relationships and perceived economic opportunities. In all cases, entry was voluntary and situational.

Managing and Building the Business

Since bookies can never be certain of the number of bets that will be made on a given game or the amount wagered, being successful involves more than simply taking the action. While several factors affect success, five appear to be common among all bookies: knowledge of sports, building a credible customer base, volatility in the market, productivity of the bookmaker, and avoiding arrest. Roger, a young former bookie who now operates a legitimate sports betting advice service, explains how knowledge of sports plays an integral role in bookmaking and succeeding in the business.

I read everything that's printed about the games and players. I did that when I was in the business and I'm doing it right now. Every morning I'm on the Internet reading what the sports writers have to say, looking at statistics, checkin' the weather conditions, going over last year's statistics, everything. I study; study so I can handicap each game. Half my day is spent digesting information. We offer 1 or 2 picks each week—and my track record is pretty good on this, mostly winners. I miss it now and then, but when you put all the facts and figures together with what the sports experts are saying, then you're more likely to come up with winners. Even though I'm on the other side of the business now, the same rules

apply. Rule number one is that you've got to pay attention to the details. When I was takin bets, the only way you could make money was if your people picked losers. They're always looking at the Internet, the paper, and checkin with friends. So if you're gonna make money, you gotta have a solid line—one that's based on the facts and figures of the game. I handicapped all my games, see some guys get their lines from the books in Vegas, but not me, I did it myself. I always prided myself on offering a precise line especially on close games—you know, 1/2 point difference or maybe a 1 point difference on a close game can make a huge difference. And 1/2 point is appealing to a lot of bettors. You gotta be firm about your line too. Once you set it, it's the same for everybody. I like to think of myself as an expert on these games—you know, Jimmy the Greek. I've studied them all my life—ask me anything and I can tell you. Bettors are emotional, so they don't think about that 1/2 point difference—they're gonna go with their hearts. If the bookie's gonna make money, you gotta get 'em to go with their favorites and you do that by a solid line. You don't have to be perfect all the time, 50% of the time will do.

The illegal status of bookmaking limits what bookmakers can do to attract customers. They cannot advertise or make cold calls from the phonebook or random digit dialing, yet a successful bookmaking business requires a customer base. How do bookies build a customer base? Jimmy, a 22 year veteran, claims that the customers are just there waiting for bookies because of the popularity of sports.

Who doesn't like sports? Everybody loves sports—and betting on them is part of the entertainment. Ever been to a Steelers game? Look around right before a game. You see all these guys on the horn talking to guys like me. Betting makes a game interesting. Every guy I ever knew from my neighborhood and school bet on games. My Dad even bet—on boxing matches. My brother bet. Since I got into the business, I never had any trouble findin' customers. Listen, they're just there waiting to give you their business. Some of my customers I've known since school, some of 'em I work with, some I've met here or there. I've even had guys give me customers. You know my brother brought customers to me. My sister's husband bets with me. My biggest problem is keeping it manageable and making sure I got good people. I don't want too many customers, you know I work and that complicates things a little. Fifty, sixty's about all I can handle at a time. It seems like if I loose one, then I pick up another one or two.

In addition to building a customer base, bookmakers assume the financial risks in their business. Like any for profit business, the focus is on the bottom line which means bookies must be attuned to the market, especially patterns and changes in betting action. Volatility of the market affects success and while the line enables the bookie to balance the action between opposing teams, it is insufficient by itself for ensuring a profit. Knowing this, bookies take other steps to minimize potential risks from dramatic swings in the action. According to Carl, who has been in the business for 18 years, limiting your customer base and setting a maximum limit on wagers are two such steps.

> If you're gonna hold everything yourself, you've got to have a minimum of 40 players. Right now I got about 100 people. You've gotta have a strong limit, say $500 and not take any more on a game from any one person because if you get 20 of those people giving you a $500 bet, you've got $10,000 you could lose.

An instrumental factor that affects success is knowing how closely specific customers fit the average customer profile. For example, bookies know that bettors tend to wager on multiple games. Multiple game bettors always work to the bookie's advantage and Tony, who has been in the business for 25 years, explains why:

> As you research this, you'll find out that the more games people play, the more chance the bookie has to make money. They gotta pick about 66% to break even. They don't realize it 'cause everybody thinks if they pick 50% they're even, but they're not, they're losing money cause you gotta pay the juice. For the customer to come out and ahead, they gotta pick about 66% of the time. There's only one way to make money betting and that's to bet one game. Of course, most people bet

a lot of games and they bet their hearts—which is good for the bookie.

Like many small businesses, bookmaking requires a high degree of commitment from the bookie. I frequently heard this level of commitment described in terms of bookmaking being a bookie's life. Bookmaking has a higher priority than other activities. To sustain a business for any length of time, bookies must set aside their other interests and be on call for their customers. Depending on the sports season this could range from every day to every weekend of that season. The bookie must learn to manage his time and balance his bookmaking responsibilities against other responsibilities. Sam describes the various ways that bookmaking consumes the bookie's life.

> You can work year round or you can work a couple seasons. See there's a sport for every season. I work year round. Football's not so bad, but basketball, baseball, and hockey'll kill you—every single day, six and seven days a week. If you're gonna work year round, you basically have no life outside of this. It's not over just cause the phones stop ringin' either. You gotta keep truck of everything and I don't know many guys who can just turn out the light after the phones stop. I have my set times see, so half hour before a game starts I'm taking action. I get all my 7 and 8:00 games startin 6:30. Now Sunday football, I've got my 1:00 games and I start around 12:40, my 4:00 games I start around 3:40, and my 9:00 game I start at 8:45. You've gotta little free time in between, but then you're watchin the game to see whether you're makin money. Some games don't end till after midnight. Then I gotta get up and go to work. My wife complains, less now than she used to. I guess she's used to it now. I think extra money helps. She likes to go out, so I get someone to take over for a day or night. I got a couple of guys who'll do that, you know, answer the phone for me. We always go away in the summer, but I don't relax, cause I'm worrying about things. I'm checking every day when I'm away. It's just part of you, it's there all the time.

With the exception of online sports books (who accept credit cards), bookmaking is strictly a cash business. Thus, having customers who are willing to pay their loses is essential for profitability. In lieu of a credit report, the bookie depends on informal means of scrutinizing customers. Typically customers are acquired in two ways: either by referral from other customers (which happened in my case) or directly by knowing the customers. What happens when a customer refuses to pay the bookie? Although proponents of criminalization argue that bookies resort to the corruption and extortion to collect gambling debts, none of the bookies I talked to engaged in these practices. Instead, they avoid potential losses by screening their customer. Ray, who has been in the business for 35 years, describes how this process works:

> You see, it's not illegal to place a bet, but you can't take bets—that's why you always gotta watch yourself; most guys know that. You gotta know who you're dealing with. I don't take on anybody I don't know. I check, check, check. Say some guy calls me and wants me to take his action. If I don't know him, I won't do it 'cause I don't need the aggravation. I had to learn this though. There was a guy—oh this was about 15 years ago—who bet with a bunch of different guys. Back then what I thought was important was the numbers. I didn't know the guy so well, but I took his bets anyway. He ran up quite a number with me—he owed me about $10,000. I'd call him and he'd tell me he'd have something for me next week. Well, he never gave me a thing—this happened over and over again. Then one night I see the guy out, and he's buying drinks for everybody, playing the big shot. The guy owes me $10,000 and he's buying people drinks. I was mad. I went up to him and said, 'hey where's my money?' The guy was a jerk; he stammered around and told me that if I didn't leave him alone he'd call the feds. He owes me and he's gonna call the feds. I don't want the feds bothering me, I'm not interested in goin' to prison and I don't want anybody checking up on what I'm doing. I knew right then I'd never see any of it and that the guy was a jerk. So what I did was put the word out that he wasn't paying up, that he wasn't good for his bets. You gotta understand, this is a disease for someone like this guy—he just has to bet and there's nothin worse you can do to somebody who likes to bet than to cut him off. When the word is out, nobody will take their action. The only way someone would take his action is if he put money up first.

As discussed earlier, the line is used to balance the action between two teams. There are times however when the betting public is not enticed by fluctuations in the line. When there is more money wagered on one team than another, the bookie could end up having to pay out more than he takes in—a situation he wants to avoid. A degree of bettor psychology is relevant in such cases. Bookies know that sentimentality motivates many bettors. Sentimentality is the tendency to bet on the hometown team—or as one bookie described it, to "bet with your heart." For example, in the recent NBA championship game between the Indiana Pacers and the Los Angeles Lakers, the majority of bettors in the L.A. area bet on the Lakers while the majority of bettors from the Midwest bet on the Pacers. Sentimentality also can explain regional variations in betting patterns, and bookies must always be aware of potential regional differences in betting patterns, and adjust accordingly. When it is impossible to balance the action, bookies can hedge their losses by selling their action to other bookmakers. This practice is called "laying off." Every bookmaker in this study had connections with a layoff bookmaker. Sam, a 17 year veteran, discusses the way sentimentality works:

> What you wanna do is get equal amounts of money on both sides. This is a Steeler town so when the line first came out at 14 then 13 and 1/2 for the Super Bowl (1995), all the money was on Pittsburgh. I lowered it to 12, but still everyone was takin' Pittsburgh. Even at 12—nobody was taking Dallas. Even though I thought Dallas would win, I was worried. You know, Super Bowl is big, big, big. I was holding a lotta money. I checked around to see what was happening and after thinkin about it, I figured I'd better get rid of it, that's called a layoff, and if I hadn't gotten rid of it, I'd a been ruined.

Layoff bookmakers charge a fee for taking unwanted action. This can be thought of in terms of 'juice.' The existence of layoff operations is essential precisely because of the sentimentality factor. If movement in the line does not affect the direction of incoming wagers on a Pittsburgh/Dallas game, then the Pittsburgh layoff bookie could layoff with a Dallas bookie. Six of the bookies in this study took layoff action from other bookies and would sometimes layoff their layoff action with larger operators. It is important to distinguish the relationships bookies have with layoff bookies from organized crime associations. Bookies choose whether to layoff their action and they have choices of who to lay it off with. Instead of reflecting organized crime connections, these relationships more closely resemble what Reuter (1983:42) calls networks. Occasionally, bookies in these networks combined their bookmaking business with social activities, but this appears to be infrequent.

Although the typical bettor is the bookie's bread and butter, there are always exceptions to typicality and in bookmaking the exception is the wiseguy. The wiseguy almost always beats the bookie. Since bookies lose money to wiseguys they must know who the wiseguy bettors are and limit the action they will accept from them. Eddy, a 52 year old who has been in the business since high school differentiates the wiseguy from the typical bettor.

> The wiseguy is someone who follows the games—they really do. They know who they wanta bet before the line comes out of Las Vegas—oh, that's the outlaw line that comes out on Sunday night. They'll bet the game with you Sunday night. They know who they want to bet. They take a chance on the line going up or down. They're sharp, but remember this is their livelihood, that's what they do, they don't do anything else. The way it works is that when the game plays, the line might be at 16 and they're locked in at 10. Being locked in they have an option, they can keep the bet at 10 or take the 16 backup with somebody else and hope it falls in between, take the game, and make twice the money. They always bet with a couple of people and they study the games. I did that myself one Saturday a couple of weeks ago, it was the playoff games. I took 17 basketball games. I bet a couple of hundred per game. I was trying to catch it in the middle. I made $17,000 that afternoon. A lot of bookies have quit basketball because you can get destroyed with basketball. If you get a wiseguy who's hot, he'll beat you to death. Those guys just don't bet $100, they want to bet your limit—whatever it is, $500, $1,000, $5,000 whatever, and if they're hot, they'll win 9 out of 10 games.

In addition to developing a reliable client base and learning how to avoid losses, sports bookies can affect their profits by the number of hours they work and the sports they will accept. If they choose to, bookies can work year round or limit their business to particular seasons. Harry, who operated one of the larger businesses, took action on all sports events and explains the relation of this to being successful.

> I got into this for the money; I figured that the more I worked, the more money I'd make. And I'm good at this. But I've had to put in hundreds of thousands of hours at this thing and that's year round. I mean I take every kind of game there is. I take horses, baseball, boxing, football, the baskets, hockey, jai lai, you name it. If there's some esoteric thing going on, I'll take it too. You should see my house, I got TVs everywhere. I got one of those dishes so I can watch what's going on all over the world and I got a computer to keep up with the scores and see what other people are doin'. If you wanta make money you've got to put the hours in and you have to know what's going on. See it's not the bookie that puts the fix in, but gamblers. You see they're lookin for the big win. But like anything else, if you wanna make money you gotta have a work ethic and you gotta know what's going on. I'll buy whatever I need that gives me an edge.

One of the biggest risks from any form of illicit activity is arrest, but unlike other forms of deviant behavior bookies are not afraid of getting arrested. In fact, none of the bookies in this study had been arrested for their bookmaking activities. Tony explains why:

> The cops don't care about guys like me. I don't cause any trouble. I don't drink or do drugs and I don't know of any guys who do, cause for one thing you gotta be alert in this business and drinking and drugs will mess up your thought processes. So what kind of a danger am I? What I'm doing is providing a service, that's all, like any other service type of business. That doesn't mean I'm out there advertising what I'm doin', no way. I just don't do anything that's gonna draw attention to me. My customers are happy. So I'm clean, my people are happy, and everything's okay. The idea that we go out and break

people's legs is something you see on TV. What sense does that make? I mean, if I broke some guy's legs, I'd get arrested. Just doin this, nobody's gonna arrest me.

Retiring From Bookmaking

Like getting into the business, retirement is a personal and voluntary decision. While not all of the bookies talked about retirement, those who did identified three factors that influenced their decision to leave the business: the desire to spend more time with family, having achieved their economic goals, and retirement from their legitimate job. Pete, who has been in the business approximately 15 years, plans to retire within the next year because he wants to spend more time with his family. Pete discusses how the business interferes with spending time with his family and the preparation involved in leaving the business.

I don't need to do this any more. Financially, things are different now, my wife is working a little and I'm working a lot of hours, I've got seniority. I'm thinking in about a year this thing will come to an end. I mean I don't have the time anymore. I'm coaching my kids, spending a lotta time with them. With the money I've made, I've put it into my house. We couldn't afford the house we're in without this. When I look back on the years, they've been good to me. I made some money where I could buy some things—my house, I got Debbie a new car, and for my kids I've been putting money away for their tuition for school. You know I don't want them going to public school. With my kids, when I'm doing this, they're with me. And to be honest with you, I don't want them growing up with this thing all the time. Everyone in my family, even my wife's family, knows what I do. The way things are now, when there's a family thing, I'm tied up with this—I can't go cause I'm answering the phone all the time. I've talked about leaving with George, he doesn't have the responsibilities I have with my family and all, so he can just step right in cause the business is running smoothly. He wants to take it over, it's a good deal for him—I've got good customers, I know them, there's no trouble, and things are real good now.

Unlike Pete, Joe's retirement plan does not involve a successor; instead he shifted his business to the Caribbean where he owns a legitimate sports book. Joe views the earnings from his legitimate enterprise as his retirement cushion. In the following excerpt, Joe discusses how his legitimate operation will provide a comfortable lifestyle without the worry that accompanies an illegal operation.

The island operation is easy. You can advertise. We do, we have a full page advertisement in Las Vegas, in fact, even in some of the local papers we've got ads with a sports book number. You can call, set up an account. You can use a credit card, wire money, or send us a certified check. Somebody puts money into their account, we give them 10% interest on their money, sometimes 15% depending on the amount they have and the type of bets they make. If somebody puts a considerable amount of money in there, we give 'em 15% right off the bat. If the person is a bettor, you have the money in an account, you know you're getting paid, and that's a big part of the business, getting paid. So we give 'em those extra bucks back to use. It works, it works great. Your regular bettors, guys that have the money and want to bet have no problem puttin' the money up front. It's a beautiful way to do business. So, I'm looking at this as my retirement plan. Cause, I'm out of everything here and I don't work any work. I've got eight people employed around the clock there and the action comes in 24 hours a day and a woman who manages the operation for me. She's real good. The business grows each week. Even though I worked all my life, I'm not depending on social security to live on—I never trusted the government, I gotta look out for myself. So by the time I'm 65, I'll have enough money saved up from this business to continue to enjoy my life. Like I told you before, me and my wife like to travel, so this way, we can do whatever we want. You know, I'm not a flamboyant guy, I drive a modest car and we live in a nice house, but nothing outrageous. People know me, I belong to the club here, me and my wife come up here a couple times a week, have dinner, watch a game here and there. I'm just beginning to enjoy all that time I've put into this thing 'cause I'll tell you I've worked hard in this business, all those hours. For years, I couldn't leave the house cause I was on the phone all day and all night. But now, I can relax now.

Frank, who is 60, is planning to retire from his legitimate job and quit bookmaking at the same time within a year.

> I plan to retire in a year and I've been savin' the money I've made so I can enjoy myself. I got this house in Florida, so as soon as I retire I'm gone. I don't have anybody to worry about, my wife left me 30 years ago, no kids, so it's just me. I saved up a little nest egg so I can go here or there. If I wanta go to Atlantic City or out to Vegas every now and then for a week or so, I'll go. You know, I like to play blackjack. Right now, I'm only working with about 45 people anyway, so when I leave they can find somebody else. Some of my people already work with a couple of bookies, so I'm not worrying, they're set up.

Although Pete, Joe, and Frank give different reasons for leaving the business, their accounts show that retirement is their choice. Pete wants to spend more time with his family and has designated a successor; Joe has worked hard and wants to enjoy his life more fully and is depending on the income from his Caribbean operation, and Frank is planning to retire from bookmaking when he retires from his legitimate job and expects his customers to find another bookie. In all three cases, the decision to retire was voluntary. The different paths to retirement further indicate that not only is the decision to retire voluntary, but also the way that one retires is a choice.

TIES TO ORGANIZED CRIME

I asked about the alleged link to organized crime and found no support in bookies' accounts. Joe offers an interesting counter perspective suggesting that the allegation is a government ploy to control competition from bookies.

> The idea that the Mafia is pullin the strings on guys like me is a myth. We keep sports honest, believe me. If you're looking for the crooked angle, then government's the real criminal cause they really don't want competition from independents like me. So they keep betting illegal. The truth is that the independents do a better job than the government. Look, our payout is higher, we

> don't use gimmicks—think about all the gimmicks for the lottery—we don't have the overhead, and we don't cheat. Remember that Nick Perry incident? Bookies have integrity. You gotta have integrity, 'cause if I cheated I'd be outta business. And remember this, we're not on every corner or in every retail store either, pushing people to gamble. But the government can't do it right cause they're greedy and most people know that. Just look at the OTB (Off Track Betting). How could you lose money with horses? Think about it. When the government's involved there are too many hands in the till. Guys like me make the government look bad 'cause we know what we're doin' and they don't want us out there. So they gotta make it seem like we're a bunch of thugs fixin' games and workin' players. The only games I ever heard about being fixed are the ones you read about in the paper. I'm not saying that it never happens, but when it does, bookies aren't doing the fixin. It's usually gamblers.

According to Sasuly (1982), all of the major sporting associations (i.e., the NCAA, NBA, NFL, and Thoroughbred Tracks of America) maintain contact with illegal bookmakers as a way of monitoring irregularities. Bookies can be thought of as barometers of irregularities for the world of sports.

Since there have been numerous reports of athletes shaving points, I asked whether bookies had first hand knowledge of professional athletes betting on games. None of the bookies I talked to knew of any instances where athletes placed bets on or against teams they played for. However, bookies did acknowledge having customers who were professional athletes and who wagered on sports events. Roger describes his experiences with a professional athlete who bets with him.

> I knew players who gambled. I took bets from_____. He didn't call me every week, but once in a while he'd bet on this game or that game. Sometimes he'd win and sometimes he'd lose. If you're askin' whether he'd bet his own game, the answer is no. In fact, he wouldn't bet on any games while he was playing. You know, he had to keep it low key because he could be kicked out of the pros if anyone ever found out. So we kept it quiet. We'd meet at_____, you know a lot

of guys worked out there, you could be sorta invisible there, and you know no one noticed, so we'd square up in the locker room. Players gamble. Believe me, all the players know what the line is on their own game and every other game. But me, personally, I've never known a player who bet his own game. I have known guys who are no longer in the pros who bet on their team. The way I look at it is, it's just part of sports.

When unusual or irregular betting patterns emerge, bookies stop taking bets on the game. This practice is referred to as "taking it off the board." It would not be in the bookie's best interests to try to rig or fix games since it is the bookie that stands to lose. Moreover, rigged games would disrupt the logic of bookmaking and the inherent parity in sports betting.

DISCUSSION AND CONCLUSIONS

Contrary to the image found in the public debate and popular culture, namely that bookies are slick operators looking for the "fast buck, the quick fix, or the desire to get something for nothing" and have ties to organized crime, this analysis shows that bookmakers work hard, are dedicated, possess skill and specialized knowledge, and operate with integrity. Bookies put in long hours and make many sacrifices in other parts of their life in order build a successful business.

Despite the maxim that "you can't beat the bookie," bookies need paying customers and since they cannot recruit or solicit their customers openly, they must devise ways that weed out the poor risks. Bookies must also be attuned to betting patterns and know when to raise or lower the line in order to balance the action. When it is not possible to balance the action, bookies need to know when to layoff with another bookmaker to avoid heavy losses. Finally, successful bookies operate with honesty and integrity. The relationship between the bookie and his customer is based on trust, and bookies know that if they cheat their customers they will loose them.

While the findings show that bookies are fully aware that what they are doing is illegal (Ray captures this by saying, "You see, it's not illegal to place a bet, but you can't take bets—that's why you always gotta watch yourself"), they do not define themselves or other bookies as deviant or criminal. Instead they perceive themselves as providing a service to a niche market. What accounts for this nondeviant identity? One possible explanation has to do with the way bookmaking is socially organized. To the lay observer deviance appears to lack order, but sociologists know that like the social world of which it is a part, deviance is highly organized. The social organization of all forms of deviance is a microcosm of society—there are distinct roles, norms of interaction, relationships, and distinct settings and contexts for its occurrence. Best and Luckenbill (1994) argue that the social organization of deviance is an endless source of knowledge because it is where the relationship between the individual and the larger society is mediated (5).

Like many service transactions, the product in the transaction (i.e., the bet) is nonmaterial—it consists of a verbal agreement about the outcome of a sports event. But unlike most illegal service transactions, the bet does not require face-to-face interaction. This fact sets bookmaking apart from other types of deviant transactions. Bookies and customers communicate by phone which means that the transaction can take place anywhere—at work, at home, in a car, in a restaurant, or even while walking down the street. This distances the bookie and the bettor from typical deviant or subcultural settings such as back alleys, street corners, or designated areas like red light districts. The deviant transaction is integrated into the rhythms of everyday actions within the context of normal everyday settings.

In addition to occurring outside a criminal milieu, the structure of the transaction is unlike other illegal transactions because it is nonhierarchical. No middlemen are involved. While it is true that most bookies get their line from oddsmakers (usually working out of Las Vegas), the line nevertheless is in the public domain; you can find it anywhere. Consequently, nothing in the transaction requires the involvement of suppliers, venders, or front men. Because the transaction is between the bookie and the bettor, bookies have autonomy, discretion, control, and independence over their business.

There are three major phases of the bookmaking process: establishing the terms for betting, taking bets, and settling accounts. Within each of these processes, the bookies maintain independence, autonomy, discretion, and control. The conditions and terms for the transaction are determined by the bookie. This involves making decisions about what the line will be, the maximum and minimum amount to be accepted, the types of bets that will be accepted (a straight bet, over/under bet, parlay, etc.), and whether to extend credit to customers. Bookies know that if their terms are not competitive with other bookies, they could lose customers. Taking the bet consummates the transaction. Bettors tell the bookie who they want to wager on, the amount of the bet, and the type of bet they want to make. Bookies repeat this information back to the customer so that there is no misunderstanding about the wager and both wait for the outcome. Settling accounts involves the transfer of money either from losing bettors to bookies or from bookies to winning bettors and is the only phase that involves face-to-face interaction. Since bookmaking is illegal, bookies have no legal right to collect monies owed them nor can bettors demand monies owed to them. Instead of being coercive, settling accounts is based on trust between the bookie and bettor. In every aspect of the bookmaking process we see that bookies are the innovators, decision makers, and risk-takers. They initiate their involvement in bookmaking and see it as an opportunity to provide a service to a growing market; they hone their skills, gather the resources, and manage the process of building their own business.

There is further evidence that bookies differ from other types of deviants when we examine their career trajectories. On average, the bookies in this study were involved in bookmaking for 17 years, which can only be thought of in terms of a lengthy career. Throughout this period, the majority held full-time legitimate jobs, were committed to their families, planned for the future of their families, set economic goals, were involved in community activities, and generally were committed to conventional values. It also is worth noting that only one of the bookies in this study had ever been arrested. Yet in the literature on deviant careers, it is argued that prolonged deviant involvement requires rejecting conventional values (Stebbins 1971:63).

Best and Luckenbill (1994:235–237) argue that the social organization of deviance affects the duration of deviant careers in at least four ways. First, most deviant careers are dependent on the support of other deviants, the more contacts they have with other deviants, the longer the career. Second, career deviants have connections to other deviant organizations because of the resources such organizations offer. Third, since illegal activities are characterized by unregulated competition, career deviants must minimize competition, and this is usually done through negotiating with other deviants and by sharing or dividing the territory or resources. The final way career length is affected is through contacts with social control agents—the more contacts the higher the risk and the shorter the career. The analysis shows that bookmakers' careers are distinct from other deviant careers in the following ways. First, while bookies may associate with other bookies, this occurs only occasionally. Most importantly, their bookmaking business does not depend on these associations or on other deviants. In short, bookies are independent of other deviants. Second, since bookies assume all of the risks (financial and otherwise) involved in the bookmaking enterprise, they do not need to divide or share their customers nor do they have ties to other deviant organizations. Third, the most effective device against competition is the bookie's honesty and integrity. Finally, bookies are buffered from contact with social control agents because the enterprise has such low criminal visibility. Gambling on sports is integrated into the rhythms of everyday life, particularly male culture, and the bookie is a feature of that culture.

This leads to the question of whether it is time that the organized crime, corruption, and sports betting link be put to rest. The findings presented here provide a compelling argument for doing so. Although there is evidence showing organized crime's involvement in bookmaking, the findings presented here show bookies to be small entrepreneurs who operate independently. They see themselves as providing a service that is in great demand. It could be argued that major media that promote sports events increase the demand for this service. The line is readily available to the public and is very much a part of the hype that accompanies sports events. The line increases public interest in sporting events and betting on events ensures that millions of fans will watch them. Given the ubiquity of these accoutrements of gambling on sports, it is not surprising that sports betting is widespread and that our anti-gambling policies against it have little impact.

Although the sample for this study is small, confined to a single geographical area, and was selected by word of mouth, the accounts show remarkable consistency. It is possible that the bookies I interviewed represent an unusual group of bookies. Since there is no way of knowing just how large the population of bookies is, it is difficult to assess the representativeness of this study's sample. More research is needed, particularly research done in other geographical areas. Nevertheless, the findings presented here show that while bookmaking is illegal, sports betting is widespread, and that bookmakers are more like small business entrepreneurs than petty criminals.

REFERENCES

Anderson, Annelies. 1979. *The Business of Organized Crime*. Stanford, CA: Hoover Institute.

Best, Joel and David Luckinbill. 1994. *Organizing Deviance*. Englewood Cliffs, NJ: Prentice Hall.

Bonanno, Bill. 1999. *Bound by Honour*. NY: St. Martin's Press.

Chambliss, William. 1978. *On the Take*. Bloomington, IN: University of Indiana Press.

Clark, Ramsey. 1971. *Crime in America*. New York: Simon and Schuster.

Congressional Hearings for H.R. 497, National Gambling Impact and Policy Commission Act, September, 1995.

Cook, James. "If Roxborough Says the Spread is 7 it's 7." Forbes, September 14, 1992, pp. 350–363.

Cressey, Donald. 1969. *Theft of the Nation*. New York: Harper & Row.

DeFina, B., Reidy, J. P. (Producers) & Scorsese, M. (Director). (1995). *Casino* [Film]. (Available from Universal Pictures).

DeFina, B., Pustin, B., and Winkler, I. (Producers) & Scorsese, M. (Director). (1990). *Goodfellas* [Film]. (Available from Wanner Bros).

Holcomb, R. & Marshall, P. D. (Directors). (1987). *Wiseguy* [Film]. (Available from Cannell Films).

McGraw, Dan. "The National Bet." U.S. News & World Report, April 7, 1997, pp. 50–55.

Moody's Industry Review. Gaming. April 6, 1995.

National Amateur and Professional Sports Protection Act, S. 474, 1992.

Nevada Gaming Commission and State Gaming Control Board, personal communication with Research Division, June 1998.

O'Brien, Timothy. 1998. *Bad Bet*. New York: Times Business.

President's Commission on Law Enforcement and the Administration of Justice. 1967. *Task Force Report: Organized Crime*. Washington, DC: US Government Printing Office.

Puzo, Mario. 1969. *The Godfather*. New York: Putnum.

Reuter, Peter. 1983. *Disorganized Crime: Illegal Markets and the Mafia*. Cambridge, MA: MIT Press.

Reuter, Peter and Jonathan Rubinstein. 1982. *Illegal Gambling in New York*. Washington, DC: National Institute of Justice.

Rosencrance, John. 1987. "Bookmaking: A Case Where Honesty is the Best Policy." *Sociology and Social Research* 72(1):7–11.

Sasuly, Richard. 1982. *Bookies and Bettors: Two Hundred Years of Gambling*. New York: Holt Rinehart & Winston.

Scarne, John. 1974. *Scarne's New Complete Guide to Gambling*. New York: Simon & Schuster.

Shapiro, Joseph, Penny Loeb, Kenan Pollack, Timothy Ito, and Gary Cohen. "America's Gambling Fever." U.S. News & World Report. January 15, 1996, pp. 52–60.

Stebbins, Robert. 1971. *Commitment to Deviance*. Westport, CT: Greenwood.

Talese, Gay. 1971. *Honour Thy Father*. New York: World Publishing.

16

Denying the Guilty Mind: Accounting for Involvement in White-Collar Crime

Michael L. Benson

In order to neutralize their behavior, white-collar criminals deny any criminal intent once they are arrested. Benson examined the reactions of offenders involved in antitrust and tax violations, as well as those convicted for fraud, making false statements, and embezzlement. In the process of denying their white-collar crimes, the author shows how excuses and justifications are utilized in an attempt to deny their involvement in any illegal behavior. Once convicted, the white-collar offender experiences an undesirable status passage, from that of law-abiding citizen to that of criminal. The process of investigation and arrest leading to prosecution resembles a public degradation ceremony whereby the morality of the offender is held up for public scrutiny. Thus, accounts are formulated by the offender in an attempt to diminish the effect of the criminal label and permit him or her to maintain a conventional moral identity.

Adjudication as a criminal is, to use Garfinkel's (1956) classic term, a degradation ceremony. The focus of this article is on how offenders attempt to defeat the success of this ceremony and deny their own criminality through the use of accounts. However, in the interest of showing in as much detail as possible all sides of the experience undergone by these offenders, it is necessary to treat first the guilt and inner anguish that is felt by many white-collar offenders even though they deny being criminals. This is best accomplished by beginning with a description of a unique feature of the prosecution of white-collar crimes.

In white-collar criminal cases, the issue is likely to be *why* something was done, rather than *who* did it (Edelhertz, 1970:47). There is often relatively little disagreement as to what happened. In the words of one Assistant U.S. Attorney interviewed for the study:

EDITOR'S NOTE: From Benson, M., "Denying the guilty mind: Accounting for involvement in white collar crime," in *Criminology, 23,* pp. 589–599. Copyright ©1985. Reprinted with permission from the American Society of Criminology.

If you actually had a movie playing, neither side would dispute that a person moved in this way and handled this piece of paper, etc. What it comes down to is, did they have the criminal intent?

If the prosecution is to proceed past the investigatory stages, the prosecutor must infer from the pattern of events that conscious criminal intent was present and believe that sufficient evidence exists to convince a jury of this interpretation of the situation. As Katz (1979:445–446) has noted, making this inference can be difficult because of the way in which white-collar illegalities are integrated into ordinary occupational routines. Thus, prosecutors in conducting trials, grand jury hearings, or plea negotiations spend a great deal of effort establishing that the defendant did indeed have the necessary criminal intent. By concentrating on the offender's motives, the prosecutor attacks the very essence of the white-collar offender's public and personal image as an upstanding member of the community. The offender is portrayed as someone with a guilty mind.

Not surprisingly, therefore, the most consistent and recurrent pattern in the interviews, though not present in all of them, was denial of criminal intent, as opposed to the outright denial of any criminal behavior whatsoever. Most offenders acknowledged that their behavior probably could be construed as falling within the conduct proscribed by statute, but they uniformly denied that their actions were motivated by a guilty mind. This is not to say, however, that offenders *felt* no guilt or shame as a result of conviction. On the contrary, indictment, prosecution, and conviction provoke a variety of emotions among offenders.

The enormous reality of the offender's lived emotion (Denzin, 1984) in admitting guilt is perhaps best illustrated by one offender's description of his feelings during the hearing at which he pled guilty.

> You know (the plea's) what really hurt. I didn't even know I had feet. I felt numb. My head was just floating. There was no feeling, except a state of suspended animation. . . . For a brief moment, I almost hesitated. I almost said not guilty. If I had been alone, I would have fought, but my family. . . .

The traumatic nature of this moment lies, in part, in the offender's feeling that only one aspect of his life is being considered. From the offender's point of view his crime represents only one small part of his life. It does not typify his inner self, and to judge him solely on the basis of this one event seems an atrocious injustice to the offender.

For some the memory of the event is so painful that they want to obliterate it entirely, as the two following quotations illustrate.

> I want quiet. I want to forget. I want to cut with the past.
>
> I've already divorced myself from the problem. I don't even want to hear the names of certain people ever again. It brings me pain.

For others, rage rather than embarrassment seemed to be the dominant emotion.

> I never really felt any embarrassment over the whole thing. I felt rage and it wasn't false or self-serving. It was really (something) to see this thing in action and recognize what the whole legal system has come to through its development, and the abuse of the grand jury system and the abuse of the indictment system. . . .

The role of the news media in the process of punishment and stigmatization should not be overlooked. All offenders whose cases were reported on by the news media were either embarrassed or embittered or both by the public exposure.

> The only one I am bitter at is the newspapers, as many people are. They are unfair because you can't get even. They can say things that are untrue, and let me say this to you. They wrote an article on me that was so blasphemous, that was so horrible. They painted me as an insidious, miserable creature, wringing out the last penny. . . .

Offenders whose cases were not reported on by the news media expressed relief at having avoided that kind of embarrassment, sometimes saying that greater publicity would have been worse than any sentence they could have received.

In court, defense lawyers are fond of presenting white-collar offenders as having suffered enough by virtue of the humiliation of public adjudication as criminals. On the other hand, prosecutors present them as cavalier individuals who arrogantly ignore the law and brush off its weak efforts to stigmatize them as criminals. Neither of these stereotypes is entirely accurate. The subjective effects of conviction on white-collar offenders are varied and complex. One suspects that this is true of all offenders, not only white-collar offenders.

The emotional responses of offenders to conviction have not been the subject of extensive research. However, insofar as an individual's emotional response to adjudication may influence the deterrent or crime-reinforcing impact of punishment on him or her, further study might reveal why some offenders stop their criminal behavior while others go on to careers in crime (Casper, 1978:80).

Although the offenders displayed a variety of different emotions with respect to their experiences, they were nearly unanimous in denying basic criminality. To see how white-collar offenders justify and excuse their crimes, we turn to their accounts. The small number of cases rules out the use of any elaborate classification techniques. Nonetheless, it is useful to group offenders by offense when presenting their interpretations.

ANTITRUST VIOLATORS

Four of the offenders had been convicted of antitrust violations, all in the same case involving the building and contracting industry. Four major themes characterized their accounts. First, antitrust offenders focused on the everyday character and historical continuity of their offenses.

> It was a way of doing business before we even got into the business. So it was like why do you brush your teeth in the morning or something. . . . It was part of the everyday. . . . It was a method of survival.

The offenders argued that they were merely following established and necessary industry practices. These practices were presented as being necessary for the well-being of the industry as a whole, not to mention their own companies. Further, they argued that cooperation among competitors was either allowed or actively promoted by the government in other industries and professions.

The second theme emphasized by the offenders was the characterization of their actions as blameless. They admitted talking to competitors and admitted submitting intentionally noncompetitive bids. However, they presented these practices as being done not for the purpose of rigging prices nor to make exorbitant profits. Rather, the everyday practices of the industry required them to occasionally submit bids on projects they really did not want to have. To avoid the effort and expense of preparing full-fledged bids, they would call a competitor to get a price to use. Such a situation might arise, for example, when a company already had enough work for the time being, but was asked by a valued customer to submit a bid anyway.

> All you want to do is show a bid, so that in some cases it was for as small a reason as getting your deposit back on the plans and specs. So you just simply have no interest in getting the job and just call to see if you can find someone to give you a price to use, so that you didn't have to go through the expense of an entire bid preparation. Now that is looked on very unfavorably, and it is a technical violation, but it was strictly an opportunity to keep your name in front of a desired customer. Or you may find yourself in a situation where somebody is doing work for a customer, has done work for many, many years and is totally acceptable, totally fair. There is no problem. But suddenly they (the customer) get an idea that they ought to have a few tentative figures, and you're called in, and you are in a moral dilemma. There's really no reason for you to attempt to compete in that circumstance. And so there was a way to back out.

Managed in this way, an action that appears on the surface to be a straightforward and conscious violation of antitrust regulations becomes merely a harmless business practice that happens to be a "technical violation." The offender can then refer to his personal history to verify his claim that, despite technical violations, he is in reality a law-abiding person. In the words of one offender, "Having been in the business for

33 years, you don't just automatically become a criminal overnight."

Third, offenders were very critical of the motives and tactics of prosecutors. Prosecutors were accused of being motivated solely by the opportunity for personal advancement presented by winning a big case. Further, they were accused of employing prosecution selectively and using tactics that allowed the most culpable offenders to go free. The Department of Justice was painted as using antitrust prosecutions for political purposes.

The fourth theme emphasized by the antitrust offenders involved a comparison between their crimes and the crimes of street criminals. Antitrust offenses differ in their mechanics from street crimes in that they are not committed in one place and at one time. Rather, they are spatially and temporally diffuse and are intermingled with legitimate behavior. In addition, the victims of antitrust offenses tend not to be identifiable individuals, as is the case with most street crimes. These characteristics are used by antitrust violators to contrast their own behavior with that of common stereotypes of criminality. Real crimes are pictured as discrete events that have beginnings and ends and involve individuals who directly and purposely victimize someone else in a particular place and a particular time.

> It certainly wasn't a premeditated type of thing in our cases as far as I can see. . . . To me it's different than _____ and I sitting down and we plan, well, we're going to rob this bank tomorrow and premeditatedly go in there. . . . That wasn't the case at all. . . . It wasn't like sitting down and planning I'm going to rob this bank type of thing. . . . It was just a common everyday way of doing business and surviving.

A consistent thread running through all of the interviews was the necessity for antitrust-like practices, given the realities of the business world. Offenders seemed to define the situation in such a manner that two sets of rules could be seen to apply. On the one hand, there are the legislatively determined rules—laws—which govern how one is to conduct one's business affairs. On the other hand, there is a higher set of rules based on the concepts of profit and survival, which are taken to define what it means to be in business in a capitalistic society. These rules do not just regulate behavior; rather, they constitute or create the behavior in question. If one is not trying to make a profit or trying to keep one's business going, then one is not really "in business." Following Searle (1969:33–41), the former type of rule can be called a regulative rule and the latter type a constitutive rule. In certain situations, one may have to violate a regulative rule in order to conform to the more basic constitutive rule of the activity in which one is engaged.

This point can best be illustrated through the use of an analogy involving competitive games. Trying to win is a constitutive rule of competitive games in the sense that if one is not trying to win, one is not really playing the game. In competitive games, situations may arise where a player deliberately breaks the rules even though he knows or expects he will be caught. In the game of basketball, for example, a player may deliberately foul an opponent to prevent him from making a sure basket. In this instance, one would understand that the fouler was trying to win by gambling that the opponent would not make the free throws. The player violates the rule against fouling in order to follow the higher rule of trying to win.

Trying to make a profit or survive in business can be thought of as a constitutive rule of capitalist economies. The laws that govern *how* one is allowed to make a profit are regulative rules, which can understandably be subordinated to the rules of trying to survive and profit. From the offender's point of view, he is doing what businessmen in our society are supposed to do—that is, stay in business and make a profit. Thus, an individual who violates society's laws or regulations in certain situations may actually conceive of himself as thereby acting more in accord with the central ethos of his society than if he had been a strict observer of its law. One might suggest, following Denzin (1977), that for businessmen in the building and contracting industry, an informal structure exists below the articulated legal structure, one which frequently supersedes the legal structure. The informal structure may define as moral and "legal" certain actions that the formal legal structure defines as immoral and "illegal."

TAX VIOLATORS

Six of the offenders interviewed were convicted of income tax violations. Like antitrust violators, tax violators can rely upon the complexity of the tax laws and an historical tradition in which cheating on taxes is not really criminal. Tax offenders would claim that everybody cheats somehow on their taxes and present themselves as victims of an unlucky break, because they got caught.

> Everybody cheats on their income tax, 95% of the people. Even if it's for ten dollars it's the same principle. I didn't cheat. I just didn't know how to report it.

The widespread belief that cheating on taxes is endemic helps to lend credence to the offender's claim to have been singled out and to be no more guilty than most people.

Tax offenders were more likely to have acted as individuals rather than as part of a group and, as a result, were more prone to account for their offenses by referring to them as either mistakes or the product of special circumstances. Violations were presented as simple errors which resulted from ignorance and poor recordkeeping. Deliberate intention to steal from the government for personal benefit was denied.

> I didn't take the money. I have no bank account to show for all this money, where all this money is at that I was supposed to have. They never found the money, ever. There is no Swiss bank account, believe me.
>
> My records were strictly one big mess. That's all it was. If only I had an accountant, this wouldn't even of happened. No way in God's creation would this ever have happened.

Other offenders would justify their actions by admitting that they were wrong while painting their motives as altruistic rather than criminal. Criminality was denied because they did not set out to deliberately cheat the government for their own personal gain. Like the antitrust offenders discussed above, one tax violator distinguished between his own crime and the crimes of real criminals.

> I'm not a criminal. That is, I'm not a criminal from the standpoint of taking a gun and doing this and that. I'm a criminal from the standpoint of making a mistake, a serious mistake. . . . The thing that really got me involved in it is my feeling for the employees here, certain employees that are my right hand. In order to save them a certain amount of taxes and things like that, I'd extend money to them in cash, and the money came from these sources that I took it from. You know, cash sales and things of that nature, but practically all of it was turned over to the employees, because of my feeling for them.

All of the tax violators pointed out that they had no intention of deliberately victimizing the government. None of them denied the legitimacy of the tax laws, nor did they claim that they cheated because the government is not representative of the people (Conklin, 1977:99). Rather, as a result of ignorance or for altruistic reasons, they made decisions which turned out to be criminal when viewed from the perspective of the law. While they acknowledged the technical criminality of their actions, they tried to show that what they did was not criminally motivated.

VIOLATIONS OF FINANCIAL TRUST

Four offenders were involved in violations of financial trust. Three were bank officers who embezzled or misapplied funds, and the fourth was a union official who embezzled from a union pension fund. Perhaps because embezzlement is one crime in this sample that can be considered *mala in se,* these offenders were much more forthright about their crimes. Like the other offenders, the embezzlers would not go so far as to say "I am a criminal," but they did say "What I did was wrong, was criminal, and I knew it was." Thus, the embezzlers were unusual in that they explicitly admitted responsibility for their crimes. Two of the offenders clearly fit Cressey's scheme as persons with financial problems who used their positions to convert other people's money to their own use.

Unlike tax evasion, which can be excused by reference to the complex nature of tax regulations or antitrust violations, which can be justified as for the good of the organization as a whole, embezzlement requires deliberate action on the part of the offender and is almost inevitably committed for personal reasons. The crime of embezzlement, therefore, cannot be accounted for by using the same techniques that tax violators or antitrust violators do. The act itself can only be explained by showing that one was under extraordinary circumstances which explain one's uncharacteristic behavior. Three of the offenders referred explicitly to extraordinary circumstances and presented the offense as an aberration in their life history. For example, one offender described his situation in this manner:

> As a kid, I never even—you know kids will sometimes shoplift from the dime store—I never even did that. I had never stolen a thing in my life and that was what was so unbelievable about the whole thing, but there were some psychological and personal questions that I wasn't dealing with very well. I wasn't terribly happily married. I was married to a very strong-willed woman and it just wasn't working out.

The offender in this instance goes on to explain how, in an effort to impress his wife, he lived beyond his means and fell into debt.

A structural characteristic of embezzlement also helps the offender demonstrate his essential lack of criminality. Embezzlement is integrated into ordinary occupational routines. The illegal action does not stand out clearly against the surrounding set of legal actions. Rather, there is a high degree of surface correspondence between legal and illegal behavior. To maintain this correspondence, the offender must exercise some restraint when committing his crime. The embezzler must be discrete in his stealing; he cannot take all of the money available to him without at the same time revealing the crime. Once exposed, the offender can point to this restraint on his part as evidence that he is not really a criminal. That is, he can compare what happened with what could have happened in order to show how much more serious the offense could have been if he was really a criminal at heart.

> What I could have done if I had truly had a devious criminal mind and perhaps if I had been a little smarter—and I am not saying that with any degree of pride or any degree of modesty whatever, [as] it's being smarter in a bad, an evil way—I could have pulled this off on a grander scale and I might still be doing it.

Even though the offender is forthright about admitting his guilt, he makes a distinction between himself and someone with a truly "devious criminal mind."

Contrary to Cressey's (1953:57–66) findings, none of the embezzlers claimed that their offenses were justified because they were underpaid or badly treated by their employers. Rather, attention was focused on the unusual circumstances surrounding the offense and its atypical character when compared to the rest of the offender's life. This strategy is for the most part determined by the mechanics and organizational format of the offense itself. Embezzlement occurs within the organization but not for the organization. It cannot be committed accidentally or out of ignorance. It can be accounted for only by showing that the actor "was not himself" at the time of the offense or was under such extraordinary circumstances that embezzlement was an understandable response to an unfortunate situation. This may explain the finding that embezzlers tend to produce accounts that are viewed as more sufficient by the justice system than those produced by other offenders (Rothman and Gandossy, 1982). The only plausible option open to a convicted embezzler trying to explain his offense is to admit responsibility while justifying the action, an approach that apparently strikes a responsive chord with judges.

Fraud and False Statements

Ten offenders were convicted of some form of fraud or false statements charge. Unlike embezzlers, tax violators, or antitrust violators, these offenders were much more likely to deny committing any crime at all. Seven of the ten claimed that they, personally, were innocent of any crime, although each admitted that fraud had occurred. Typically, they claimed to have been set up by associates and to have been wrongfully convicted by the U.S. Attorney handling the case. One might call this the scapegoat strategy. Rather than admitting technical wrong doing and then justifying or excusing it, the offender attempts to paint himself as a victim by shifting the blame entirely to another party. Prosecutors were presented as being either ignorant or politically motivated.

The outright denial of any crime whatsoever is unusual compared to the other types of offenders studied here. It may result from the nature of the crime of fraud. By definition, fraud involves a conscious attempt on the part of one or more persons to mislead others. While it is theoretically possible to accidentally violate the antitrust and tax laws, or to violate them for altruistic reasons, it is difficult to imagine how one could accidentally mislead someone else for his or her own good. Furthermore, in many instances, fraud is an aggressively acquisitive crime. The offender develops a scheme to bilk other people out of money or property, and does this not because of some personal problem but because the scheme is an easy way to get rich. Stock swindles, fraudulent loan scams, and so on are often so large and complicated that they cannot possibly be excused as foolish and desperate solutions to personal problems. Thus, those involved in large-scale frauds do not have the option open to most embezzlers of presenting themselves as persons responding defensively to difficult personal circumstances.

Furthermore, because fraud involves a deliberate attempt to mislead another, the offender who fails to remove himself from the scheme runs the risk of being shown to have a guilty mind. That is, he is shown to possess the most essential element of modern conceptions of criminality: an intent to harm another. His inner self would in this case be exposed as something other than what it has been presented as, and all of his previous actions would be subject to reinterpretation in light of this new perspective. For this reason, defrauders are most prone to denying any crime at all. The cooperative and conspiratorial nature of many fraudulent schemes makes it possible to put the blame on someone else and to present oneself as a scapegoat. Typically, this is done by claiming to have been duped by others.

Two illustrations of this strategy are presented below.

I figured I wasn't guilty, so it wouldn't be that hard to disprove it, until, as I say, I went to court and all of a sudden they start bringing in these guys out of the woodwork implicating me that I never saw. Lot of it could be proved that I never saw.

Inwardly, I personally felt that the only crime that I committed was not telling on these guys. Not that I deliberately, intentionally committed a crime against the system. My only crime was that I should have had the guts to tell on these guys, what they were doing, rather than putting up with it and then trying to gradually get out of the system without hurting them or without them thinking I was going to snitch on them.

Of the three offenders who admitted committing crimes, two acted alone and the third acted with only one other person. Their accounts were similar to others presented earlier and tended to focus on either the harmless nature of their violations or on the unusual circumstances that drove them to commit their crimes. One claimed that his violations were only technical and that no one besides himself had been harmed.

First of all, no money was stolen or anything of that nature. The bank didn't lose any money.... What I did was a technical violation. I made a mistake. There's no question about that, but the bank lost no money.

Another offender who directly admitted his guilt was involved in a check-kiting scheme. In a manner similar to embezzlers, he argued that his actions were motivated by exceptional circumstances.

I was faced with the choice of all of a sudden, and I mean now, closing the doors or doing something else to keep that business open.... I'm not going to tell you that this wouldn't have happened if I'd had time to think it over, because I think it probably would have. You're sitting there with a dying patient. You are going to try to keep him alive.

In the other fraud cases more individuals were involved, and it was possible and perhaps necessary for each offender to claim that he was not really the culprit.

Discussion: Offenses, Accounts, and Degradation Ceremonies

The investigation, prosecution, and conviction of a white-collar offender involves him in a very undesirable status passage (Glaser and Strauss, 1971). The entire process can be viewed as a long and drawn-out degradation ceremony with the prosecutor as the chief denouncer and the offender's family and friends as the chief witnesses. The offender is moved from the status of law-abiding citizen to that of convicted felon. Accounts are developed to defeat the process of identity transformation that is the object of a degradation ceremony. They represent the offender's attempt to diminish the effect of his legal transformation and to prevent its becoming a publicly validated label. It can be suggested that the accounts developed by white-collar offenders take the forms that they do for two reasons: (1) the forms are required to defeat the success of the degradation ceremony, and (2) the specific forms used are the ones available given the mechanics, history, and organizational context of the offenses.

Three general patterns in accounting strategies stand out in the data. Each can be characterized by the subject matter on which it focuses: the event (offense), the perpetrator (offender), or reduced. Although there are overlaps in the accounting strategies used by the various types of offenders, and while any given offender may use more than one strategy, it appears that accounting strategies and offenses correlate.

References

Casper, Jonathan D. 1978. *Criminal Courts: The Defendant's Perspective.* Washington, D.C.: U.S. Department of Justice.

Conklin, John E. 1977. *Illegal But Not Criminal: Business Crime in America.* Englewood Cliffs, N.J.: Prentice Hall.

Cressey, Donald. 1953. *Other People's Money.* New York: Free Press.

Denzin, Norman K. 1977. "Notes on the criminogenic hypothesis: A case study of the American liquor industry." *American Sociological Review* 42:905–920.

——. 1984. *On Understanding Emotion.* San Francisco: Jossey-Bass.

Edelhertz, Herbert. 1970. *The Nature, Impact, and Prosecution of White Collar Crime.* Washington, D.C.: U.S. Government Printing Office.

Garfinkel, Harold. 1956. "Conditions of successful degradation ceremonies." *American Journal of Sociology* 61:420–424.

Glaser, Barney G. and Anselm L. Strauss. 1971. *Status Passage.* Chicago: Aldine.

Katz, Jack. 1979. "Legality and equality: Plea bargaining in the prosecution of white-collar crimes." *Law and Society Review* 13:431–460.

Rothman, Martin and Robert F. Gandossy. 1982. "Sad tales: The accounts of white-collar defendants and the decision to sanction." *Pacific Sociological Review* 4:449–473.

Searle, John R. 1969. *Speech Acts.* Cambridge: Cambridge University Press.

Part V

GANGS AND CRIME

The existence of gangs historically is not a new phenomenon, but it was not until the late 1980s and 1990s that youth street gangs began to grow in number and sustain themselves in urban America by their involvement in drug trafficking and sales, together with the publicized violence and use of weapons that fueled a national public outcry. For many urban youth from economically deprived neighborhoods, gang membership became a refuge from a violent and stressful home life, chaotic neighborhoods, other gang members, and a lack of social outlets. Gang membership provided a sense of belonging, another type of family that was accepting of marginalized and impoverished adolescents. Gangs do engage in various types of criminal behavior, as the articles in this section will illustrate. However, the gang offers a haven for the multiple problems that inner city youth face in their communities. Besides the monetary gains some gang members experience from membership, they also experience social ones in the way of peer acceptance and family-like unity.

The most dangerous part of being a gang member is the culture of violence. Decker explores gang violence and finds that, like other gang activities, the use of violent acts defined the gang's organizational and normative structure. Gang violence serves multiple functions; most important, it causes continued violence through threat and retaliation. Further, it temporarily unites gang members and increases solidarity of the group, escalating the need to stick together against a perceived common enemy, which results in a greater dependency on one another. Last, the sustained use of violence by a gang can result in the fracturing of gangs into various groups and results in some members' decision to leave gang life because the violent activities exceed tolerable limits.

Hagedorn portrays another type of gang activity than that of Decker's violent gang. He addresses gang members' relationship with drugs and their drug dealing as an integral part of their membership activity. He found that the majority of gang members did not have a great commitment to drug dealing and discovered that they move in and out of legitimate and illegitimate employment and sell drugs on an irregular basis, depending on what employment opportunities exist at the time. Most gang members see drug dealing as a short-term method to make money due to their lack of skills to earn a decent wage by way of legitimate employment. Hagedorn concludes that those gang members he had contact with internalized a conventional working-class value system and desired to lead a more conventional lifestyle but could not procure steady employment. Due to their inability to climb out of the impoverished economic circumstances that they find themselves trapped in, the hopelessness of their situation sustains their gang membership via drug use and the street sales of narcotics, which offers one of the limited choices gang members have to make money due to their socioeconomic circumstances.

Drawing on a telephone survey of municipal and county police agencies in non-metropolitan counties in the United States, Weisheit and Wells examined the pressure of gang membership in rural America. By focusing their phone interviews with law enforcement respondents in these small communities, the authors were able to elicit the perceived definition of gangs, gang-related problems, and their official responses to gangs. The authors found that gangs have been reported to be present in some small communities, but by no means is their existence universal nor are they considered to pose a serious problem in these rural areas. They further report that many of the non-metropolitan communities that indicated gang existence in a prior national study now claim that they presently do not have a gang problem. An explanation for the decrease in the number of gangs in small towns may be attributable to the fact that rural gangs are historically small in membership and apparently short lived. Another factor in their demise is rural law enforcement's ability to rapidly recognize gang existence and apply resources to the problem with effective suppression by arresting suspected gang members for even the smallest law violations. Such swift reaction on the part of rural law enforcement authorities played a large part in breaking up local gangs.

In her analysis of gender and victimization in gangs, Miller suggests that the research needs to focus on ways in which belonging to a gang can increase the risks for becoming the recipient of violent interactions. By studying mixed-gender gangs, Miller found that participation in delinquent lifestyles plays a big part in gang life and often places female gang members at a high degree of risk for experiencing violent victimization. Merely joining a gang places a new female initiate in a position of submission by becoming a target for physical abuse at the hands of her gang peers. Miller found that gender may at times act to insulate a female gang member from some forms of assaultive victimization and decrease her exposure to violence from other gangs, but it also has the tendency to make her even more vulnerable to other forms of violence, mainly frequent victimization by her own male gang members through sexual exploitation and sexual assault.

17

Collective and Normative Features of Gang Violence

Scott H. Decker

Based on a 3-year field study of gang members who engage in a great deal of violent activity, Decker examined those situations and circumstances where violence occurs. He found that the use of violence, like other gang activities, reflected the type of organizational and normative structure of the gang itself. That is, violence is a result of the gangs' loose organizational structure, with multiple goals, very few members acting as leaders, and a low degree of loyalty among the members of the gang. In addition, Decker concludes that most gang violence is retaliatory in nature, a response to violence, real or perceived, against the gang, which ends up producing more violence. Such violence, notes the author, has causal effects of life in the gang; it precludes violence by outside threats, and it tends to increase the solidarity of the membership by a collective perception of a common enemy, which increases dependency on each other. But violent activity can also exceed limits of toleration for some gang members and result in their leaving the gang.

In 1927 Frederic Thrasher observed that gangs shared many of the properties of mobs, crowds, and other collectives, and engaged in many forms of collective behavior. Despite the prominent role of his work in gang research, few attempts have been made to link the behavior of gangs to theories of collective behavior. This omission is noteworthy because, despite disagreements about most other criteria—turf, symbols, organizational structure, permanence, criminality—all gang researchers include "group" as a part of their definition of gangs. Gang members are individuals with diverse motives, behaviors, and socialization experiences. Their *group* membership, behavior, and values, however, make them interesting to criminologists who study gangs.

In this paper we explore the mechanisms and processes that result in the spread and escalation of gang violence. In particular, we focus on contagion as an aspect of collective behavior that produces expressive gang violence. Collective behavior explanations provide insights into gang processes, particularly the escalation of violence, the spread of gangs from one community to another, and increases in gang membership in specific communities.

EDITOR'S NOTE: From Decker, S., "Collective and normative features in gang violence," in *Justice Quarterly, 13,* pp. 243–264. Copyright © 1996. Reprinted with permission of the Academy of Criminal Justice Sciences.

GANG VIOLENCE

Violence is integral to life in the gang, as Klein and Maxson (1989) observed, and gang members engage in more violence than other youths. Thrasher (1927) noted that gangs developed through strife and flourish on conflict. According to Klein (1972:85), violence is a "predominant 'myth system'" among gang members and is constantly present.

Our analysis of gang violence focuses on the role of *threat,* actual or perceived, in explaining the functions and consequences of gang violence. We define threat as the potential for transgressions against or physical harm to the gang, represented by the acts or presence of a rival group. Threats of violence are important because they have consequences for future violence. Threat plays a role in the origin and growth of gangs, their daily activities, and their belief systems. In a sense, it helps to define them to rival gangs, to the community, and to social institutions.

Katz (1988) argues that gangs are set apart from other groups by their ability to create "dread," a direct consequence of involvement in and willingness to use violence. Dread elevates these individuals to street elites through community members' perceptions of gang members as violent. In many neighborhoods, groups form for protection against the threat of outside groups (Suttles 1972). Sometimes these groups are established along ethnic lines, though territorial concerns often guide their formation. Sanders (1993), in a 10-year study of gangs in San Diego, argued that the mix of conventional values with underclass values—spiced by the realities of street culture—was a volatile combination. Hagedorn (1988) found that conflicts between the police and young men "hanging out" on the corner led to more formalized structures, and ultimately to gangs. Both Suttles (1972:98) and Sullivan (1989) underscored the natural progression from a neighborhood group to a gang, particularly in the face of "adversarial relations" with outside groups. The emergence of many splinter gangs can be traced to the escalation of violence within larger gangs, and to the corresponding threat that the larger gang comes to represent to certain territorial or age-graded subgroups. Sullivan (1989)

documented the expressive character of most gang violence, and described the role of fighting in the evolution of cliques into street gangs. Because this occurs at a young age, the use of group violence attains a normative character.

Threat also may contribute to the growth of gangs. This mechanism works in two ways: through building cohesiveness and through contagion. Threats of physical violence increase the solidarity or cohesiveness of gangs within neighborhoods as well as across neighborhoods. Klein (1971) identified the source of cohesion in gangs as primarily external—the result of intergang conflict; Hagedorn (1988) also made this observation. According to Klein, cohesion within the gang grows in proportion to the perceived threat represented by rival gangs. Padilla (1992) reported a similar finding, noting that threat maintains gang boundaries by strengthening the ties among gang members and increasing their commitment to each other, thus enabling them to overcome any initial reluctance about staying in the gang and ultimately engaging in violence. Thus the threat of a gang in a geographically proximate neighborhood increases the solidarity of the gang, motivates more young men to join their neighborhood gang (see Vigil 1988), and enables them to engage in acts of violence that they might not have committed otherwise.

The reciprocal nature of gang violence explains in part how gangs form initially, as well as how they increase in size and strength. Klein and Maxson (1989:223) demonstrated that fear of retaliation was three times more likely to characterize gang homicides than other homicides involving juveniles. The perceived need to engage in retaliatory violence also helps to explain the increasing sophistication of weapons used by gang members. As Horowitz (1983) observed, gang members arm themselves in the belief that their rivals have guns; they seek to increase the sophistication of their weaponry in the hope that they will not find themselves in a shootout with less firepower than their rival. This process was documented by Block and Block (1993) in their explanation of the increase in street gang homicides in Chicago.

As gangs and gang members engage in acts of violence and create "dread" (Katz 1988:135), they are viewed as threatening by other (gang and non-gang) groups and individuals. Also, over time, the threats that gang members face and pose isolate them from legitimate social institutions such as schools, families, and the labor market. This isolation, in turn, prevents them from engaging in the very activities and relationships that might reintegrate them into legitimate roles and reduce their criminal involvement. It weakens their ties to the socialization power and the controlling norms of such mainstream institutions, and frees them to commit acts of violence.

DATA AND METHODS

We contacted gang members directly on the street and conducted interviews at a neutral site. This procedure was consistent with our goal of learning about gang activities in the words and terms used by gang members to describe them. Our working definition of a gang includes age-graded peer groups that exhibit permanence, engage in criminal activity, and have symbolic representations of membership. Field contacts with active gang members were made by a street ethnographer, an ex-offender himself, who had built a reputation as "solid" on the street through his work with the community and previous fieldwork. The street ethnographer had been shot several years earlier, and now used a wheelchair. The combination of his reputation in the community and his experience in contacting and interviewing active offenders enhanced the validity of the responses. Using snowball sampling procedures (Biernacki and Waldorf 1981; Wright et al. 1992), the research team made initial field contacts with gang members, verified membership, and built the sample to include more subjects.

We interviewed 99 active gang members representing 29 different gangs. Sixteen of these gangs, accounting for 67 of our 99 subjects, were affiliated with the Crips. Gang members affiliated with the Bloods accounted for the remainder of our sample, and included 13 gangs and 32 members. Field techniques cannot provide a representative sample, but our subjects varied considerably in age, gang affiliation, and activities, thus assuring that we received

information from a variety of gang members. As a result, it was unlikely that our respondents revealed information about only a narrow segment of gang activity.

Ages for members of our sample ranged from 13 to 29, with an average age of 17. The majority were black (96%) and male (93%). On average, the gang members we interviewed had been active members for three years. More than three-quarters of our subjects told us that their gang existed before they joined; the average age of gangs in our sample was six years. An average of 213 gang members were involved in the larger gang; subgroups ranged from six to ten members. Ninety percent of our sample reported that they had participated in violent crime; 70% reported that they had committed a property crime. Thus it is not surprising that our subjects also had extensive experience with the criminal justice system: 80 percent reported an arrest, and the average number of arrests was eight.

COLLECTIVE VIOLENCE PROCESSES WITHIN THE GANG

Gang violence includes a number of acts and is most likely to involve assaults and the use of weapons. Although the motives for these acts are diverse, much gang violence is retaliatory.

As further evidence of the importance of violence, nine of our 99 subjects have been killed since the study began in 1990; several showed us bullet wounds during the interview. As stated earlier, this group had extensive arrest histories: 80 percent had been arrested at least once, the mean number of arrests per subject was eight, and one-third reported that their most recent arrest was for assault or weapons violations.

The research reported here attempts to provide a framework for understanding the peaks and valleys of gang violence. As Short and Strodtbeck (1974) observed, efforts to understand gang violence must focus both on process variables (such as interactions) and on situational characteristics (such as neighborhood structure, age, race, and sex). For these reasons we concentrate on stages in the gang process that illustrate important aspects of gang violence, and we examine such violence in the context of five spheres of gang activity: (1) the role of violence in defining life in the gang, (2) the role of violence in the process of joining the gang, (3) the use of violence by the gang, (4) staging grounds for violence, and (5) gang members' recommendations for ending their gang.

The Role of Violence in Defining Life in the Gang

A fundamental way to demonstrate the centrality of violence to life in the gang is to examine how gang members defined a gang. Most answers to this question included some mention of violence. Our subjects were able to distinguish between violence within the gang and that which was unrelated to the gang.

INT: What is a gang to you?

007: A gang is, I don't know, just a gang where people hang out together and get into fights. A lot of members of your group will help you fight.

INT: So if you just got into a fight with another girl because you didn't like her?

007: Then it would be a one-on-one fight, but then like if somebody else jump in, then somebody would come from my side.

INT: Why do you call the group you belong to a gang?

047: Violence, I guess. There is more violence than a family. With a gang it's like fighting all the time, killing, shooting.

INT: What kind of things do members of your organization do together?

085: We have drive-bys, shootings, go to parties, we even go to the mall. Most of the things we do together is dealing with fighting.

Most often the violence was protective, reflecting the belief that belonging to a gang at least would reduce the chance of being attacked.

INT: Are you claiming a gang now?

046: I'm cool with a gang, real cool.

INT: What does that mean to be cool?

046: You don't got to worry about nobody jumping you. You don't got to worry about getting beat up.

Other subjects found the violence in their gang an attractive feature of membership. These individuals were attracted not so much by protection as by the opportunity to engage in violence.

INT: Why did you start to call that group a gang?

009: It's good to be in a gang cause there's a lot of violence and stuff.

INT: So the reason you call it a gang is basically why?

101: Because I beat up on folks and shoot them. The last person I shot I was in jail for five years.

INT: What's good about being in a gang?

101: You can get to fight whoever you want and shoot whoever you want. To me, it's kind of fun. Then again, it's not . . . because you have to go to jail for that shit. But other than that, being down for who you want to be with, it's kind of fun.

INT: What's the most important reason to be in the gang?

057: Beating Crabs. If it wasn't for beating Crabs, I don't think I would be in a gang right now.

Whether for protection or for the opportunity to engage in violence, the members of our sample attached considerable importance to the role of violence in their definition of a gang. Many of the comments evoke what Klein (1971) termed "mythic violence"—discussions of violent activities between gangs that reinforce the ties of membership and maintain boundaries between neighborhood gangs and those in "rival" neighborhoods. In this sense, violence is a central feature of the normative system of the gang; it is the defining feature and the central value of gang life.

Violence in Joining the Gang

Most gangs require an initiation process that includes participation in violent activities. This ritual fulfills a number of important functions. First, it determines whether a prospective member is tough enough to endure the level of violence he or she will face as a gang member. Equally important, the gang must learn how tough a potential member is because they may have to count on this individual for support in fights or shootings. The initiation serves other purposes as well. Most important, it increases solidarity among gang members by engaging them in a collective ritual. The initiation reminds active members of their earlier status, and gives the new member something in common with other gang members. In addition, a violent initiation provides a rehearsal for a prospective member for life in the gang. In short, it demonstrates the centrality of violence to gang life.

Three-quarters of our subjects were initiated into their gangs through the process known as "beating in." This ritual took many forms; in its most common version a prospective gang member walked between lines of gang members or stood inside a circle of gang members who beat the initiate with their fists.

> **020:** I had to stand in a circle and there was about ten of them. Out of these ten there was just me standing in the circle. I had to take six to the chest by all ten of them. Or I can try to go to the weakest one and get out. If you don't get out, they are going to keep beating you. I said "I will take the circle."

One leader, who reported that he had been in charge of several initiations, described the typical form:

001: They had to get jumped on.

INT: How many guys jump on em?

001: Ten.

INT: And then how long do they go?

001: Until I tell em to stop.

INT: When do you tell em to stop?

001: I just let em beat em for bout two or three minutes to see if they can take a punishment.

The initiation also communicates information about the gang and its activities.

099: I fought about four people at one time.

INT: Fought who?

099: I fought some old Gs.

INT: How long did you have to fight them?

099: It seemed like forever.

INT: So they beat you down or you beat them down?

099: It went both ways because I knocked that one motherfucker out.

INT: So that was your initiation?

099: Yeah. And then they sat down and blessed me and told me the sixteen laws and all that. But now in the new process there is a seventeenth and eighteenth law.

Other gang members reported that they had the choice of either being beaten in or "going on a mission." On a mission, a prospective member had to engage in an act of violence, usually against a rival gang member on rival turf. Initiates often were required to confront a rival gang member face-to-face.

041: You have to fly your colors through enemy territory. Some step to you; you have to take care of them by yourself; you don't get no help.

084: To be a Crip, you have to put your blue rag on your head and wear all blue and go in a Blood neighborhood—that is the hardest of all of them—and walk through the Blood neighborhood and fight Bloods. If you come out without getting killed, that's the way you get initiated.

Every gang member we interviewed reported that his or her initiation involved participating in some form of violence. This violence was rarely directed against members of other gangs; most often it took place within the gang. Then, in each successive initiation, recently initiated members participated in "beating in" new members. Such violence always has a group context and a normative purpose: to reinforce the ties between members while reminding them that violence lies at the core of life in the gang.

The Use of Gang Violence

To understand gang violence more clearly, it is critical to know when such violence is used. In the four following situations, gang members did not regard themselves as initiating violence; rather, its purpose was to respond to the violent activities of a rival gang. Retaliatory violence corresponds to the concept of contagion (Loftin 1984) as well as to the principle of crime as social control (Black 1983). According to this view, gang violence is an attempt to enact private justice for wrongs committed against the gang, one of its members, or a symbol of the gang. These wrongs may be actual or perceived; often the perceived threat of impending violence is as powerful a motivator as violence itself.

This view of gang violence helps to explain the rapid escalation of intergang hostilities that lead to assaults, drive-by shootings, or murders between gangs. Such actions reflect the collective behavior processes at work, in which acts of violence against the gang serve as the catalyst that brings together subgroups within the gang and unites them against a common enemy. Such violent events are rare, but are important in gang culture. Collective violence is one of the few activities involving the majority of gang members, including fringe members. The precipitation of such activities pulls fringe members into the gang and increases cohesion.

When violence comes to the gang. We asked gang members when they used violence. Typically they claimed that violence was seldom initiated by the gang itself, but was a response to "trouble" that was "brought" to them. In these instances, the object of violence was loosely defined and was rarely identified; it represented a symbolic enemy against whom violence would be used. These statements, however, indicate an attempt to provide justifications for gang violence.

INT: How often do gang members use violence?

005: When trouble comes to them.

INT: When do you guys use violence?

018: When people start bringing violence to us. They bring it to us and set it up. We take it from there.

INT: When do members of the gang use violence?

037: When somebody approaches us. We don't go out looking for trouble. We let trouble come to us.

INT: When do you guys use violence?

042: Only when it's called for. We don't start trouble. That's the secret of our success.

The view of gang members passively sitting back and waiting for violence to come to them is inconsistent with much of what we know about gang life. After all, many gang members reported that they joined the gang expressly for the opportunity to engage in violence; many lived in neighborhoods where acts of violence occurred several times each day; and most had engaged in violence before joining the gang. Even so, unprovoked violence against another gang is difficult to justify; retaliatory actions against parties that wronged them can be justified more easily. Also, such actions are consistent with the view of the gang as a legitimate social organization serving the legitimate purpose of protecting its members—a central value in the gang's normative structure.

Retaliation. A number of gang members told us that they used violence to even the score with a specific group or individual. Unlike the subjects above, who reported generalized responses, these individuals identified a specific target for their violence: someone who had committed a violent act against them or their gang in the past.

002: I had on a blue rag and he say what's up cuz, what's up blood, and I say uh, what's up cuz, just like that, and then me and him got to arguin' and everything, and teachers would stop it, and then me and him met up one day when nobody was round. We got to fightin. Naw, cause I told Ron, my cousin, my cousin and em came up to the school and beat em up. And the next day when he seen me, he gonna ask me where my cousin and em at. I say I don't need my cousin and em for you. They just came up there cause they heard you was a Blood. And they whooped em. Then me and him had a fight the next day, yeah. And then I had to fight some other dudes that was his friends and I beat em up. Then he brought some boys up to the school and they, uh, pulled out a gun on me and I ran up in the school. And then I brought my boys up the next day and we beat on em.

INT: What happened yesterday?

039: This dude had beat up one of our friends. He was cool with one of my friends but he had beat up another one of my friends before. They came back and busted one of my friends' head. We was going to get him.

Specific examples of retaliation against rival gangs were mentioned less frequently than was general gang violence. This point underscores the important symbolic function of gang violence, a value that members must be ready to support. The idea that rival gangs will "bring violence" to the gang is an important part of the gang belief system; it is pivotal in increasing cohesion among members of otherwise loosely confederated organizations.

Graffiti. A third type of gang violence occurred in response to defacing gang graffiti. Organizational symbols are important to all groups, and perhaps more so to those whose members are adolescents. The significance of graffiti to gangs has been documented by a number of observers in a variety of circumstances (Block and Block 1993; Hagedorn 1988; Moore 1978; Vigil 1988). In particular, graffiti identify gang territory, and maintaining territory is an important feature of gang activity in St. Louis and other cities. As Block and Block observed in Chicago, battles over turf often originated in attempts by rival gangs to "strike out" graffiti. Several gang members told us that attempts to paint over their graffiti by rival gangs were met with a violent response, but no gang members could recall a specific instance. Claiming to use violence in response to such insults again reflects the mythic character of gang violence; it emphasizes the symbolic importance of violence for group processes such as cohesion, boundary maintenance, and identity. Further, such responses underscore the threat represented by rivals who would

encroach on gang territory to strike out gang graffiti.

INT: What does the removal of graffiti mean?

043: That's a person that we have to go kill. We put our enemies up on the wall. If there is a certain person, we "X" that out and know who to kill.

INT: What if somebody comes and paints a pitchfork or paints over your graffiti? What does that mean to your gang?

046: First time we just paint it back up there, no sweat. Next time they come do it, we go find out who did it and go paint over theirs. If they come back a third time, it's like three times you out. Obviously that means something if they keep painting over us. They telling us they ready to fight.

Territory. Most gang members continued to live in the neighborhood where their gang started. Even for those who had moved away, it retained a symbolic value. Protecting gang turf is viewed as an important responsibility, which extends well beyond its symbolic importance as the site where the gang began. Our subjects' allegiance to the neighborhood was deeply embedded in the history of neighborhood friendship groups that evolved into gangs. Thus, turf protection was an important value.

When we asked gang members about defending their turf, we received some generalized responses about their willingness to use violence to do so.

INT: If someone from another gang comes to your turf, what does your gang do?

019: First try to tell him to leave.

INT: If he don't leave?

019: He'll leave one way or the other—carry him out in a Hefty bag.

INT: What was your interest in it (the gang)?

036: We started out, we didn't want nobody coming out and telling us, walking through our neighborhood cause we grew up in this hood and we was going to protect it even if it did mean us fighting every day, which we done. We fought every day. If you walked through the neighborhood and we didn't know you or you didn't know where you was going in the neighborhood, we would rush you.

In other instances, however, the responses identified an individual or an incident in which the gang used violence to protect its turf.

INT: What kind of things does the gang have to do to defend its turf?

013: Kill. That's all it is, kill.

INT: Tell me about your most recent turf defense. What happened, a guy came in?

013: A guy came in, he had the wrong colors on, he got to move out. He got his head split open with a sledge-hammer, he got two ribs broken, he got his face torn up.

INT: Did he die from that?

013: I don't know. We dropped him off on the other side of town. If he did die, it was on the other side of town.

INT: If someone from another gang came to you-all turf, what happens? What do your gang do?

068: We shoot. If it's a lot, we gonna get organized and we shoot.

Staging Grounds for Violence

Gang members expect that when they go to certain locations they will be the targets of violence from other gangs or will be expected by members of their own gang to engage in violence. In some cases, large-scale violence will occur. Other encounters result only in "face-offs." These encounters highlight the role of situational characteristics in gang violence. Most often the staging grounds are public places such as a restaurant.

INT: Do they ever bring weapons to school?

011: No, cause we really don't have no trouble. We mainly fight up at the White Castle. That's where our trouble starts, at the White Castle.

In other instances, the encounters may take place at the skating rink.

INT: What kind of fights have you guys had lately?

057: Yeah, last Saturday at Skate King.

INT: Do you go there to skate or were you just hanging?

057: We used to skate a long time ago, but we just all the sudden went [crazy]. Crabs started hanging out there. Usually all Bloods up there but the Crabs started hanging out there so we had to get rid of them.

Dances are not new locations for youthful violence. Members of our sample identified them as locations that produced violent encounters between rival gangs.

INT: Do you go to dances?

017: Yeah. That's when we mostly get into the gang fights. Yeah, we go to dances.

INT: What about the last fight? What was that about?

031: That was at a dance. It was some Slobs there. They was wanting to show they colors and just didn't know who they was around. They weren't really paying attention.

INT: How often do you guys use violence?

033: Only if we go out to a dance or something.

The expectation of violence at certain locations was so strong that some members avoided going to those places.

INT: Do you go to dances or parties?

047: I don't. I stay away from house parties. Too many fights come out of there.

According to another gang member, violence at house parties had reached such a level that many hosts searched their guests for weapons.

074: Sometimes people wait until they get out of the party and start shooting. Now at these parties they have people at the door searching people, even at house parties.

In general, gang members reported that they "hung out" in small cliques or subgroups and that it was rare for the entire gang to be together. This reflects the general character of social organization in the gangs we studied. An external threat—usually from another gang—was needed to strengthen cohesion among gang members and to bring the larger gang together. Many members of our sample reported that they did not go skating, to the mall, or to dances alone or in small groups because they knew that gang violence was likely to erupt at such locations. Thus the gang went *en masse* to these locations, prepared to start or respond to violence. These expectations contributed to the eventual use of violence. In this way, the gang's belief system contributed to the likelihood of violent encounters.

Ending Gangs

When we asked for gang members' perspectives on the best way to end gangs, we expected to find a variety of recommendations targeted at fundamental causes (racism, unemployment, education) as well as more proximate solutions (detached workers, recreation centers, job training). Instead the modal response reflected the centrality of violence in the gang. Twenty-five of our 99 subjects told us that the only way to get rid of their gang would be to use violence to get rid of the members. This response was confirmed by gang members in their conversations with the field ethnographer. For many gang members, life in the gang had become synonymous with violence; for one respondent, even job offers were not sufficient to end the gang.

INT: What would be the best way to get rid of your gang, the Rolling Sixties?

033: Smoke us all.

INT: Kill you all?

033: Yeah.

INT: We couldn't give you guys jobs?

033: No, just smoke us.

Others recommended using extreme violence to get rid of their gang.

INT: What would it take to get rid of your gang?

035: Whole lot of machine guns. Kill us all. We just going to multiply anyway cause the Pee Wees gonna take over.

INT: What would be the best way to get rid of the Sixties?

042: Kill us all at once. Put them in one place and blow them up.

Violence is so central a part of gang culture that even the members' recommendations about ending gangs include elements of violence.

CONCLUSION

Gang violence, like other gang activities, reflects the gang's organizational and normative structure. Such violence, especially retaliatory violence, is an outgrowth of a collective process that reflects the loose organizational structure of gangs with diffuse goals, little allegiance among members, and few leaders.

If gangs are composed of diffuse subgroups, how is violence organized? Our answer to this question is "Not very well and not very often," because most gang violence serves important symbolic purposes within the gang. In addition, most gang violence is retaliatory, a response to violence—real or perceived—against the gang.

Gang violence serves many functions in the life of the gang. First, and most important, it produces more violence through the processes of threat and contagion. These mechanisms strongly reflect elements of collective behavior. Second, it temporarily increases the solidarity of gang members, uniting them against a common enemy by heightening their dependence on each other. When gang violence exceeds tolerable limits, a third function may be evident: the splintering of gangs into subgroups and the decision by some individuals to leave the gang.

REFERENCES

Biernacki, P. and D. Waldorf. 1981. "Snowball Sampling: Problems and Techniques of Chain Referral Sampling." *Sociological Methods and Research* 10:141–63.

Black, D. 1983. "Crime as Social Control." *American Sociological Review* 43:34–45.

Block, C.R. and R. Block. 1993. "Street Gang Crime in Chicago." *Research in Brief* (December). Washington, DC: National Institute of Justice.

Hagedorn, J. 1988. *People and Folks.* Chicago: Lakeview Press.

Horowitz, R. 1983. *Honor and the American Dream.* New Brunswick, NJ: Rutgers University Press.

Katz, J. 1988. *The Seductions of Crime.* New York: Basic Books.

Klein, M. 1971. *Street Gangs and Street Workers.* Englewood Cliffs, NJ: Prentice-Hall.

Klein, M. and C. Maxson, 1989. "Street Gang Violence." Pp. 198–234 in *Violent Crimes, Violent Criminals,* edited by N. Weiner. Beverly Hills: Sage.

Loftin, C. 1984. "Assaultive Violence as Contagious Process." *Bulletin of the New York Academy of Medicine* 62:550–55.

Moore, J. 1978. *Homeboys.* Philadelphia: Temple University Press.

Padilla, F. 1992. *The Gang as an American Enterprise.* New Brunswick, NJ: Rutgers University Press.

Sanders, W. 1993. *Drive-Bys and Gang Bangs: Gangs and Grounded Culture.* Chicago: Aldine.

Short, J. and F. Strodtbeck. 1974. *Group Process and Gang Delinquency.* Chicago: University of Chicago Press.

Sullivan, M. 1989. *Getting Paid: Youth Crime and Work in the Inner City.* Ithaca: Cornell University Press.

Suttles, G. 1972. *The Social Construction of Communities.* Chicago: University of Chicago Press.

Thrasher, F. 1927. *The Gang.* Chicago: University of Chicago Press.

Vigil, D. 1988. *Barrio Gangs.* Austin: University of Texas Press.

Wright, R., S.H. Decker, A. Redfern, and D. Smith. 1992. "A Snowball's Chance in Hell: Doing Field Work with Active Residential Burglars." *Journal of Research in Crime and Delinquency* 29:148–61.

18

Homeboys, Dope Fiends, Legits, and New Jacks

John M. Hagedorn

In this study of gangs and drugs, Hagedorn found that the majority of gang members he interviewed were not committed to selling drugs. He found gang members moving in and out of legitimate and illegal employment on a sporadic basis dependent upon the types of opportunities that existed. Drug dealing was seen as a short-term method to earn money that they could not make as fast in the type of unskilled labor they could attain. Hagedorn notes that most gang members did not reject the values and aspirations of working-class people and, if they were able, would take full-time jobs and lead a conventional lifestyle. Due to their lack of legitimate, steady employment opportunities and finding themselves trapped in a life of poverty, using and selling drugs on an occasional basis was perceived as a limited but acceptable alternative. The author concludes that they deal in order to survive.

This paper addresses issues that are controversial in both social science and public policy. First, what happens to gang members as they age? Do most gang members graduate from gangbanging to drug sales, as popular stereotypes might suggest? Is drug dealing so lucrative that adult gang members eschew work and become committed to the drug economy? Have changes in economic conditions produced underclass gangs so deviant and so detached from the labor market that the only effective policies are more police and more prisons?

Second, and related to these questions, are male adult gang members basically similar

kinds of people, or are gangs made up of different types? Might some gang members be more conventional, and others less so? What are the implications of this "continuum of conventionality" within drug-dealing gangs for public policy? Data from a Milwaukee study on gangs and drug dealing shed some light on these issues.

The interview picks up the lives of the founding members since 1987, when we conducted our original study, and asks them to recount their careers in the drug business to discuss their pursuit of conventional employment, and to reflect on their personal lives. The respondents also were asked to describe the current status of their fellow gang members.

A Typology of Male Adult Gang Members

We developed four ideal types on a continuum of conventional behaviors and values: (1) those few who had gone *legit,* or had matured out of the gang; (2) *homeboys,* a majority of both African American and Latino adult gang members, who alternately worked conventional jobs and took various roles in drug sales; (3) *dope fiends,* who were addicted to cocaine and participated in the dope business as a way to maintain access to the drug; and (4) *new jacks,* who regarded the dope game as a career.

Some gang members, we found, moved over time between categories, some had characteristics of more than one category, and others straddled the boundaries (see Hannerz, 1969:57). Thus a few homeboys were in the process of becoming legit, many moved into and out of cocaine addiction, and others gave up and adopted a new jack orientation. Some new jacks returned to conventional life; others received long prison terms or became addicted to dope. Our categories are not discrete, but our typology seemed to fit the population of gang members we were researching. Our "member checks" (Lincoln and Guba, 1985:314–316) of the constructs with gang members validated these categories for male gang members.

Legits

Legits were those young men who had walked away from the gang. They were working or may have gone on to school. Legits had not been involved in the dope game at all, or not for at least five years. They did not use cocaine heavily, though some may have done so in the past. Some had moved out of the old neighborhood; others, like our project staff, stayed to help out or "give back" to the community. These are prime examples of Whyte's "college boys" or Cloward and Ohlin's Type I, oriented to economic gain and class mobility. The following quote is an example of a young African American man who "went legit" and is now working and going to college.

Q: Looking back over the past five years, what major changes took place in your life—things that happened that really made things different for you?

R#105:   I had got into a relationship with my girl, that's one thing. I just knew I couldn't be out on the streets trying to hustle all the time. That's what changed me, I just got a sense of responsibility.

Today's underclass gangs appear to be fundamentally different from those in Thrasher's or Cloward and Ohlin's time, when most gang members "matured out" of the gang. Of the 236 Milwaukee male founders, only 12 (5.1 percent) could be categorized as having matured out: that is, they were working full time *and* had not sold cocaine in the past five years. When these data are disaggregated by race, the reality of the situation becomes even clearer. We could verify only two of 117 African Americans and one of 87 Latino male gang founders who were currently working and had not sold dope in the past five years. One-third of the white members fell into this category.

Few African American and Latino gang founders, however, were resigned to a life of crime, jail, and violence. After a period of rebellion and living the fast life, the majority of gang founders, or "homeboys," wanted to settle down and go legit, but the path proved to be very difficult.

Homeboys

"Homeboys" were the majority of all adult gang members. They were not firmly committed to the drug economy, especially after the early thrill of fast money and "easy women" wore off. They had reached an age, the mid-twenties, when criminal offenses normally decline (Gottfredson and Hirschi, 1990). Most of these men were unskilled, lacked education, and had had largely negative experiences in the secondary labor market. Some homeboys were committed more strongly to the streets, others to a more conventional life. Most had used cocaine, some heavily at times, but their use was largely in conjunction with selling from a house or corner with their gang "homies." Most homeboys either were married or had a "steady" lady. They also had strong feelings of loyalty to their fellow gang members.

Here, two different homeboys explain how they had changed, and how hard that change was:

Q: Looking back over the past five years, what major changes took place in your life—things that happened that really made things different for you?

R#211: The things that we went through wasn't worth it, and I had a family, you know, and kids, and I had to think about them first, and the thing with the drug game was, that money was quick, easy, and fast, and it went like that, the more money you make the more popular you was. You know, as I see it now it wasn't worth it because the time that I done in penitentiaries I lost my sanity. To me it feels like I lost a part of my kids, because, you know, I know they still care, and they know I'm daddy, but I just lost out. Somebody else won and I lost.

Q: Is she with somebody else now?

R#211: Yeah. She hung in there about four or five months after I went to jail.

Q: It must have been tough for her to be alone with all those kids.

R#211: Yeah.

Q: What kind of person are you?

R#217: Mad. I'm a mad young man. I'm a poor young male. I'm a good person to my kids and stuff, and given the opportunity to have something nice and stop working for this petty-ass money I would try to change a lot of things . . .

. . . I feel I'm the type of person that given the opportunity to try to have something legit, I will take it, but I'm not going to go by the slow way, taking no four, five years working at no chicken job and trying to get up to a manager just to start making six, seven dollars. And then get fired when I come in high or drunk or something. Or miss a day or something because I got high smoking weed, drinking beer, and the next day come in and get fired; then I'm back in where I started from. So I'm just a cool person, and if I'm given the opportunity and if I can get a job making nine, ten

dollars an hour, I'd let everything go; I'd just sit back and work my job and go home. That kind of money I can live with. But I'm not going to settle for no three, four dollars an hour, know what I'm saying?

Homeboys present a more confused theoretical picture than legits. Cloward and Ohlin's Type III delinquents were rebels, who had a "sense of injustice" or felt "unjust deprivation" at a failed system (1960:117). Their gang delinquency is a collective solution to the failure of institutional arrangements. They reject traditional societal norms; other, success-oriented illegitimate norms replace conventionality.

Others have questioned whether gang members' basic outlook actually rejects conventionality. Matza (1964) viewed delinquents' rationalizations of their conduct as evidence of techniques meant to "neutralize" deeply held conventional beliefs. Cohen (1955:129–137) regarded delinquency as a nonutilitarian "reaction formation" to middle-class standards, though middle-class morality lingers, repressed and unacknowledged. What appears to be gang "pathological" behavior, Cohen points out, is the result of the delinquent's striving to attain core values of "the American way of life." Short and Strodtbeck (1965), testing various gang theories, found that white and African American gang members, and lower- and middle-class youths, held similar conventional values.

Our homeboys are older versions of Cohen's and Matza's delinquents, and are even more similar to Short and Strodtbeck's study subjects. Milwaukee homeboys shared three basic characteristics: (1) They worked regularly at legitimate jobs, although they ventured into the drug economy when they believed it was necessary for survival. (2) They had very conventional aspirations; their core values centered on finding a secure place in the American way of life. (3) They had some surprisingly conventional ethical beliefs about the immorality of drug dealing. To a man, they justified their own involvement in drug sales by very Matza-like techniques of "neutralization."

Homeboys are defined by their in-and-out involvement in the legal and illegal economies. Recall that about half of our male respondents had sold drugs no more than 12 of the past

36 months. More than one-third never served any time in jail. Nearly 60 percent had worked legitimate jobs at least 12 months of the last 36, with a mean of 14.5 months. Homeboys' work patterns thus differed both from those of legits, who worked solely legal jobs, and new jacks who considered dope dealing a career.

To which goal did homeboys aspire, being big-time dope dealers or holding a legitimate job? Rather than having any expectations of staying in the dope game, homeboys aspired to settling down, getting married, and living at least a watered-down version of the American dream. Like Padilla's (1992:157) Diamonds, they strongly desired to "go legit." Although they may have enjoyed the fast life for a while, it soon went stale. Listen to this homeboy, the one who lost his lady when he went to jail:

Q: Five years from now, what would you want to be doing?

R#211: Five years from now? I want to have a steady job, I want to have been working that job for about five years, and just with a family somewhere.

Q: Do you think that's gonna come true?

R#211: Yeah, that's basically what I'm working on. I mean, this bullshit is over now, I'm twenty-five, I've played games long enough, it don't benefit nobody. If you fuck yourself away, all you gonna be is fucked, I see it now.

Others had more hopeful or wilder dreams, but a more sobering outlook on the future. The other homeboy, who said he wouldn't settle for three or four dollars an hour speaks as follows:

Q: Five years from now, what would you want to be doing?

R#217: Owning my own business. And rich. A billionaire.

Q: What do you realistically expect you'll be doing in five years?

R#217: Probably working at McDonald's. That's the truth.

Homeboys' aspirations were divided between finding a steady full-time job and setting up their own business. Their strivings pertained less to being for or against "middle-class status" than to finding a practical, legitimate occupation that could support them (see Short and Strodtbeck, 1965). Many homeboys believed that using skills learned in selling drugs to set up a small business would give them a better chance at a decent life than trying to succeed as an employee.

Most important, homeboys "grew up" and were taking a realistic look at their life chances. This homeboy spoke for most:

Q: Looking back over the past five years, what major changes have taken place in your life—things that made a difference about where you are now?

R#220: I don't know, maybe maturity. . . . Just seeing life in a different perspective . . . realizing that from 16 to 23, man, just shot past. And just realizing that it did, shucks, you just realizing how quick it zoomed past me. And it really just passed me up without really having any enjoyment of a teenager. And hell, before I know it I'm going to hit 30 or 40, and I ain't going to have nothing to stand on. I don't want that shit. Because I see a lot of brothers out here now, that's 43, 44 and ain't got shit. They's still standing out on the corner trying to make a hustle. Doing this, no family, no stable home and nothing. I don't want that shit. . . . I don't give a fuck about getting rich or nothing, but I want a comfortable life, a decent woman, a family to come home to. I mean, everybody needs somebody care for. This ain't where it's at.

Finally, homeboys were characterized by their ethical views about selling dope. As a group, they believed dope selling was "unmoral"—wrong, but necessary for survival. Homeboys' values were conventional, but in keeping with Matza's findings, they justified their conduct by neutralizing their violation of norms. Homeboys believed that economic necessity was the overriding reason why they could not live up to their values (see Liebow, 1967:214). They were the epitome of ambivalence, ardently believing that dope selling was both wrong and absolutely necessary. One longtime dealer expressed this contradiction:

Q: Do you consider it wrong or immoral to sell dope?

R#129: Um-hum, very wrong.

Q: Why?

R#129: Why, because it's killing people.

Q: Well how come you do it?

R#129: It's also a money maker.

Q: Well how do you balance those things out? I mean, here you're doing something that you think is wrong, making money. How does that make you feel when you're doing it, or don't you think about it when you're doing it?

R#129: Once you get a (dollar) bill, once you look at, I say this a lot, once you look at those dead white men [*presidents' pictures on currency*], you care about nothing else, you don't care about nothing else. Once you see those famous dead white men. That's it.

Q: Do you ever feel bad about selling drugs, doing something that was wrong?

R#129: How do I feel? Well a lady will come in and sell all the food stamps, all of them. When they're sold, what are the kids gonna eat? They can't eat the dope cause she's gonna go smoke that up, or do whatever with it. And then you feel like "wrong." But then, in the back of your mind, man, you just got a hundred dollars worth of food stamps for thirty dollars worth of dope, and you can sell them at the store for seven dollars on ten, so you got seventy coming. So you get seventy dollars for thirty dollars. It is not wrong to do this. It is not wrong to do this!

Homeboys also refused to sell to pregnant women or to juveniles. Contrary to Jankowski's (1991:102) assertion that in gangs "there is no ethical code that regulates business ventures," Milwaukee homeboys had some strong moral feelings about how they carried out their business:

R#109: I won't sell to no little kids. And, ah, if he gonna get it, he gonna get it from someone else besides me. I won't sell to no pregnant woman. If she gonna kill her baby, I want to sleep not knowing that I had anything to do with it. Ah, for anybody else, hey, it's their life, you choose your life how you want.

Q: But how come—I want to challenge you. You know if kids are coming or a pregnant woman's coming, you know they're going to get it somewhere else, right? Someone else will make their money on it; why not you?

R#109: 'Cause the difference is I'll be able to sleep without a guilty conscience.

Homeboys were young adults living on the edge. On the one hand, like most Americans, they had relatively conservative views on social issues and wanted to settle down with a job, a wife, and children. On the other hand, they were afraid they would never succeed, and that long stays in prison would close doors and lock them out of a conventional life. They did not want to continue to live on the streets, but they feared that hustling might be the only way to survive.

Dope Fiends

Dope fiends are gang members who are addicted to cocaine. Thirty-eight percent of all African American founders were using cocaine at the time of our interview, as were 55 percent of Latinos and 53 percent of whites. African Americans used cocaine at lower rates than white gang members, but went to jail twice as often. The main focus in a dope fiend's life is getting the drug. Asked what they regretted most about their life, dope fiends invariably said "drug use," whereas most homeboys said "dropping out of school."

Most Milwaukee gang dope fiends, or daily users of cocaine, smoked it as "rocks." More casual users, or reformed dope fiends, if they used cocaine at all, snorted it or sprinkled it on marijuana (called a "primo") to enhance the high. Injection was rare among African Americans but more common among Latinos. About one-quarter of those we interviewed, however, abstained totally from use of cocaine. A majority of the gang members on our rosters had used cocaine since its use escalated in Milwaukee in the late 1980s.

Of 110 gang founders who were reported to be currently using cocaine, 37 percent were reported to be using "heavily" (every day, in our data), 44 percent, "moderately" (several times per week), and 19 percent "lightly" (sporadically). More than 70 percent of all founders on our rosters who were not locked up were currently using cocaine to some extent. More than one-third of our male respondents considered themselves, at some time in their lives, to be "heavy" cocaine users.

More than one-quarter of our respondents had used cocaine for seven years or more, roughly the total amount of time cocaine has dominated the illegal drug market in Milwaukee. Latinos had used cocaine slightly longer than African Americans, for a mean of 75 months compared with 65. Cocaine use followed a steady pattern in our respondents' lives; most homeboys had used cocaine as part of their day-to-day life, especially while in the dope business.

Dope fiends were quite unlike Cloward and Ohlin's "double failures," gang members who used drugs as part of a "retreatist subculture." Milwaukee dope fiends participated regularly in conventional labor markets. Of the 110 founders who were reported as currently using cocaine, slightly more were working legitimate jobs than were not working. Most dope fiends worked at some time in their homies' dope houses or were fronted an ounce or an "eightball" (3.5 grams) of cocaine to sell. Unlike Anderson's "wine-heads," gang dope fiends were not predominantly "has-beens" and did not "lack the ability and motivation to hustle" (Anderson, 1978: 96–97). Milwaukee cocaine users, like heroin users (Johnson et al., 1985; Moore, 1978; Preble and Casey, 1969), played an active role in the drug-selling business.

Rather than spending their income from drug dealing on family, clothes, or women, dope fiends smoked up their profits. Eventually many stole dope belonging to the boss or "dopeman" and got into trouble. At times their dope use made them so erratic that they were no longer trusted and were forced to leave the neighborhood. Often,

however, the gang members who were selling took them back and fronted them cocaine to sell to put them back on their feet. Many had experienced problems in violating the cardinal rule, "Don't get high on your own supply," as in this typical story:

> R#131: . . . if you ain't the type that's a user, yeah, you'll make fabulous money but if you was the type that sells it and uses it and do it at the same time, you know, you get restless. Sometimes you get used to taking your own drugs. . . . I'll just use the profits and just do it . . . and then the next day if I get something again, I'd just take the money to pay up and keep the profits. . . . You sell a couple of hundred and you do a hundred. That's how I was doing it.

Cocaine use was a regular part of the lives of most Milwaukee gang members engaged in the drug economy. More than half of our respondents had never attended a treatment program; more than half of those who had been in treatment went through court-ordered programs. Few of our respondents stopped use by going to a treatment program. Even heavy cocaine use was an "on-again, off-again" situation in which most gang members alternately quit by themselves and started use again (Waldorf et al., 1991).

Alcohol use among dope fiends and homeboys (particularly 40-ounce bottles of Olde English 800 ale) appears to be even more of a problem than cocaine use. Like homeboys, however, most dope fiends aspired to have a family, to hold a steady job, and to find some peace. The wild life of the dope game had played itself out; the main problem was how to quit using.

New Jacks

Whereas homeboys had a tentative relationship with conventional labor markets and held some strong moral beliefs, new jacks had chosen the dope game as a career. They were often loners, strong individualists like Jankowski's (1991) gang members, who cared little about group norms. Frequently they posed as the embodiment of media stereotypes. About one-quarter of our interview respondents could be described as new jacks: they had done nothing in the last 36 months except hustle or spend time in jail.

In some ways, new jacks mirror the criminal subculture described by Cloward and Ohlin. If a criminal subculture is to develop, Cloward and Ohlin argued, opportunities to learn a criminal career must be present, and close ties to conventional markets or customers must exist. This situation distinguishes the criminal from the violent and the retreatist subcultures. The emergence of the cocaine economy and a large market for illegal drugs provided precisely such an opportunity structure for this generation of gang members. New jacks are those who took advantage of the opportunities, and who, at least for the present, have committed themselves to a career in the dope game.

Q: Do you consider it wrong or immoral to sell dope?

R#203:   I think it's right because can't no motherfucker live your life but you.

Q: Why?

R#203:   Why? I'll put it this way . . . I love selling dope. I know there's other niggers out here love the money just like I do. And ain't no motherfucker gonna stop a nigger from selling dope . . . I'd sell to my own mother if she had the money.

New jacks, like other gang cocaine dealers, lived up to media stereotypes of the "drug dealer" role and often were emulated by impressionable youths. Some new jacks were homeboys from Milwaukee's original neighborhood gangs, who had given up their conventional dreams; others were members of gangs that were formed solely for drug dealing (see Klein and Maxson, 1993). A founder of one new jack gang described the scene as his gang set up shop in Milwaukee. Note the strong mimicking of media stereotypes:

> **R#126:** . . . it was crime and drug problems before we even came into the scene. It was just controlled by somebody else. We just came on with a whole new attitude, outlook, at the whole situation. It's like, have you ever seen the movie *New Jack City,* about the kid in New York? You see, they was already there. We just came out with a better idea, you know what I'm saying?

New jacks rejected the homeboys' moral outlook. Many were raised by families with long traditions of hustling or a generation of gang affiliations, and had few hopes of a conventional future. They are the voice of the desperate ghetto dweller, those who live in Carl Taylor's (1990:36) "third culture" made up of "underclass and urban gang members who exhibit signs of moral erosion and anarchy" or propagators of Bourgois's (1990:631) "culture of terror." New jacks fit the media stereotype of all gang members, even though they represent fewer than 25 percent of Milwaukee's adult gang members.

DISCUSSION: GANGS, THE UNDERCLASS, AND PUBLIC POLICY

Our study was conducted in one aging postindustrial city, with a population of 600,000. How much can be generalized from our findings can be determined only by researchers in other cities, looking at our categories and determining, whether they are useful. Cloward and Ohlin's opportunity theory is a workable general theoretical framework, but more case studies are needed in order to recast their theory to reflect three decades of economic and social changes. We present our typology to encourage others to observe variation within and between gangs, and to assist in the creation of new taxonomies and new theory.

Our paper raises several empirical questions for researchers: Are the behavior patterns of the founding gang members in our sample representative of adult gang members in other cities? In larger cities, are most gang members now new jacks who have long given up the hope of a conventional life, or are most still homeboys? Are there "homeboy" gangs and "new jack" gangs, following the "street gang/drug gang" notion of Klein and Maxson (1993)? If so, what distinguishes one from the other? Does gang members' orientation to conventionality vary by ethnicity or by region? How does it change over time? Can this typology help account for variation in rates of violence between gang members? Can female gang members be typed in the same way as males?

Our data also support the life course perspective of Sampson and Laub (1993:255), who ask whether present criminal justice policies "are producing unintended criminogenic effects." Milwaukee gang members are like the persistent, serious offenders in the Gluecks' data (Glueck and Glueck, 1950). The key to their future lies in building social capital that comes from steady employment and a supportive relationship, without the constant threat of incarceration (Sampson and Laub, 1993:162–168). Homeboys largely had a wife or a steady lady, were unhappily enduring "the silent, subtle humiliations" of the secondary labor market (Bourgois, 1990:629), and lived in dread of prison. Incarceration for drug charges undercut their efforts to find steady work and led them almost inevitably back to the drug economy.

Long and mandatory prison terms for use and intent to sell cocaine lump those who are committed to the drug economy with those who are using or are selling in order to survive. Our prisons are filled disproportionately with minority drug offenders (Blumstein, 1993) like our homeboys, who in essence are being punished for the "crime" of not accepting poverty or of being addicted to cocaine. Our data suggest that jobs, more accessible drug treatment, alternative sentences, or even decriminalization of nonviolent drug offenses would be better approaches than the iron fist of the war on drugs (see Hagedorn, 1991; Reinarman and Levine, 1990; Spergel and Curry, 1990).

Finally, our typology raises ethical questions for researchers. Wilson (1987:8) called the underclass "collectively different" from the poor of the past, and many studies focus on underclass deviance. Our study found that some underclass gang members had embraced the drug economy and had forsaken conventionality, but we also found that the *majority* of adult gang members are still struggling to hold onto a conventional orientation to life.

Hannerz (1969:36) commented more than two decades ago that dicthotomizing community residents into "respectables" and "disrespectables" "seems often to emerge from social science writing about poor black people or the lower classes in general." Social science that emphasizes differences within poor communities, without noting commonalities, is one-sided and often distorts and demonizes underclass life.

Our data emphasize that there is no Great Wall separating the underclass from the rest of the central-city poor and working class. Social research should not build one either. Researchers who describe violent and criminal gang actions without also addressing gang members' orientation to conventionality do a disservice to the public, to policy makers, and to social science.

REFERENCES

Anderson, Elijah. 1978. *A Place on the Corner.* Chicago: University of Chicago Press.

Bourgois, Phillippe. 1990. "In search of Horatio Alger: Culture and ideology in the crack economy." *Contemporary Drug Problems,* 16:619–649.

Cloward, Richard and Lloyd Ohlin. 1960. *Delinquency and Opportunity.* Glencoe, Ill.: Free Press.

Cohen, Albert. 1955. *Delinquent Boys.* Glencoe, Ill.: Free Press.

Glueck, Sheldon and Eleanor Glueck. 1950. *Unraveling Juvenile Delinquency.* New York: Commonwealth Fund.

Gottfredson, Michael and Travis Hirschi. 1990. *A General Theory of Crime.* Stanford: Stanford University Press.

Hagedorn, John M. 1991. "Gangs, neighborhoods, and public policy." *Social Problems,* 38:529–542.

Hannerz, Ulf. 1969. *Soulside: Inquiries into Ghetto Culture and Community.* New York: Columbia University Press.

Jankowski, Martin Sanchez. 1991. *Islands in the Street: Gangs and American Urban Society.* Berkeley: University of California Press.

Johnson, Bruce D., Terry Williams, Kojo Dei, and Harry Sanahria. 1985. *Taking Care of Business: The Economics of Crime by Heroin Abusers.* Lexington, Mass.: Heath.

Klein, Malcolm W. and Cheryl L. Maxson. 1993. "Gangs and cocaine trafficking." In Craig Uchida and Doris Mackenzie (eds.), *Drugs and the Criminal Justice System.* Newbury Park: Sage.

Liebow, Elliot. 1967. *Tally's Corner.* Boston: Little, Brown.

Lincoln, Yvonna S. and Egon G. Guba. 1985. *Naturalistic Inquiry.* Beverly Hills: Sage.

Matza, David. 1964. *Delinquency and Drift.* New York: Wiley.

Moore, Joan W. 1978. *Homeboys: Gangs, Drugs, and Prison in the Barrios of Los Angeles.* Philadelphia: Temple University Press.

Padilla, Felix. 1992. *The Gang as an American Enterprise.* New Brunswick: Rutgers University Press.

Preble, Edward and John H. Casey. 1969. "Taking care of business: The heroin user's life on the street." *International Journal of the Addictions,* 4:1–24.

Sampson, Robert J. and John H. Laub. 1993. *Crime in the Making: Pathways and Turning Points through Life.* Cambridge: Harvard University Press.

Short, James F. and Fred L. Strodtbeck. 1965. *Group Process and Gang Delinquency.* Chicago: University of Chicago Press.

Spergel, Irving A. and G. David Curry. 1990. "Strategies and perceived agency effectiveness in dealing with the youth gang problem." In C. Ronald Huff (ed.), *Gangs in America.* Beverly Hills: Sage.

Taylor, Carl. 1990. *Dangerous Society.* East Lansing: Michigan State University Press.

Waldorf, Dan, Craig Reinarman, and Sheigla Murphy. 1991. *Cocaine Changes: The Experience of Using and Quitting.* Philadelphia: Temple University Press.

Wilson, William Julius. 1987. *The Truly Disadvantaged.* Chicago: University of Chicago.

19

THE PERCEPTION OF GANGS AS A PROBLEM IN NONMETROPOLITAN AREAS

RALPH A. WEISHEIT AND L. EDWARD WELLS

The existence and sustained number of gangs in rural areas has received little attention by both the media and researchers in recent times. Weisheit and Wells studied the emergence and existence of gangs in smaller communities by utilizing telephone interviews with 216 law enforcement agencies that had previously reported the prevalence of gang activity in their jurisdictions. The authors focused their research on issues involving the definition of gangs, problems related to gangs, and the types of strategies that police and sheriff's departments utilize in their responses to gangs. Weisheit and Wells conclude that there is wide disagreement among rural police concerning the actual existence of gangs, and that past gang organizations were small in number and short-lived, due to active law enforcement practices of arresting suspected members for any legal infractions they could find.

Traditionally, youth gangs or "street gangs" in the U.S. have been regarded as urban phenomena—the products of large, crowded, disorganized metropolitan communities. This view has been amply detailed in ethnographic studies focused on a few of the largest metropolitan centers. However, in recent decades the problem of youth gangs has visibly grown, spreading to smaller cities and to less metropolitan communities across the United States. This trend has been widely reported in the media in sometimes sensational stories about "gangs in the heartland" or "gangs invading small town America" (Coates & Blau, 1989; "Gangs in the Heartland," 1996; A. Miller, 1996; Poe, 1998). This pattern has also been reported by scholars who study gangs (Caldarella, Sharpnack, Loosli, & Merrell, 1996; Curry, Ball, & Decker, 1996; Hagedorn, 1999; Howell, 1998; Klein, 1995; Maxson, 1998; Short, 1998), although the empirical basis for these claims has been rather limited.

EDITOR'S NOTE: From Weisheit, R. & Wells, L. E., "The perception of gangs as a problem in nonmetropolitan areas," in *Criminal Justice Review, 26*(2), pp. 170-192. Copyright © 2001. Reprinted with permission from the Criminal Justice Review

Although we now have more extensive information about gangs in more types of places, our knowledge about gang dynamics in smaller cities, towns, and rural communities is still incomplete. Although little is known about nonmetropolitan gangs, there are reasons to believe that our understanding of urban gangs will not automatically apply in nonmetropolitan areas. Thus, urban conceptions of gangs are a useful starting point for an analysis of nonmetropolitan gangs, but the applicability of urban conceptions of gangs to nonmetropolitan settings is very much an open question.

This study is based on the premise that the police as the officially designated agency in a community for dealing with troublesome and order-threatening groups are the most influential social audience for officially defining gang problems. The presumption is supported by the work of McCorkle and Miethe (1998), who found that police were the primary source of information about gangs for the local media and for local politicians. Thus, police play a key role in shaping media images of gangs, which, in turn, shape public opinion. This research examines how police define, perceive, and identify gang phenomena. Police respond to objective characteristics of groups, and the focus here is on the nature of that response. Our interest is in how police perceive and define gang problems. Thus, our focus is on perceptual rather than organizational issues and our study is comparative rather than focused on single cases. Utilizing descriptions and explanations generated by police respondents, this study seeks to better understand what police mean by the term "gang" in community contexts outside the traditional large urban centers and to better understand the nature of gang problems in nonmetropolitan areas.

The analysis is framed by several distinct but related questions. What do police respondents in nonmetropolitan agencies mean when they report "gangs" or "gang problems" in their community? How do police as peacekeeping law enforcers in smaller nonurban communities recognize or define persons as "gang members?" How do police in smaller jurisdictions deal with the presence of gangs or gang members? How closely do these nonurban responses correspond to conventional urban descriptions of gang phenomena?

METHOD

For this study, the county in which the agency was located had to be classified as nonmetropolitan by the U.S. Bureau of the Census. This meant that there could be no urbanized center of 50,000 or more in the county, plus contiguous areas having strong economic and social ties to that urban center.

Telephone interviews were conducted from November 1999 through February 2000. The length of the interviews varied, depending on the nature and extent of gang problems in a community and on the extent to which subjects were talkative. Repeated efforts were made to reach respondents, and cases were only dropped if the respondent explicitly refused, or if after more than a dozen tries it was not possible to reach the respondent.

On average the interviews lasted about 20 minutes. All interviews were tape recorded and transcribed. The transcriptions were then analyzed through a series of discrete steps. First, responses to general topic areas (e.g., How do you know that you have a gang?) were extracted. These responses were then examined to determine whether patterns emerged, and responses were then grouped according to any perceived patterns. These patterns were then reexamined to determine whether there were additional subcategories for each category.

FINDINGS

Of agencies reporting the presence of at least one youth gang in 1997, only 40 percent ($n = 86$) still reported the presence of a gang when we called them three years later. These numbers are surprising and are substantially lower than would be expected if gangs were pervasive, well defined, and persistent in nonmetropolitan areas. Further, of the nonmetropolitan agencies that reported gangs in 1997, the percentage that also reported them in our 2000 interview declined dramatically as the county in which the agency was located became more rural (that is, the percentage declined across county size) (see Table 19.1). In the most rural counties, only 14 percent of the agencies that reported gangs in 1997 still reported gangs three years later.

Table 19.1   Number and Percentage of Nonmetropolitan Agencies Still Reporting Gangs in 2000 After First Reporting Gangs in 1997

Rurality of county in which jurisdiction is located	Number reporting gangs	Percentage reporting gangs
Urban population of 20,000 or more	51 of 88 agencies	58
Urban population of 2,500 to 19,999	31 of 97 agencies	32
Completely rural or less than 2,500 urban population	4 of 28 agencies	14

Note. Number of cases = 213. All agencies had reported the presence of gangs in 1997.

Respondents reported that, although they had no gangs or resident gang members in their communities, they occasionally did have "gang problems" or "gang situations" due to gangs in neighboring communities dropping in or transient gangs passing through the community:

> In the last maybe four years we have had various problems with gang members from other communities. We don't really have a text-book type definition of gangs here. I mean we do have gang members here but we don't have a large number of them. Most of them and most of their activity and most gangs are affiliated with our neighboring communities both to the north and west. They come to obviously commit various offenses and go back to their home communities. (Agency 146)

> The other thing that is a headache is that some drug dealers are from out of town, from [names large cities within driving distance]. We do feel strongly that these people have gang associations, so they are a problem. But again, they don't have a lot of success at creating a gang organization in this part of the state. In that way gangs are a headache for us, but not organized. (Agency 065)

> Periodically we will discover an organized gang member from another part of the state operating in this area primarily in drug distribution, but they're not actually gang banging here. They are up here making money, selling drugs to the kids. We have had several gang members . . . come up here and infiltrate the students at [local college] and conduct their business. But, they are not actually banging while they're here. (Agency 024)

> We don't really have any gangs that are centered here in our community, because we just don't have that large of a community. But we have some that are members of gangs in surrounding communities and, occasionally, they come over here. (Agency 212)

For purposes of this study, such communities were categorized as not having gangs, because they explicitly reported that they did not have any resident groups identified as gangs.

Our attention in the telephone interviews focused on four primary questions or issues: (a) How did the respondent's agency know that there was a gang or gang members in its jurisdiction? (b) What kinds of problems did the agency have from gangs? (c) How did gangs emerge in the community? (d) How did the agency respond to gangs?

Identifying Gangs

Respondents generally used several items as indicators of a gang presence. The most frequent indicator was a juvenile's self-identification as a gang member. Also frequently used were the presence of graffiti, tattoos, a youth's affiliation with others thought to be gang members, and the wearing of gang colors. In a number of jurisdictions, any one of these indicators might, by itself, be used as evidence of the presence of a gang. Other jurisdictions were more selective, requiring several of these indicators. When asked "How have you determined that you have an organized gang in your jurisdiction? What are some of the things you used as indicators?" some agencies reported using relatively concrete indicators:

We have six criteria . . . The officer can check as many of those as apply. Obviously, if they check the one about whether they fit their style of dress or tattoos, I mean if they check that one box and that one box only, we don't classify them as being a gang member solely on how they dress. (Agency 039)

Other jurisdictions used criteria that were more vague and impressionistic:

Just by their names, with the colors they wear, the things they are doing . . . And, well, I don't know. I just look at them. (Agency 186)

Of course dress, tattoos, the gang signs—you know they'll wear certain colors, or they used to, jewelry that they wear with six and five point stars. Just their appearance. (Agency 129)

Through our school resource officers that have developed a relationship with a lot of the kids, and they talk to them and the kids are opening up to them and saying, yeah, I was approached by so and so to be a member of this gang or that gang, and that's basically the way we've determined what gangs we have in the area. (Agency 046)

Our local community doesn't allow them to fly colors. We don't allow that. We have an ordinance against it. So the only way we know that they are part of a gang is the actual corners they hang on, and everyone of them has a different area of town that they sort of claim as their own. (Agency 043)

Mostly it's the fights between youth at school. (Agency 198)

Relying on such indicators as colors, tattoos, and signs has become more problematic in communities in which gangs are attempting to keep a low profile—something that many agencies thought was becoming more common:

Most of the gangs I understand are getting away from indicators, tattoos, colors, or whatever. They still have them but they don't display them for police, ever since the gang laws and everything came into effect. Basically, the people they hang out with are the people that they don't find

offensive. In other words, when you talk with a person and he hates the people in this side of town and he won't go over there, it's an indicator that he's probably affiliated with some group that hates another here. (Agency 038)

A few jurisdictions used guidelines established by their states. These guidelines listed specific criteria and required that a certain proportion of these criteria be met before someone could be called a gang member (e.g., 4 of 12 criteria). For most states, these criteria are quite demanding. One Minnesota official described their system:

We have identified members through a criteria system that we follow through the Minnesota Gang Strike Force, which is basically a ten-point criteria system. If someone has been convicted of a gross misdemeanor or higher gross misdemeanor felony and also meets three out of these ten criteria, we can then identify them as a known gang member. (Agency 109)

Generally, however, these criteria were used to decide whether an individual could be labeled as a gang member and were not used as proof of the existence of a gang. This system could be problematic in rural jurisdictions in which there were reported to be gang members but no gangs, a situation noted earlier in the discussion.

The Nature of Gang Problems

Questions about the types of problems associated with gangs led to a wide range of responses. In some jurisdictions, having a gang problem meant nothing more than finding graffiti, whereas in others there were reports of murders committed by gang members.

Respondents were asked to describe the types of problems that they experienced as a result of gangs, without being provided specific topics or other prompting. Among those who reported specific problems, the most frequent responses were drugs (69 percent), assaults (52 percent), theft (32 percent), and burglary (31 percent). Despite reports of drugs, assaults, drive-by shootings, and even homicides, only 42 percent of those reporting gangs described the gang problems in their community as "serious," and some of those who described the problem as serious qualified their rating:

> I consider any gang activity to be serious. (Agency 052)
>
> In a small town like this our little gangs, to the people, are serious. But, to the big city, this would be minor. (Agency 179)
>
> Well, again, the problem is significant for us, but I suppose if you were comparing it to an urban environment it would be minimal. (Agency 151)

Although drug use and drug sales were common among gang members, and violence was occasionally seen, most of the observed gang crime problems were of a relatively minor nature, what might be described as general delinquency:

> Most of the problems would be graffiti, parties, and alcohol consumption. Occasionally we have had problems in the past with some shootings and some fights, and weapon violations. (Agency 133)
>
> For the most part, our problems from a criminal standpoint involve minor property crime, auto burglaries, residential burglaries, and then of course we have the assaults and nothing, at least in our community, nothing has exceeded the point beyond maybe an aggravated assault. We have had I think three drive-by shootings that we've actually been able to trace back to a gang. We have had a couple of gang members involved in a homicide but that wasn't a crime that was in furtherance of the gang per se. I think that was something outside that realm. So for the most part of I would say property crimes, graffiti, auto burglaries, residential burglaries that kind of stuff. (Agency 014)

It is clear from these interviews that "gang problems" include activities that vary greatly in seriousness from one community to the next. Some rural jurisdictions have problems that are serious by any standard, whereas for others the problems are rather minor. It is important to recognize this variability across jurisdictions when describing rural gangs and when establishing policies for responding to them. Asking police agencies whether they have "gang problems" in their jurisdictions (as some surveys have done) does not provide much specific information about the presence of gangs or the nature of gang activity in those communities.

The Emergence of Gangs

Some have assumed that gangs spread from urban to rural areas through a process in which urban gang members themselves migrate to rural areas (Donnermeyer, 1994), while others have argued that only the symbols and culture of the gang are exported to rural communities (Hagedorn, 1988, 1998). We asked those in jurisdictions reporting gangs how many of the gang members seemed to come from outside the local area. The results were mixed. There were a few jurisdictions in which all gang members were reportedly from other areas, but this was not the most typical circumstance, and, although it is sometimes assumed that rural gang problems are almost entirely imported, there were a few jurisdictions in which the gang problem was completely home-grown:

> [How many local gang members came from somewhere else?] Almost none. I would say that any that did probably grew up here and went off somewhere for whatever reason, and wound up back here with it. I mean it actually did come from somewhere else, but percentage wise it is almost non-existent. It all somehow started here. (Agency 019)
>
> [How many youth gang members came from somewhere else?] I'd say most of these came straight from here. I can't say none, but I bet it would be close to none. (Agency 247)
>
> We primarily have local kids who have for some reason got the idea that having a gang would be cool, would be the thing to do. Over the years we have had a few people actually come up from places like L.A. and Salt Lake that were gang members in those areas and have started groups, but they've been arrested and sent away and are no longer in our area. (Agency 051)

Further, even when outsiders moved into the area, continued gang activity was frequently reported to depend on the cooperation of local youth:

> [Gang members are] primarily local residents. We found that even if you're from another city and

Table 19.2   Reasons Why Gang Members Moved Into the Nonmetropolitan Community

Reason	Percentage
Moved for social reasons	86
Moved to avoid the police	46
Moved to expand drug markets	41
Moved to engage in other illegal activities	33
Moved to get away from gangs	30

Note. $n = 80$. Multiple responses were allowed. Percentages reflect the percentage of agencies reporting gangs in which the respondent believed that this reason had been true for at least some gang members in their community.

you come up here to set up business, you have to work through local residents to do business. (Agency 246)

Focusing on gang members who moved into the community, officials gave a variety of reasons why these youth moved into the area, but they also were specifically asked about five particular reasons: moving for social reasons (e.g., their family moved there), moving to expand drug markets, moving specifically to engage in other illegal activities, moving to avoid the police, and moving to get away from gang influences. These reasons include a mix of the sinister and the more benign. Table 19.2 shows the frequency of response for each reason on this more focused probe.

Although urban gang members often moved into these rural areas for more than one of these reasons, Table 19.2 shows that most gang youth move into the area for social reasons, that is, to accompany their family or to move in with relatives. Family moves were generally precipitated by changing jobs or by the availability of subsidized housing, consistent with the speculations of Maxson (1998). Other reasons occurred frequently enough to suggest that a single model or explanation for the migration of urban gang members into rural areas will not suffice.

Displacement: In response to our structured question about gang migration, 46 percent of the respondents believed that gang members moved into their rural area to avoid the police. Several examples of this process were provided on the open-ended query:

We have a lot of drug activity and I think a lot of the problem is that the gangers from Washington and Oregon, you know, head over this way to evade the law over there. (Agency 143)

The ones that I've interviewed, which has been about 50 percent of them, have either been cooling off from the area that they're from, or avoiding trouble with California laws. Three strike laws and stuff like that. (Agency 059)

However, although it was commonly mentioned, crime displacement was not the dominant issue reported by police respondents. Nearly half agreed that some gang members moved into the area to avoid the police, but displacement was mentioned as the primary reason for the move in only a few jurisdictions.

Branch office: In response to our structured question, 41 percent agreed that urban gang youth moved into their area to expand drug markets because the area was seen as a lucrative market where drugs would sell for a higher price and enforcement would be weaker:

Nearly all of them have done that. Some of them have recruited the local kids from different high schools and communities but our problem gangs have been all based out of Chicago and Minneapolis. And they've just reached out to the rural area because there's less pressure and they're able to obviously get more money for their narcotics . . . There's more of a demand you know in the area of the state where you're not real close to metropolitan cities. There is less law enforcement in those areas and a less likelihood of being detected by law enforcement . . . A gang member will move in from Chicago and he'll talk to his brother, or he'll talk to his sister, or relative or friend and they'll say come up to [his community], you know there is very little law enforcement and there's a need for us to come up here to set up our illegal activities. (Agency 091)

They move from place to place. They don't really set up any permanent residence in this community. They operate either quick crack houses that they don't keep in operation very long at that location or they move to motel rooms. They jump around on us quite a bit. (Agency 084)

Although the establishment of a branch office did occur in some of these jurisdictions, it was not the most common situation reported by these agencies. This finding seems consistent with Klein's argument (1995) that urban gangs do not primarily spread through the entrepreneurial expansion of drug distribution networks. Further, we would expect such branch offices to be relatively easily neutralized by local police.

Franchise: There was little evidence of local drug dealers reaching out to forge business links with urban gangs. Although we are aware of communities in which local dealers have claimed ties to urban gangs, our police respondents generally believed that those claims were questionable:

> For the most part they may claim to be a faction of the "Bloods" or the "Crips" or the "Latin Kings," but they're not really closely associated. I mean they're not taking any marching orders. They're not funneling any profits to anybody in particular, so I would say they're kind of autonomous groups that claim national affiliation for extra power and prestige. (Agency 136)

Social learning: Though it was not a frequent response, respondents in some agencies did report that local youth learned about gangs and gang activities from urban youth with whom they had contact outside the local community:

> A couple of juveniles that came back from JDC (juvenile detention center) started with a basic interest in the gang and then they recruited, oh,

probably 30 some members at the height of their glory. (Agency 139)

You know a lot of them go to boot camp or they go to the job corps and they bring that stuff back. Some have relatives who leave and they come back and they learn it from the relatives. (Agency 041)

Right now I don't have any migrant gang members at all. All of these are local folks. They went off somewhere and brought it back home, but they're our local folks. (Agency 079)

This quote suggests that youth may leave the community and return for a variety of reasons, not simply for being institutionalized, as was suggested in earlier research (Donnermeyer, 1994; Weisheit et al., 1996, 1999).

Urban flight: About 30 percent of respondents to our structured question reported that gang members (or their families) had migrated to their area in an attempt to get away from urban gangs. On our open-ended query, a number of jurisdictions reported the presence of individual gang members whose families had moved them from the city to get them away from gang influences. Instead of leaving the gang, these youth often initiated or otherwise became involved in gang activity with local youth:

> One gang has its origins from [a large city] and has gravitated here when some of the kids who were involved with gangs have been redistributed by families to get them out of that environment. But, rather than getting them out of that environment they bring to these rural communities that gang culture. (Agency 026)

> Actually we have run into quite a few who have said that their parents moved them to get away from gangs. We hear that a lot from the few kids from the Chicago area that live here that we have a pretty consistent contact with. (Agency 014)

> They left [the city] because of the gangs there. Every single parent says the same thing. We left Chicago because of the gangs, or we left Detroit because of the gangs, and then they end up starting their own here. (Agency 090)

> I have yet to find a parent that didn't say I was trying to move them and get them into a new environment. See our gang members are sophisticated kids from Tacoma, Los Angeles, Phoenix, or where

ever. They weren't a gang member but they were around it enough that when they came here they were able to talk the talk and walk the walk enough to impress the bumpkin locals who have a leaning that way. That's kind of where our people have sprung up. If you were to take these people and drop them into Compton, California they would all disappear within a matter of five minutes. These people have no real idea of what a true gang-banger is. That's what we see here. (Agency 050)

A lot of times it's the parents. The parents will move back south because they feel like their kids are getting too out of hand up north, and actually we're suffering the same problems they do up there, we just get it later. A lot of them decide to come home just to be home. Most of their homes, their relatives, or their parents, whatever, were probably originally from here and they wanted to come back home. Usually a sick grandparent or something like that brings most of the parents home and that's how they wind up working themselves back in. (Agency 099)

Although the notion of urban flight was not the most frequent response to our more structured question, it generated the most comment from police respondents in their open-ended responses.

Thus, according to police respondents, gang members often move into rural areas for reasons completely unrelated to their involvement in gangs, their fear of urban police, or their interest in establishing drug markets.

Responding to Gangs

Several researchers have argued that rural policing involves a very different style of policing. In particular, research has suggested that rural police rely more on interpersonal skills and diplomacy, commanding respect because of who they are as persons rather than because of the uniform that they are wearing (Weisheit, Wells, & Falcone, 1994). If this were true, it was expected that rural police might also adopt very different styles of responding to gangs and might be less effective in dealing with organized gangs. This expectation was not generally borne out by the interviews, in which the most frequently described agency response to gang activity was suppression through strict

enforcement, a style usually associated with urban police:

> It is our philosophy that we don't give warnings for tobacco, we don't give warnings for alcohol, we don't give warnings for trespass or truancy, and we have adopted a zero tolerance policy on possession. And then, the second prong of the strategy is that when we code an offense a particular way that indicates a gang or a group activity—we've gone to our judicial partners, the judicial system, and demanded maximum penalties or guilty verdicts . . . And being creative in enforcement, looking at zoning, looking at parked cars, looking at animal control laws, and having zero tolerance on possession at an address or on a group. Whenever we receive a complaint there's an arrest made. There's no negotiating, no discussing, no officer discretion, and that has actually driven out the last five houses that we've had to deal with that were identified as being used by a gang. They weren't even residents. People were just using them to hang out. We were able to close those up and get the landlord to do different things with it through the zoning and through ordinance violations. (Agency 171)

> We try to keep the pressure up with the youth gangs. We had one several years ago that aligned itself with the Crips. Every time we'd see several of them together we'd talk to them. We also talked to their parents and sent letters to their parents to advise them. (Agency 133)

> We did have a gang trying to move in and we just kept constant pressure on them. The kids that started getting involved, once we identified them and their groups we had constant pressure on those groups. Any time we saw anything, even smoking under age or anything like that, we'd be on them for it. And, of course, the juveniles would get the parents involved and we would go that route with them . . . But it's just constant pressure on them. We just try to make it a pain in the butt to be a gang member. (Agency 187)

Our interview subjects also suggested that zero tolerance practices were easier to apply in smaller communities where gang members and outsiders of any sort stand out and in which individual police officers, prosecutors, probation officers, and judges may have a closer working relationship:

We are fortunate in that we are still small enough that once these individuals are identified, every law enforcement officer is aware of them and their criminal activity. And they do take a zero tolerance approach, and if it warrants felony charges we can get these individuals out of the community and into the prison system . . . I think it would [work in other communities] but when you're talking a metropolitan area there is no way it can be enforced. You're not going to as easily recognize one individual in your community who may be identified as a gang member and target them for the criminal activity they commit. (Agency 039)

For many agencies, strict enforcement against individuals perceived to be gang members was also accompanied by a more tempered approach to handling potential gang members, or gang members who have not yet become involved in serious crime. Many stressed the importance of prevention and of working with the community:

> . . . talking, communicating with them. If you sit and talk to them you may find out they're on a borderline as to whether they are even a member of a gang or not. If you can carry on a conversation with them and find out why they are in gangs, maybe you can address that, get them involved in some other activity, and pull them out of that gang. (Agency 084)
> When we have targeted gang members, especially, we try to communicate with them before there's any criminal activity. We've got a basketball league and we try to get them involved with other things, give them alternatives . . . That's through most of the church organizations that we work with and we try to give our assistance, so we have a couple of officers involved heavily in that. I guess it's been successful in isolated instances where we know kids are about to become involved or have recently become involved with gangs. We target those folks and try to give them an alternative. (Agency 093)
> I handle them by talking to each individual one. [For example, there was a fight between two groups who identified themselves as gangs.] I dealt with it individually. I knew the parents, called them all in and dealt with it. There weren't even charges filed. And I think the gangs pretty much broke up. These were pretty much wanna

be's, because they didn't have anything else to do. We don't have movie theaters, bowling alleys, or nothing here in this community. (Agency 046)

Thus it appeared that for outsiders engaged in gang activity, or for insiders deemed to be beyond redemption, harsh criminal penalties were seen as appropriate. However, for youth with stronger bonds to the local community, for whom there was some hope, the emphasis shifted to community and family pressure and to prevention. It is likely that police in these kinds of nonmetropolitan communities have a greater familiarity with their citizens, including troubled youth, and have more information than urban police have to make judgments about a particular juvenile's ties to the community and likely prospects for reform.

Discussion

The data suggest that many of the nonmetropolitan gang members were from the local area, although gang members who moved in or visited may have brought the symbols and ideas of urban gangs with them. Gang influences, as opposed to gang members, were reported to have come from a variety of sources. In some cases local youth left the community and returned with knowledge about urban gangs. In other cases urban gang members moved into the area, accompanying family members, following employment opportunities, or through any one of a number of other avenues. In many cases, gang youth were moved to the area by their families for the express purpose of getting them away from urban gang influences.

Having identified a gang problem, the typical rural agency responded through suppression, often arresting suspected gang youth for even the smallest infraction. The intent was to make suspected gang members feel uncomfortable and unwelcome, with the expectation that they would then leave. In the smallest jurisdictions continuous and conspicuous monitoring of suspected gang members was possible at a level that would probably not be practical in most large jurisdictions. By most police accounts, this approach was effective.

REFERENCES

Caldarella, P., Sharpnack, J., Loosli, T., & Merrell, K. W. (1996). The spread of youth gangs into rural areas: A survey of school counselors. *Rural Special Education Quarterly, 15*, 18–27.

Coates, J., & Blau, R. (1989, September 13). Big-city gangs fuel growing crack crisis. *Chicago Tribune*, pp. 1, 8.

Curry, G. D., Ball, R. A., & Decker, S. H. (1996). Estimating the national scope of gang crime from law enforcement data. In C. R. Huff, R. A. Ball, & S. H. Decker (Eds.), *Gangs in America* (2nd ed.) (pp. 21–36). Newbury Park, CA: Sage.

Donnermeyer, J. F. (1994, March). *Crime and violence in rural communities*. Paper presented at the annual meeting of the Academy of Criminal Justice Sciences, Chicago, IL.

Gangs in the heartland. (1996, May 25). *The Economist, 339*, 29–30.

Goldstein, A. P. (1991). *Delinquent gangs: A psychological perspective*. Champaign, IL: Research Press.

Hagedorn, J. M. (1988). *People and folks: Gangs, crime, and the underclass in a rustbelt city*. Chicago: Lakeview Press.

Hagedorn, J. M. (1998). Gang violence in the post-industrial era. In M. Tonry & M. H. Moore (Eds.), *Youth violence* (pp. 365–419). Chicago: University of Chicago Press.

Hagedorn, J. M. (1999). *People and folks: Gangs, crime, and the underclass in a rustbelt city* (2nd ed.). Chicago: Lakeview Press.

Klein, M. W. (1995). *The American street gang: Its nature, prevalence, and control*. New York: Oxford University Press.

Maxson, C. L. (1998). *Gang members on the move*. Washington, DC: Office of Juvenile Justice and Delinquency Prevention, Office of Justice Programs, U.S. Department of Justice.

McCorkle, R. C., & Miethe, T. D. (1998). The political and organizational response to gangs: An examination of a "moral panic" in Nevada. *Justice Quarterly, 15*, 41–64.

Miller, A. (1996, February 22). Gang murder in the heartland. *Rolling Stone*, pp. 48–54, 72.

Short, J. F. (1998). The level of explanation problem revisited—The American Society of Criminology 1997 presidential address. *Criminology, 36*, 3–36.

Weisheit, R. A., Falcone, D. N., & Wells, L. E. (1996). *Crime and policing in rural and small-town America*. Prospect Heights, IL: Waveland Press.

Weisheit, R. A., Falcone, D. N., & Wells, L. E. (1999). *Crime and policing in rural and small-town America* (2nd ed.). Prospect Heights, IL: Waveland Press.

Weisheit, R. A., & Wells, L. E. (1996). Rural crime and justice: Implications for theory and research. *Crime and Delinquency, 42*, 379–397.

Weisheit, R. A., Wells, L. E., & Falcone, D. N. (1994). Community policing in small town and rural America. *Crime and Delinquency, 40*, 549–567.

20

Gender and Victimization Risk Among Young Women in Gangs

Jody Miller

Many studies have been conducted concerning the criminal effects of gang involvement, but there is only sparse literature to be found on gang member victimization. Miller explores the extent that female gang membership in mixed gender gangs shapes young women's risk of victimization by male members. She conducted interviews with active female and male gang members, and she points out that merely being a member increases one's risk of assaultive victimization and that these risks are greater for females. She suggests that being a woman in a gang may offer protection from outside gangs, but it results in routine violent victimization on the part of fellow gang members, which causes women in gangs to be vulnerable to sexual exploitation and sexual assault.

Gang participation can be recognized as a delinquent lifestyle that is likely to involve high risks of victimization (see Huff 1996:97). Although research on female gang involvement has expanded in recent years and includes the examination of issues such as violence and victimization, the oversight regarding the relationship between gang participation and violent victimization extends to this work as well.

Based on in-depth interviews with female gang members, this article examines the ways in which gender shapes victimization risk within street gangs.

Girls, Gangs, and Crime

Until recently, however, little attention was paid to young women's participation in serious and violent gang-related crime. Most traditional gang research emphasized the auxiliary and peripheral nature of girls' gang involvement and often resulted in an almost exclusive emphasis on girls' sexuality and sexual activities with male gang members, downplaying their participation in delinquency (for critiques of gender bias in gang research, see Campbell 1984, 1990; Taylor 1993).

However, recent estimates of female gang involvement have caused researchers to pay greater attention to gang girls' activities. This evidence suggests that young women approximate anywhere from 10 to 38 percent of gang members (Campbell 1984; Chesney-Lind 1993; Esbensen 1996; Fagan 1990; Moore 1991), that female gang participation may be increasing (Fagan 1990; Spergel and Curry 1993; Taylor 1993), and that in some urban areas, upward of one-fifth of girls report gang affiliations (Bjerregaard and Smith 1993; Winfree et al., 1992). As female gang members have become recognized as a group worthy of criminologists' attention, we have garnered new information regarding their involvement in delinquency in general, and violence in particular.

Few would dispute that when it comes to serious delinquency, male gang members are involved more frequently than their female counterparts. However, this evidence does suggest that young women in gangs are more involved in serious criminal activities than was previously believed and also tend to be more involved than nongang youths—male or female. As such, they likely are exposed to greater victimization risk than nongang youths as well.

In addition, given the social contexts described above, it is reasonable to assume that young women's victimization risk within gangs is also shaped by gender. Gang activities (such as fighting for status and retaliation) create a particular set of factors that increase gang members' victimization risk and repeat victimization risk. Constructions of gender identity may shape these risks in particular ways for girls. For instance, young women's adoption of masculine attributes may provide a means of participating and gaining status within gangs but may also lead to increased risk of victimization as a result of deeper immersion in delinquent activities. On the other hand, experiences of victimization may contribute to girls' denigration and thus increase their risk for repeat victimization through gendered responses and labeling—for example, when sexual victimization leads to perceptions of sexual availability or when victimization leads an individual to be viewed as weak. In addition, femaleness is an individual attribute that has the capacity to mark young women as "safe" crime victims (e.g.,

easy targets) or, conversely, to deem them "off limits." My goal here is to examine the gendered nature of violence within gangs, with a specific focus on how gender shapes young women's victimization risk.

METHODOLOGY

Data presented in this article come from survey and semistructured in-depth interviews with 20 female members of mixed-gender gangs in Columbus, Ohio. The interviewees ranged in age from 12 to 17; just over three-quarters were African American or multiracial (16 of 20), and the rest (4 of 20) were white.

Girls who admitted gang involvement during the survey participated in a follow-up interview to talk in more depth about their gangs and gang activities. The goal of the in-depth interview was to gain a greater understanding of the nature and meanings of gang life from the point of view of its female members.

The in-depth interviews were open-ended and all but one were audiotaped. They were structured around several groupings of questions. We began by discussing girls' entry into their gangs—when and how they became involved, and what other things were going on in their lives at the time. Then we discussed the structure of the gang—its history, size, leadership, and organization, and their place in the group. The next series of questions concerned gender within the gang; for example, how girls get involved, what activities they engage in and whether these are the same as the young men's activities, and what kind of males and females have the most influence in the gang and why. The next series of questions explored gang involvement more generally—what being in the gang means, what kinds of things they do together, and so on. Then, I asked how safe or dangerous they feel gang membership is and how they deal with risk. I concluded by asking them to speculate about why people their age join gangs, what things they like, what they dislike and have learned by being in the gang, and what they like best about themselves. This basic guideline was followed for each interview subject, although when additional topics arose in the context of the interview we often deviated from the interview guide to pursue them. Throughout the interviews, issues related to violence emerged; these issues form the core of the discussion that follows.

SETTING

The young women I interviewed described their gangs in ways that are very much in keeping with these findings. All 20 are members of Folks, Crips, or Bloods sets. All but 3 described gangs with fewer than 30 members, and most reported relatively narrow age ranges between members. Half were in gangs with members who were 21 or over, but almost without exception, their gangs were made up primarily of teenagers, with either one adult who was considered the OG ("Original Gangster," leader) or just a handful of young adults. The majority (14 of 20) reported that their gangs did not include members under the age of 13.

Although the gangs these young women were members of were composed of both female and male members, they varied in their gender composition, with the vast majority being predominantly male. Six girls reported that girls were one-fifth or fewer of the members of their gang; 8 were in gangs in which girls were between a quarter and a third of the overall membership; 4 said girls were between 44 and 50 percent of the members; and 1 girl reported that her gang was two-thirds female and one-third male. Over-all, girls were typically a minority within these groups numerically, with 11 girls reporting that there were 5 or fewer girls in their set.

This structure—male-dominated, integrated mixed-gender gangs—likely shapes gender dynamics in particular ways. Much past gang research has assumed that female members of gangs are in auxiliary subgroups of male gangs, but there is increasing evidence—including from the young women I spoke with—that many gangs can be characterized as integrated, mixed-gender groups.

GENDER, GANGS, AND VIOLENCE

Gangs as Protection and Risk

An irony of gang involvement is that although many members suggest one thing they get out of the gang is a sense of protection (see also Decker 1996; Joe and Chesney-Lind 1995; Lauderback et al., 1992), gang membership itself means exposure to victimization risk and even a willingness to be victimized. These contradictions are apparent when girls talk about what they get out of the gang, and what being in the gang means in terms of other members' expectations of their behavior. In general, a number of girls suggested that being a gang member is a source of protection around the neighborhood. Erica, a 17-year-old African American, explained, "It's like people look at us and that's exactly what they think, there's a gang, and they respect us for that. They won't bother us. . . . It's like you put that intimidation in somebody." Likewise, Lisa, a 14-year-old white girl, described being in the gang as empowering: "You just feel like, oh my God, you know, they got my back. I don't need to worry about it." Given the violence endemic in many inner-city communities, these beliefs are understandable, and to a certain extent, accurate.

In addition, some young women articulated a specifically gendered sense of protection that they felt as a result of being a member of a group that was predominantly male. Gangs operate within larger social milieus that are characterized by gender inequality and sexual exploitation. Being in a gang with young men means at least the semblance of protection from, and retaliation against, predatory men in the social environment. Heather, a 15-year-old white girl, noted. "You feel more secure when, you know, a guy's around protectin' you, you know, than you would a girl." She explained that as a gang member, because "you get protected by guys . . . not as many people mess with you." Other young women concurred and also described that male gang members could retaliate against specific acts of violence against girls in the gang. Nikkie, a 13-year-old African American girl, had a friend who was raped by a rival gang member, and she said, "It was a Crab [Crip] that raped my girl in Miller Ales, and um, they was ready to kill him." Keisha, an African American 14-year-old, explained, "if I got beat up by a guy, all I gotta do is go tell one of the niggers, you know what I'm sayin'? Or one of the guys, they'd take care of it."

At the same time, members recognized that they may be targets of rival gang members and were expected to "be down" for their gang at those times even when it meant being physically hurt. In addition, initiation rites and internal rules were structured in ways that required individuals to submit to, and be exposed to, violence. For example, young women's descriptions of the qualities they valued in members revealed the extent to which exposure to violence was an expected element of gang involvement. Potential members, they explained, should be tough, able to fight and to engage in criminal activities, and also should be loyal to the group and willing to put themselves at risk for it. Erica explained that they didn't want "punks" in her gang: "When you join something like that, you might as well expect that there's gonna be fights. . . . And, if you're a punk, or if you're scared of stuff like that, then don't join." Likewise, the following dialogue with Cathy, a white 16-year-old, reveals similar themes. I asked her what her gang expected out of members and she responded, "to be true to our gang and to have our backs." When I asked her to elaborate, she explained,

Cathy: Like, uh, if you say you're a Blood, you be a Blood. You wear your rag even when you're by yourself. You know, don't let anybody intimidate you and be like, 'Take that rag off.' You know, 'you better get with our set.' Or something like that.

JM: Ok. Anything else that being true to the set means?

Cathy: Um. Yeah, I mean, just, just, you know, I mean it's, you got a whole bunch of people comin', up in your face and if you're by yourself they ask you what's your claimin', you tell 'em. Don't say 'nothin.'

JM: Even if it means getting beat up or something?

Cathy: Mmhmm.

One measure of these qualities came through the initiation process, which involved the individual submitting to victimization at the hands of the gang's members. Typically this entailed either taking a fixed number of "blows" to the head and/or chest or being "beaten in" by members for a given duration (e.g., 60 seconds). Heather described the initiation as an important event for determining whether someone would make a good member:

> When you get beat in if you don't fight back and if you just like stop and you start cryin' or somethin' or beggin' 'em to stop and stuff like that, then, they ain't gonna, they'll just stop and they'll say that you're not gang material because you gotta be hard, gotta be able to fight, take punches.

In addition to the initiation, and threats from rival gangs, members were expected to adhere to the gang's internal rules (which included such things as not fighting with one another, being "true" to the gang, respecting the leader, not spreading gang business outside the gang, and not dating members of rival gangs). Breaking the rules was grounds for physical punishment, either in the form of a spontaneous assault or a formal "violation," which involved taking a specified number of blows to the head. For example, Keisha reported that she talked back to the leader of her set and "got slapped pretty hard" for doing so. Likewise, Veronica, an African American 15-year-old, described her leader as "crazy, but we gotta listen to 'im. He's just the type that if you don't listen to 'im, he gonna blow your head off. He's just crazy."

It is clear that regardless of members' perceptions of the gang as a form of "protection," being a gang member also involves a willingness to open oneself up to the possibility of victimization. Gang victimization is governed by rules and expectations, however, and thus does not involve the random vulnerability that being out on the streets without a gang might entail in high-crime neighborhoods. Because of its structured nature, this victimization risk may be perceived as more palatable by gang members. For young women in particular, the gendered nature of the streets may make the empowerment available through gang involvement an appealing alternative to the individualized vulnerability they otherwise would face. However, as the next sections highlight, girls' victimization risks continue to be shaped by gender, even within their gangs, because these groups are structured around gender hierarchies as well.

Gender and Status, Crime and Victimization

Status hierarchies within Columbus gangs, like elsewhere, were male dominated (Bowker et al., 1980; Campbell 1990). Again, it is important to highlight that the structure of the gangs these young women belonged to—that is, male-dominated, integrated mixed-gender gangs—likely shaped the particular ways in which gender dynamics played themselves out. Autonomous female gangs, as well as gangs in which girls are in auxiliary subgroups, may be shaped by different gender relations, as well as differences in orientations toward status, and criminal involvement.

All the young women reported having established leaders in their gang, and this leadership was almost exclusively male. While LaShawna, a 17-year-old African American, reported being the leader of her set (which had a membership that is two-thirds girls, many of whom resided in the same residential facility as her), all the other girls in mixed-gender gangs reported that their OG was male. In fact, a number of young women stated explicitly that only male gang members could be leaders. Leadership qualities, and qualities attributed to high-status members of the gangs—being tough, able to fight, and willing to "do dirt" (e.g., commit crime, engage in violence) for the gang—were perceived as characteristically masculine. Keisha noted, "The guys, they just harder." She explained, "Guys is more rougher. We have our G's back but, it ain't gonna be like the guys, they just don't give a fuck. They gonna shoot you in a minute."

For the most part, status in the gang was related to traits such as the willingness to use serious violence and commit dangerous crimes and, though not exclusively, these traits were viewed primarily as qualities more likely and more intensely located among male gang members.

Because these respected traits were characterized specifically as masculine, young women actually may have had greater flexibility in their gang involvement than young men. Young women had fewer expectations placed on them—by both their male and female peers—in regard to involvement in criminal activities such as fighting, using weapons, and committing other crimes. This tended to decrease girls' exposure to victimization risk compared to male members, because they were able to avoid activities likely to place them in danger. Girls could gain status in the gang by being particularly hard and true to the set. Heather, for example, described the most influential girl in her set as "the hardest girl, the one that don't take no crap, will stand up to anybody." Likewise, Diane, a white 15-year-old, described a highly respected female member in her set as follows:

> People look up to Janeen just 'cause she's so crazy. People just look up to her 'cause she don't care about nothin'. She don't even care about makin' money. Her, her thing is, 'Oh, you're a Slob [Blood]? You're a Slob? You talkin' to me? You talkie' shit to me?' Pow, pow! And that's it. That's it.

However, young women also had a second route to status that was less available to young men. This came via their connections—as sisters, girlfriends, cousins—to influential, high-status young men. In Veronica's set, for example, the girl with the most power was the OG's "sister or his cousin, one of 'em." His girl-friend also had status, although Veronica noted that "most of us just look up to our OG." Monica, a 16-year-old African American, and Tamika, a 15-year-old African American, both had older brothers in their gangs, and both reported getting respect, recognition, and protection because of this connection. This route to status and the masculinization of high-status traits functioned to maintain gender inequality within gangs, but they also could put young women at less risk of victimization than young men. This was both because young women were perceived as less threatening and thus were less likely to be targeted by rivals, and because they were not expected to prove themselves in the ways that young men were, thus decreasing their participation in those delinquent activities likely to increase exposure to violence. Thus, gender inequality could have a protective edge for young women.

Young men's perceptions of girls as lesser members typically functioned to keep girls from being targets of serious violence at the hands of rival young men, who instead left routine confrontations with rival female gang members to the girls in their own gang. Diane said that young men in her gang "don't wanna waste their time hittin' on some little girls. They're gonna go get their little cats [females] to go get 'em." Lisa remarked, "girls don't face as much violence as [guys]. They see a girl, they say, 'we'll just smack her and send her on.' They see a guy—'cause guys are like a lot more into it than girls are, I've noticed that—and they like, 'well, we'll shoot him.'" In addition, the girls I interviewed suggested that, in comparison with young men, young women were less likely to resort to serious violence, such as that involving a weapon, when confronting rivals. Thus, when girls' routine confrontations were more likely to be female on female than male on female, girls' risk of serious victimization was lessened further.

Also, because participation in serious and violent crime was defined primarily as a masculine endeavor, young women could use gender as a means of avoiding participation in those aspects of gang life they found risky, threatening, or morally troubling. Of the young women I interviewed, about one-fifth were involved in serious gang violence: A few had been involved [in] aggravated assaults on rival gang members, and one admitted to having killed a rival gang member, but they were by far the exception. Most girls tended not to be involved in serious gang crime, and some reported that they chose to exclude themselves because they felt ambivalent about this aspect of gang life. Angie, an African American 15-year-old explained,

> I don't get involved like that, be out there goin' and just beat up people like that or go stealin', things like that. That's not me. The boys, mostly the boys do all that, the girls we just sit back and chill, you know.

Likewise, Diane noted,

> For maybe a drive-by they might wanna have a bunch of dudes. They might not put the females in that. Maybe the females might be weak inside, not strong enough to do something like that, just on the insides. . . . If a female wants to go forward

and doin' that, and she wants to risk her whole life for doin' that, then she can. But the majority of the time, that job is given to a man.

Diane was not just alluding to the idea that young men were stronger than young women. She also inferred that young women were able to get out of committing serious crime, more so than young men, because a girl shouldn't have to "risk her whole life" for the gang. In accepting that young men were more central members of the gang, young women could more easily participate in gangs without putting themselves in jeopardy—they could engage in the more routine, everyday activities of the gang, like hanging out, listening to music, and smoking bud (marijuana). These male-dominated mixed-gender gangs thus appeared to provide young women with flexibility in their involvement in gang activities. As a result, it is likely that their risk of victimization at the hands of rivals was less than that of young men in gangs who were engaged in greater amounts of crime.

Girls' Devaluation and Victimization

In addition to girls choosing not to participate in serious gang crimes, they also faced exclusion at the hands of young men or the gang as a whole (see also Bowker et al., 1980). In particular, the two types of crime mentioned most frequently as "off-limits" for girls were drug sales and drive-by shootings. LaShawna explained, "We don't really let our females [sell drugs] unless they really wanna and they know how to do it and not to get caught and everything." Veronica described a drive-by that her gang participated in and said, "They wouldn't let us [females] go. But we wanted to go, but they wouldn't let us." Often, the exclusion was couched in terms of protection. When I asked Veronica why the girls couldn't go, she said, "so we won't go to jail if they was to get caught. Or if one of 'em was to get shot, they wouldn't want it to happen to us." Likewise, Sonita, a 13-year-old African American, noted, "If they gonna do somethin' bad and they think one of the females gonna get hurt they don't let 'em do it with them. . . . Like if they involved with shooting or whatever, [girls] can't go."

Although girls' exclusion from some gang crime may be framed as protective (and may reduce their victimization risk vis-à-vis rival gangs), it also served to perpetuate the devaluation of female members as less significant to the gang—not as tough, true, or "down" for the gang as male members. When LaShawna said her gang blocked girls' involvement in serious crime, I pointed out that she was actively involved herself. She explained, "Yeah, I do a lot of stuff 'cause I'm tough. I likes, I likes messin' with boys. I fight boys. Girls ain't nothin' to me." Similarly, Tamika said, "girls, they little peons."

Some young women found the perception of them as weak a frustrating one. Brandi, an African American 13-year-old, explained, "Sometimes I dislike that the boys, sometimes, always gotta take charge and they think sometimes, that the girls don't know how to take charge 'cause we're like girls, we're females, and like that." And Chantell, an African American 14-year-old, noted that rival gang members "think that you're more of a punk." Beliefs that girls were weaker than boys meant that young women had a harder time proving that they were serious about their commitment to the gang. Diane explained,

A female has to show that she's tough. A guy can just, you can just look at him. But a female, she's gotta show. She's gotta go out and do some dirt. She's gotta go whip some girl's ass, shoot somebody, rob somebody or something. To show that she is tough.

In terms of gender-specific victimization risk, the devaluation of young women suggests several things. It could lead to the mistreatment and victimization of girls by members of their own gang when they didn't have specific male protection (i.e., a brother, boyfriend) in the gang or when they weren't able to stand up for themselves to male members. This was exacerbated by activities that led young women to be viewed as sexually available. In addition, because young women typically were not seen as a threat by young men, when they did pose one, they could be punished even more harshly than young men, not only for having challenged a rival gang or gang member but also for having overstepped "appropriate" gender boundaries.

Monica had status and respect in her gang, both because she had proven herself through fights and criminal activities, and because her older brothers were members of her set. She contrasted her own treatment with that of other young women in the gang:

They just be puttin' the other girls off. Like Andrea, man. Oh my God, they dog Andrea so bad. They like, 'Bitch, go to the store.' She like, 'All right, I be right back.' She will go to the store and go and get them whatever they want and come back with it. If she don't get it right, they be like, 'Why you do that bitch?' I mean, and one dude even smacked ha. And, I mean, and, I don't, I told my brother once. I was like, 'Man, it ain't even like that. If you ever see someone tryin' to disrespect me like that or hit me, if you do not hit them or at least say somethin' to them. . . .' So my brothers, they kinda watch out for me.

However, Monica put the responsibility for Andrea's treatment squarely on the young woman: "I put that on her. They ain't gotta do her like that, but she don't gotta let them do her like that either." Andrea was seen as "weak" because she did not stand up to the male members in the gang; thus, her mistreatment was framed as partially deserved because she did not exhibit the valued traits of toughness and willingness to fight that would allow her to defend herself.

An additional but related problem was when the devaluation of young women within gangs was sexual in nature. Girls, but not boys, could be initiated into the gang by being "sexed in"—having sexual relations with multiple male members of the gang. Other members viewed the young women initiated in this way as sexually available and promiscuous, thus increasing their subsequent mistreatment. In addition, the stigma could extend to female members in general, creating a sexual devaluation that all girls had to contend with.

The dynamics of "sexing in" as a form of gang initiation placed young women in a position that increased their risk of ongoing mistreatment at the hands of their gang peers. According to Keisha, "If you get sexed in, you have no respect. That means you gotta go ho'in' for 'em; when they say you give 'em the pussy, you gotta give it to 'em. If you don't, you gonna get your ass beat. I ain't down for that." One girl in her set was sexed in and Keisha said the girl "just do everything they tell her to do, like a dummy." Nikkie reported that two girls who were sexed into her set eventually quit hanging around with the gang because they were harassed so much. In fact, Veronica said the young men in her set purposely tricked girls into believing they were being sexed into the gang and targeted girls they did not like:

> If some girls wanted to get in, if they don't like the girl they have sex with 'em. They run trains on 'em or either have the girl suck their thang. And then they used to, the girls used to think they was in. So, then the girls used to just just come try to hang around us and all this little bull, just 'cause, 'cause they thinkin' they in.

Young women who were sexed into the gang were viewed as sexually promiscuous, weak, and not "true" members. They were subject to revictimization and mistreatment, and were viewed as deserving of abuse by other members, both male and female. Veronica continued, "They [girls who are sexed in] gotta do whatever, whatever the boys tell 'em to do when they want 'em to do it, right then and there, in front of whoever. And, I think, that's just sick. That's nasty, that's dumb." Keisha concurred, "She brought that on herself, by bein' the fact, bein' sexed in." There was evidence, however, that girls could overcome the stigma of having been sexed in through their subsequent behavior, by challenging members that disrespect them and being willing to fight. Tamika described a girl in her set who was sexed in, and stigmatized as a result, but successfully fought to rebuild her reputation:

> Some people, at first, they call her 'little ho' and all that. But then, now she startin' to get bold. . . . Like, like, they be like, 'Ooh, look at the little ho. She fucked me and my boy.' She be like, 'Man, forget y'all. Man, what? What?' She be ready to squat [fight] with 'em. I be like, 'Ah, look at her!' Uh huh. . . . At first we looked at her like, 'Ooh, man, she a ho, man.' But now we look at her like she just our kickin' it partner. You know, however she got in that's her business.

The fact that there was such an option as "sexing in" served to keep girls disempowered, because they always faced the question of how they got in and of whether they were "true" members. In addition, it contributed to a milieu in which young women's sexuality was seen as exploitable. This may help explain why young women were so harshly judgmental of those girls who were sexed in. Young women who were privy to male gang members' conversations reported that male members routinely disrespect girls in the gang by disparaging them sexually. Monica explained,

> I mean the guys, they have their little comments about 'em [girls in the gang] because, I hear more because my brothers are all up there with the guys and everything and I hear more just sittin' around, just listenin'. And they'll have their little jokes about 'Well, ha I had her,' and then and everybody else will jump in and say, 'Well, I had her, too.' And then they'll laugh about it.

In general, because gender constructions defined young women as weaker than young men, young women were often seen as lesser members of the gang. In addition to the mistreatment these perceptions entailed, young women also faced particularly harsh sanctions for crossing gender boundaries—causing harm to rival male members when they had been viewed as nonthreatening. One young woman participated in the assault of a rival female gang member, who had set up a member of the girl's gang. She explained, "The female was supposingly goin' out with one of ours, went back and told a bunch of [rivals] what was goin' on and got the [rivals] to jump my boy. And he ended up in the hospital." The story she told was unique but nonetheless significant for what it indicates about the gendered nature of gang violence and victimization. Several young men in her set saw the girl walking down the street, kidnapped her, then brought her to a member's house. The young woman I interviewed, along with several other girls in her set, viciously beat the girl, then to their surprise the young men took over the beating, ripped off the girl's clothes, brutally gang-raped her, then dumped her in a park. The interviewee noted, "I don't know what happened to her. Maybe she died.

Maybe, maybe someone came and helped her. I mean, I don't know." The experience scared the young woman who told me about it. She explained,

> I don't never want anythin' like that to happen to me. And I pray to God that it doesn't. 'Cause God said that whatever you sow you're gonna reap. And like, you know, beatin' a girl up and then sittin' there watchin' somethin' like that happen, well, Jesus that could come back on me. I mean, I felt, I really did feel sorry for her even though my boy was in the hospital and was really hurt. I mean, we coulda just shot her. You know, and it coulda been just over. We coulda just taken her life. But they went farther than that.

This young woman described the gang rape she witnessed as "the most brutal thing I've ever seen in my life." While the gang rape itself was an unusual event, it remained a specifically gendered act that could take place precisely because young women were not perceived as equals. Had the victim been an "equal," the attack would have remained a physical one. As the interviewee herself noted, "we coulda just shot her." Instead, the young men who gang-raped the girl were not just enacting revenge on a rival but on a young woman who had dared to treat a young man in this way. The issue is not the question of which is worse—to be shot and killed, or gang-raped and left for dead. Rather, this particular act sheds light on how gender may function to structure victimization risk within gangs.

DISCUSSION

Gender dynamics in mixed-gender gangs are complex and thus may have multiple and contradictory effects on young women's risk of victimization and repeat victimization. My findings suggest that participation in the delinquent lifestyles associated with gangs clearly places young women at risk for victimization. The act of joining a gang involves the initiate's submission to victimization at the hands of her gang peers. In addition, the rules governing gang members' activities place them in situations in which they are vulnerable to assaults that are specifically gang related. Many acts of violence that girls described would not have occurred had they not been in gangs.

It seems, though, that young women in gangs believed they have traded unknown risks for known ones—that victimization at the hands of friends, or at least under specified conditions, was an alternative preferable to the potential of random, unknown victimization by strangers. Moreover, the gang offered both a semblance of protection from others on the streets, especially young men, and a means of achieving retaliation when victimization did occur.

Lauritsen and Quinet (1995) suggest that both individual-specific heterogeneity (unchanging attributes of individuals that contribute to a propensity for victimization, such as physical size or temperament) and state-dependent factors (factors that can alter individuals' victimization risks over time, such as labeling or behavior changes that are a consequence of victimization) are related to youths' victimization and repeat victimization risk. My findings here suggest that, within gangs, gender can function in both capacities to shape girls' risks of victimization.

Girls' gender, as an individual attribute, can function to lessen their exposure to victimization risk by defining them as inappropriate targets of rival male gang members' assaults. The young women I interviewed repeatedly commented that young men were typically not as violent in their routine confrontations with rival young women as with rival young men. On the other hand, when young women are targets of serious assault, they may face brutality that is particularly harsh and sexual in nature because they are female—thus, particular types of assault, such as rape, are deemed more appropriate when young women are the victims.

Gender can also function as a state-dependent factor, because constructions of gender and the enactment of gender identities are fluid. On the one hand, young women can call upon gender as a means of avoiding exposure to activities they find risky, threatening, or morally troubling. Doing so does not expose them to the sanctions likely faced by male gang members who attempt to avoid participation in violence. Although these choices may insulate young women from the risk of assault at the hands of rival gang members, perceptions of female gang members—and of women in general—as weak may contribute to more routinized victimization at the hands of the male members of their gangs. Moreover, sexual exploitation in the form of "sexing in" as an initiation ritual may define young women as sexually available, contributing to a likelihood of repeat victimization unless the young woman can stand up for herself and fight to gain other members' respect.

Finally, given constructions of gender that define young women as nonthreatening, when young women do pose a threat to male gang members, the sanctions they face may be particularly harsh because they not only have caused harm to rival gang members but also have crossed appropriate gender boundaries in doing so. In sum, my findings suggest that gender may function to insulate young women from some types of physical assault and lessen their exposure to risks from rival gang members, but also to make them vulnerable to particular types of violence, including routine victimization by their male peers, sexual exploitation, and sexual assault.

REFERENCES

Bjerregaard, Beth and Carolyn Smith. 1993. "Gender Differences in Gang Participation, Delinquency, and Substance Use." *Journal of Quantitative Criminology* 4:329–355.

Bowker, Lee H., Helen Shimota Gross, and Malcolm W. Klein. 1980. "Female Participation in Delinquent Gang Activities." *Adolescence* 15(59): 509–519.

Campbell, Anne. 1984. *The Girls in the Gang.* New York: Basil Blackwell.

——. 1990. "Female Participation in Gangs." Pp. 163–182 in *Gangs in America,* edited by C. Ronald Huff. Beverly Hills, CA: Sage.

Chesney-Lind, Meda. 1993. "Girls, Gangs and Violence: Anatomy of a Backlash." *Humanity & Society* 17(3):321–344.

Decker, Scott H. 1996. "Collective and Normative Features of Gang Violence." *Justice Quarterly* 13 (2):243–264.

Esbensen, Finn-Aage. 1996. Comments presented at the National Institute of Justice/Office of Juvenile Justice and Deliquency Prevention Cluster Meetings, June, Dallas, TX.

——. 1990. "Social Processes of Deliquency and Drug Use among Urban Gangs." Pp. 183–219 in *Gangs in America,* edited by C. Ronald Huff. Newbury Park, CA: Sage.

Huff, C. Ronald. 1996. "The Criminal Behavior of Gang Members and Nongang At-Risk Youth." Pp. 75–102 in *Gangs in America,* 2nd ed., edited by C. Ronald Huff. Thousand Oaks, CA: Sage.

Joe, Karen A. and Meda Chesney-Lind. 1995. "Just Every Mother's Angel: An Analysis of Gender and Ethnic Variations in Youth Gang Membership." *Gender & Society* 9(4):408–430.

Lauderback David, Joy Hansen, and Dan Waldorf. 1992. "'Sisters Are Doin' It for Themselves': A Black Female Gang in San Francisco." *The Gang Journal* 1(1):57–70.

Lauritsen, Janet L. and Kenna F. Davis Quinet. 1995. "Repeat Victimization Among Adolescents and Young Adults." *Journal of Quantitative Criminology* 1(2):143–166.

Moore, Joan. 1991. *Going Down to the Barrio: Homeboys and Homegirls in Change.* Philadelphia: Temple University Press.

Spergel, Irving A. and G. David Curry. 1993. "The National Youth Gang Survey: A Research and Development Process." Pp. 359–400 in *The Gang Intervention Handbook,* edited by Arnold P. Goldstein and C. Ronald Huff. Champaign, IL: Research Press.

Winfree, L. Thomas, Jr., Kathy Fuller, Teresa Vigil, and G. Larry Mays. 1992. "The Definition and Measurement of 'Gang Status': Policy Implications for Juvenile Justice." *Juvenile and Family Court Journal* 43:29–37

Part VI

DRUGS AND CRIME

Having experienced a recent "law and order" campaign in this country with the 1980s' "war on drugs," federal and state convictions for drug offenses have increased over 100% as compared with drug-related convictions prior to the 1980s. Many states have enacted mandated sentences for the possession of narcotics, which has resulted in our nation's prisons, both federal and state systems, experiencing drastic increases in prisoner population growth in recent years. This has led to prison overcrowding and placed financial burdens on state governments for the costs of maintaining and constructing new prison facilities. Much of this correctional population growth is a direct result of punitive policies formulated by state and federal legislation to get tough on narcotic violators.

The disproportionate drug arrest rate for cocaine use between blacks and whites is a consequence of laws against possession and use of crack cocaine, a narcotic used mainly by African Americans. Penalties against crack are more stringent than those judicial sentences handed out to white offenders for cocaine violations. These legal disparities in the laws for drug offenses, as the example of cocaine illustrates, are indicative of just how disparate the laws are for the differing sanctions between black and white drug users.

The selected studies in this part of the book offer a penetrating look into the world of narcotic addict lifestyles as well as the risks and hazards they face from life on the street as well as detection from law enforcement agencies. In the first article, Barbara Lex provides an insightful view of hustling strategies employed by heroin addicts. She found that the most successful hustle is one that delivers the largest amount of quality heroin with the least amount of effort on the part of the hustling addict. The use of deception is an integral part of hustling. A major element in hustling is to convey an illusion of reality so that those being hustled remain naive while payoffs are gained in the form of money or other services that facilitate opportunities for the purchase of heroin. Lex points out that the cultural context for successful hustling that must be employed is the use of impression management techniques in the hustler's interpersonal interactions with others, along with a solid knowledge of human behavior and a rich repertoire of role-taking skills and intuitive timing for their actions. These social and interpersonal skills closely resemble those of actors in the theater. Furthermore, Lex concludes that successful narcotic hustlers have developed acute observational, analytic, and evaluative skills comparable to schooled social scientists, all of which are utilized in the addict-hustler's deceptive game to attain drugs.

On a somewhat different perspective from that of hustling, Mieczkowski describes street-level crack buying and selling. He informs us that the past methods of street sales in urban areas that were heroin-dominated in past years have not survived in the newer world of crack cocaine. Mieczkowski found that the sale of crack is operated by what he terms a "crack house system," as opposed to the old-style crew system of street selling. This change in narcotic sales came about because customers prefer to buy their crack from fixed locales and from known sellers. Finally, Mieczkowski concludes that open street sales are considered the most risky for detection and that at a fixed location a buyer could hold the seller accountable for the quality of the crack purchased.

Waldorf and Murphy describe and analyze the perceived risks and pressures faced by middle-class cocaine dealers and what types of risks they try to avoid. The possibility of arrest for sales of cocaine was minimized by sellers through restricting their sales to small numbers of buyers that they were acquainted with and trusted. Sellers perceived a great risk in dealing drugs to strangers. Middle-class sellers were more concerned about disgruntled buyers, informants, and possible robbers via violent encounters than they were of law enforcement agents. For approximately half of those who stopped selling cocaine, criminal justice pressures did not play a part in their quitting the business of sales. For those who did experience criminal justice pressures to stop dealing, the arrest of a member of their drug supply network proved to be the most important factor in their desisting from this illegal behavior.

By examining the position of women in the drug economy, Maher and Daly focused their research on the issue of whether women who were currently involved in selling drugs had higher distribution roles in the crack cocaine markets of the 1980s and 1990s than in the drug markets of previous years. The authors studied female drug users in New York City and found that, contrary to the belief that crack cocaine markets resulted in new economic opportunities for women, these markets were in reality more profitable for men. Maher and Daly conclude that those researchers who perceived an increased change in women's roles in the drug economy due to the rapid increase in crack cocaine use in an expanded drug market economy were incorrect in their assumptions. The authors point out that temporary opportunities for an expanded role for women to participate in the distribution of crack at the street level was short-lived and did very little to change male dealers' perceptions of women as unreliable, untrustworthy, and incapable of handling violent situations that can arise in the business of illegal drug dealing.

21

NARCOTICS ADDICTS' HUSTLING STRATEGIES: CREATION AND MANIPULATION OF AMBIGUITY

BARBARA W. LEX

For this article, Lex collected data from 20 male heroin addicts who were part of a rehabilitation program that was testing a narcotic antagonist. She interviewed these 20 addicts for purposes of identifying tactics used in the process of hustling and to analyze the ways in which decisions hustlers made related to their interactions with those persons that they were hustling. The author's major contention is that hustling by addicted heroin users is an economic activity that deliberately creates a manipulated ambiguity best characterized as open to multiple interpretations. The purpose of the hustling process is to attain money, goods, or services from naive persons in the course of social interaction for the payoff of procuring heroin. Many tactics and strategies are analyzed here that generally explain addicts' hustling behavior.

The vast majority of heroin addicts exhibit characteristics of a highly specialized subculture. Predominantly male, as "junkies," "dope fiends," or "street addicts," they share certain information, experiences, skills, interests, terminology, beliefs, values, attitudes, social statuses and roles, standards of behavior, and interaction patterns. For these individuals, the dishonest or illegal means used to obtain drugs or money for drugs ("hustling") are a major focus of knowledge and skills that occupy most of their time (Burr, 1983; Carlson, 1976; Fields and Walters, 1985; Sutker, 1974; Walters, 1985).

The aim of this article is to identify and describe tactics used in the process of hustling and to analyze the ways in which the decisions and actions made by hustlers reflect at least tacit recognition of behaviors appropriate and inappropriate to specific social interactions. My major contention in this analysis is that hustling by heroin addicts is a type of economic activity that follows specific rules and thrives in an atmosphere of deliberately created and

EDITOR'S NOTE: From Lex, B., "Narcotics addicts' hustling strategies: Creation and manipulation of ambiguity," in *Journal of Contemporary Ethnography, 18,* pp. 388-415. Copyright © 1990. Reprinted with permission from Sage Publications, Inc.

manipulated ambiguity. By ambiguity, I refer to a state in which two or more meanings can be ascribed to an event, that is, a series of behaviors that constitute a sequence of interactions that are vague, unclear, and open to alternative interpretations. I begin by providing a brief description of the respondents, the context of the study, and the interviews used to obtain data about the process of hustling. I then present four rules that emerged from the interviews. Last, I use the interview data pertinent to hustling strategies to show how each of the rules operate, and show how basic concepts of economic exchange illuminate hustling tactics.

METHODS

Respondents

Data were collected from 20 male heroin addicts who were subjects in a research and treatment program testing a narcotic antagonist. Subjects were either in-patients participating in 42-day studies or outpatients residing in the community during the aftercare phase of the program. Each subject was a volunteer, over the age of 22, with history of at least two unsuccessful rehabilitation attempts, who had given his informed consent on a form administered by an attorney. All potential candidates for admission to the program were screened over a four-to six-week period and, apart from heroin addiction, had no major psychological disorders, and were medically healthy.

Respondents' average age was 25.6, with a duration of addiction from 5.4 to 7.8 years. Three respondents were Black, with the remainder second- or third-generation white Irish-Americans or Italian-Americans. All 20 men were inhabitants of the Boston metropolitan area, and the majority had been lifelong residents of this locality. All respondents had a history of narcotics-related arrests, but as volunteers in a program testing an investigational new drug none were stipulated to this form of treatment.

HUSTLING STRATEGIES

Addicts agreed that the "best hustle" is one that delivers the greatest amount and best quality of

heroin with the least amount of investment of effort. The heroin market shifts continuously, so that current information about who has what, and where, is a high-ranking need (see Agar, 1973). However, it is not always possible for an addict to have optimal information, nor is it always possible to obtain heroin by resorting to deception.

Some addicts pull "cowboy jobs" (make swift use of physical force). For example, some "throw bricks." In this strategy, addicts may steal a car, drive it into a store display window, grab as much merchandise as possible, flee in another car, and obtain drugs or money for drugs with the stolen merchandise.

At other times, addicts "burn" or "beat" (swindle) fellow addicts. A burn involves a drug transaction in which the quality of heroin (or other drugs) is less than stated, or nil. Burns, however, typically engender anger, hostility, and embarrassment. Being burnt is a stigmatizing condition to be avoided if one wishes to retain a "good" street reputation and sustain a "successful" self-image.

Rule 1: Role Playing, Time, and Space

Implicit in hustling transactions are the social roles played by the hustler and by the victim of the "game"—the "sucker" or "mark." The victim may be another addict, as in drug dealing (Adler, 1985; Burr, 1983) and in some cases of stealing (Goldstein, 1981), or a "straight" individual in instances of pimping and confidence games. These categories of hustles are groupings of subtypes that may empirically shade into one another.

Several respondents detailed their strategies for successful deception and manipulation. Paul, for example, opened his discussion by remarking (with some braggadocio), that in his various social roles he exploited potentials for ambiguity by having a "good way with people":

> I can get them to trust me. I've got good connections—just fall into them. I'm lucky—I can always get better dope than the next guy and there are very few people I don't get along with. Everyone I meet, it's like meeting a dollar sign.

As one 24-year-old Black hustler stated, he exploited both his Black identity and knowledge of office behavior while hustling. For three months Conrad had worked for a messenger service in downtown Boston and had ample opportunity to observe the behavior of office workers. His friend, Benny, had worked for a cleaning company that serviced some of the same buildings. Benny and Conrad sometimes worked together and shared their knowledge of places, people, and times to hustle. Conrad said that he and Benny were able to "creep" (steal wallets or portable equipment from large, impersonal office buildings) because, as Conrad said:

> 99 percent of people—especially White people— are asleep [are not aware of the role of ambiguity and deception in hustles]. I can play on that—get them to do what I want them to do. Some people [especially in Boston, where racial tensions have been strong] want to appear liberal when they talk to Blacks. I take advantage of that, 'cuz they don't want to insult me, certainly not when I'm togged out in a suit and tie and attache case. They'll do anything to help me—even leave their desk and show me the "right" place that I ask for. Then my partner creeps in on 'em. And, by then, it's too late.

Thus while Conrad created a seemingly benign distraction, Benny, who was dressed as a manual laborer, committed the crime. From his knowledge of middle-class white office workers, Conrad believed that no one would dare to confront him by connecting the two as accomplices because of the apparent social class differences between two Black men, each of whom wore different types of clothing and comported themselves according to class stereotypes. Such an implication might lead to an allegation of prejudice and a hostile confrontation.

Thus almost any social setting can have potential for hustling, but hustlers must first learn the prevailing norms and then discover how to manipulate them.

Rule 2: Information and Emotion

Hustlers deliberately exploit ambiguity in human communication. Both information and appropriate emotional facades are necessary to perform a specific hustle successfully. Karen was not only playing the part of a young woman who accidentally slipped and fell on a wet floor in a public place, she also claimed to be pregnant—a condition that would both arouse the sympathy and concern of bystanders and additionally threaten responsible restaurant workers.

Ideally, Paul claimed, the perfect victim for a hustle was "someone with their guard down" or a naive person "proud of never having been conned." Paul reported that he memorized information about everyone whom he met to use at a later time. Even when gainfully employed he deliberately set out to identify coworkers' interests and to develop a fictional "common bond" with potential victims. To create a seemingly genuine persona he drew on "memory and experience," and "mixed the lies with the truth." In Paul's terms, "the truth would be the highlights; lies something [the victim] had never previously heard." According to Paul, within a few weeks of starting a new job or meeting a new person, he knew that individual's background and had sufficient data to hustle. Especially discerning was his statement that he could "tell if the person was lonely and if he needed a friend." Employing flattery, he "sat with the guy and laughed at his jokes."

Another respondent, Sandy, distilled this sort of knowledge about human behavior, and observed that "everybody has a weakness." In Sandy's estimation, successful hustling depended on discovering and exploiting such weaknesses. This generalization suggests that hustling operates fundamentally through recognition and manipulation of ambiguity, in which the dynamics of human emotions play an integral part. Sandy stated this point through reflection on the emotional factors inherent in a jewelry hustle. A hustler and an accomplice realized a 500% profit by selling jewelry at inflated prices to recent widows. In this hustle, widows, whose names were obtained from obituary notices, were led to believe that just before their deaths their recently deceased husbands had planned to make a large purchase for a gift. The approach required not only "being cold," to use Sandy's term, in the face of another's grief but it also gave evidence that hustlers recognize that it is necessary to wait a suitable interval after a funeral in order to appear at a time when a widow is still bereaved and rendered vulnerable and uncritical by her recent loss. Sandy's sensitivity to bereavement parallels that of experts.

Rule 3: The Illusion of Benign Intent

Maintaining an illusion: Keith, a 29-year-old white addict, found himself "dope sick" (experiencing withdrawal symptoms). None of his usual resources (girlfriend, family, connections) would lend him any money or advance him any heroin to "get straight" (relieve abstinence symptoms). Keith found a bottle in his girlfriend's kitchen. Hoping that alcohol would relieve some of his discomfort, he went into the basement of her apartment building to drink. Nora, Keith's girlfriend, had recently purchased an expensive coffee maker. While drinking the wine, Keith worried intensely about what he would do next. In his words:

I don't know if I'd've thought of this if I hadn't been desperate and about to panic. I saw the empty box from the new coffee pot and the old coffee pot was sitting right next to it. I gulped down the wine and went upstairs to see if I could find the receipt. I was trying not to get her suspicious, so I couldn't just ask where it was. I couldn't find it, but I did find some "super glue," and the bag from the store.

I went back down into the cellar and made some noise like I was looking for tools to work on my car. I put the old pot in the new one's box and glued it all up. You couldn't get that sucker [coffee pot box] open again without tearing it all apart. I stuck it in the bag and drove off to the Cathedral [verbal shorthand for the name of a local public housing project that is mainly occupied by Blacks and a major site for heroin dealing]. I parked the car and walked down the block. I walked confident. I was determined to get some dope. Somebody was dealing on the corner. "What you got there?" I showed him the bag and the box. I had hoped to get $15 for it, but this guy gave me two bags of dope. I got back into the car. I drove to an alley and "cooked" the doped and shot it.

I forgot about the whole thing. About two weeks later I was with somebody else and we were lookin' to cop. We drove down to Cathedral and pretty soon I saw my man. He took one look at me and began pounding on the car, yelling and hollering.

It turned out that he didn't open it [the box] right away. If he had, he would've known he was ripped off when he saw the old cracked and yellow pot. Seems he didn't: "Hey, man, you know what you *did?* Made a *fool* a me in front of my *sister!* I'm gonna kick your white ass!" I got the car started and drove away, fast.

It was months before I went there again. I found out that he'd been looking for something new that somebody had boosted [shoplifted]. He thought I had just the right thing and he had his girlfriend wrap it up and he gave it to his sister for a wedding present. I could just imagine her face when she opened that box.

Testing for "weakness": If it is axiomatic in the hustler's world view that "everybody has a weakness," and one that can serve as a matrix for creating ambiguity, then an important aspect of evaluating the potential resources to be gained from a victim in a hustle is a sequence of diagnostic tests for weakness.

Perhaps a corollary of the axiom "everybody has a weakness" was the statement by a key respondent that "junkies confuse kindness with weakness." Thus the potential for exploitation may depend on the quality of the personal relationship that the hustler has developed. Bobby, a 22-year-old Black addict, needed a supply of heroin so that he could "lay up" (remain out of sight and avoid "street" contacts) after a robbery, and avoid any encounters with police. Bobby contrived a scene in which his girlfriend saw him threatened by "thugs" demanding immediate payment of an $800 debt. Convinced that Bobby would be injured or killed if she failed to provide him with the money, she willingly parted with the entire amount in cash. Bobby used the $800 to pay off an existing debt for heroin and to buy more. He gave each of his accomplices two bags of heroin and kept the rest.

Randy, a 25-year-old white addict, also was physically attractive and charming to women. He "played on" a girlfriend's affection for him by persuading her that he would serve a long prison sentence unless a lawyer were "paid off" $1,500. In a scene staged for her benefit, his accomplice (who was not an attorney but carried an attache case) took the woman's check and later divided the money with Randy.

Each woman was led to believe that her reward for "saving" her boyfriend was continued receipt of his gratitude and affection as well as the implicit promise that such an event would never recur. Several months later, however, Randy used a comparable strategy a second time. When he found himself in debt to a dealer, Randy told his girlfriend a similar story. On this occasion he took the money, told his girlfriend to wait for him in her double-parked car, and entered a busy downtown office building. He returned 10 minutes later, but the money was concealed in his sock. Randy expressed profuse thanks, accepted a ride home, and immediately went out to buy drugs.

Other respondents reported that a well-timed "sob story" told to a friend or kinsman gained them enough money to obtain sufficient heroin to avoid severe withdrawal symptoms, and thus enable them to resume more lucrative forms of hustling.

Rule 4: Breaking the Rules

From the foregoing it should be clear that hustling is complicated by the fact that most hustles are based on deception. A hustler must refrain from emitting contradictory or disquieting behaviors until the transaction is completed and he has safely left the scene. A hustler must maintain a constant image or deception with each victim, as well as make stringent efforts to keep his victims separate and in ignorance of each other. In some cases a hustler must continue his ruse indefinitely if the same victim is to be manipulated repeatedly. In two simple examples, Howard was apprehended when he tried to cash a worthless check at the same bank in which he had previously cashed a bogus check and had become known as a forger.

According to one respondent, the chances of discovery, failure, or apprehension—and potential penalties—could be minimized if the hustler worked only one sort of game at a time. In other words, a basic rule is that hustling strategies must be kept unambiguous in order to maximize the possibility of success. For example, a 25-year-old Italian-American addict attempted to purchase about $500 worth of stereo equipment with a stolen credit card. Planning to fence the equipment to obtain money for drugs, Joe

entered a busy suburban store the weekend before Christmas. He was neatly groomed and wore expensive clothing, thus presenting himself as an affluent professional purchasing a holiday gift for his "wife." However, when the clerk stepped into the stockroom to fill the order, Joe could not resist opening the drawer of a cash register.

Joe was caught in the act of theft, and obtained neither money nor goods. He reported that his "cover was blown" as an "affluent professional," for he was discovered "behaving like a common thief." He argued noisily when the suspicious clerk refused to accept the credit card, but fled when the store manager threatened to call the police. This example demonstrates that a hustler must first choose a specific strategy, then establish and maintain an unambiguous image and suitable information flow appropriate to the role that he has selected to play. Ideally, the strategy should reassure and lull a victim into acquiescence and compliance. If not, a hustler needs an escape plan.

CONCLUSION

Some observers have reported that the hustling process involves spontaneous or "opportunistic" responses to anticipated events (Agar, 1973; Fields and Walters, 1985; Goldstein, 1981). Like other cultural domains, however, the characteristics of the social contexts in which successful hustles occur can be described and the cultural rules for hustling can be identified (see Spradley, 1980). Hustling tactics appear to take advantage of expectations of behaviors associated with conventional social reality. This list is not exhaustive, but cultural differences are ascribed to individuals in terms of their age, sex, race, style of dress and speech, use of etiquette, and association with the symbolic meaning of material objects. For persons who are performing their usual social roles, conventional expectations also have a quality of inertia. Ambiguities exploited in these instances appear to stem, in part, from the impersonal qualities of social interaction in American public contexts, coupled with the human propensities to be reassured by commonplace events or distracted by surprise events. Hustling addicts, however, appear exquisitely sensitive to alternative interpretations of events. In general, the hustling process appears most successful in those contexts in which the potential for ambiguous outcomes can be manipulated to the advantage of the hustler.

Efforts at impression management in interpersonal relations are exacting and strenuous. Successful hustling requires that an individual command sound intuitive knowledge about human behavior and social dynamics as well as a rich repertoire of skills in role taking and the precise "timing" of actions. This set of knowledge and skills resembles and overlaps with that commanded by successful professional actors in the legitimate theater. But in contrast to that generated by members of the theatrical profession, this form of drama is reminiscent of "adventure," as described by Lyman and Scott (1975). The hustler employs deception in order to convey an apparently spontaneous yet unambiguous impression of "reality," while the victim of the hustle remains naive, and the rewards or "payoffs" are reaped in the form of money, goods, or services that facilitate continuation of the hustler's narcotics addiction.

Success in hustling also requires awareness of the emotional state of the victim as well as awareness of the ways in which ordinary social scenes are enacted by naive role players (see Letkemann, 1973). Hustlers' considerable ability to manipulate others is striking. Ideally, it appears that successful hustlers must continually monitor role performances of all actors in a social setting, including themselves, to ensure that their own actions appear to be unambiguous forms of generalized or balanced reciprocity rather than patent instances of negative reciprocity (burns or rip-offs).

Eventually, however, the effects of heroin and other drugs (including alcohol), and the tension of maintaining a heroin habit appear to distort judgment. The hustler is usually either apprehended by the criminal justice system, or

family members become so distressed about the hustling behavior that their sentiments impel an addict to seek treatment.

REFERENCES

Adler, P. A. (1985) Wheeling and Dealing: An Ethnography of an Upper-Level Drug Dealing and Smuggling Community. New York: Columbia Univ. Press.

Agar, M. (1973) Ripping and Running: A Formal Ethnography of Urban Heroin Addicts. New York: Seminar Press.

Burr, A. (1983) "The Piccadilly drug scene." British J. of the Addictions 78:5–19.

Carlson, K. (1976) "Heroin, hassle, and treatment: the importance of perceptual differences." Addictive Diseases 2: 569–584.

Fields, A. and J. M. Walters (1985) "Hustling: supporting a heroin habit," pp. 49–73 in B. Hanson, G. Beschner, J. M. Walters, and E. Bovelle (eds.) Life with Heroin: Voices from the Inner City. Lexington, MA: Lexington Books.

Goldstein, P. J. (1981) "Getting over: economic alternatives to predatory crime among street drug users," pp. 67–84 in J. A. Inciardi (ed.) The Drugs-Crime Connection, Sage Annual Reviews of Drug and Alcohol Abuse, Vol. 5. Beverly Hills, CA: Sage.

Letkemann, P. (1973) Crime as Work. Englewood Cliffs, NJ: Prentice-Hall.

Lyman, S. and M. Scott (1975) The Drama of Social Reality. New York: Oxford Univ. Press.

Spradley, J. P. (1980) Participant Observation. New York: Holt, Rinehart and Winston.

Sutker, P. (1974) "Field observations of a heroin addict: a case study." Amer. J. of Community Psychology 2: 35–42.

Walters, J. M. (1985) "Taking care of business updated; a fresh look at the daily routine of the heroin user," pp. 31–48 in B. Hanson, G. Beschner, J. M. Walters and E. Bovelle (eds.) Life with Heroin: Voices from the Inner City. Lexington, MA: Lexington Books.

22

CRACK DEALING ON THE STREET: THE CREW SYSTEM AND THE CRACK HOUSE

TOM MIECZKOWSKI

The objective of Mieczkowski's research is to report street-level crack buying and selling. The author compares street-level sales techniques that occurred in the heroin-dominated era of the 1960s and 1970s with the more current sales methods of the crack house system. The crack house is preferred by the majority of crack buyers because it permits greater control over the danger and uncertainty of being observed as compared with the openness and vulnerability for detection of open street sales. Mieczkowski explains the importance of knowing the differences between the heroin and cocaine markets in order to conceptualize the dangers in explaining how the different sales methods developed.

Widespread public attention has been focused on the use of crack cocaine and the social problems associated with it. Yet many aspects of the crack phenomenon remain unexplored. Some researchers have described the current media images of the crack culture as "caricatures" (Reinarman and Levine 1989). Others examining the scientific literature available on crack and crack distribution have noted the tendency to characterize crack as more dangerous or more threatening than other forms of cocaine, even though the empirical data do not support such a view (Fagan and Chin 1989).

Reports on the organization of crack selling are limited. Johnson and his associates (1990, 1991) have developed a model of crack distribution. Mieczkowski (1989a, 1989b, 1990a, 1990b) has written on crack distribution. Hamid (1990) and Bourgois (1990) have published ethnographic descriptions of crack sellers in New York. Bourgois deals mainly with social psychological aspects of crack culture among Hispanics and its relation to the ethos of the barrio. Hamid examines the impact of crack on existing dealing systems; he reports on West Indian dealers who made a transition from

EDITOR'S NOTE: From Mieczkowski, T., "Crack dealing on the street: The crew system and the crack house," in *Justice Quarterly, 9,* pp. 151–163. Copyright © 1992. Reprinted with permission of the Academy of Criminal Justice Sciences.

marijuana dealing to crack dealing, with emphasis on the aspects of crack that are destructive for their dealing systems.

Some good descriptive research on powder cocaine sales has been conducted in recent years. Patricia Adler's (1985) work focuses on middle-class cocaine users and dealers. Terry Williams's (1989) narrative *The Cocaine Kids* is a description of Hispanic powder cocaine sellers. Carl Taylor (1990) has produced a descriptive study of Detroit drug-dealing gangs with some ethnographic description. It focuses on the Young Boys, Incorporated and their successors, however, without detailing the types of drugs sold or discussing specific organizational mechanisms of crack selling. Detailed economic information and careful empirical data—such as the landmark work by Johnson et al. (1985) on the economics of heroin distribution—have not been assembled for cocaine, either powder or crack. It has been noted, however, that the lack of ethnography on drug distribution groups hampers our ability to understand this phenomenon (Reuter and Haaga 1989; Reuter, Macoun, and Murphy 1990).

The objective of this article is to report on street-level crack buying and selling. In presenting this description, I evaluate a strategic prediction made in 1986 regarding future styles of street drug sales (Mieczkowski 1986). This material is important for two reasons. First, although ethnographic descriptions of the crack culture are important in their own right, ethnography also creates the basis for interpretative analysis of other sorts of information, such as statistical data (Anderson 1978). Second, effective social responses such as the evaluation of treatment strategies and social services require accurate qualitative data to provide interpretation of the worldview of those for whom the services are provided.

The data reported here come from two sources. The first is the Detroit Crack Ethnography Project (DCEP), conducted in 1988 and 1989. The DCEP studied 100 active crack dealers in Detroit; the data base consists of in-depth interviews with crack dealers. These interviews, which were tape-recorded and transcribed into manuscripts, consisted of a structured questionnaire as well as open-ended conversation. Both statistical and text DCEP data are presented here.

The second source is data taken from the Detroit Drug Use Forecast (DUF) Crack Supplement. This information was gathered from arrestees processed in the Detroit DUF, an ongoing drug survey conducted by the National Institute of Justice. The Detroit staff administered a supplemental set of questions on crack use and distribution as an addendum to the national questionnaire. During the structured interviews, the interviewers also made notes and observations. The data in this article are based on 212 DUF cases.

THE DRUG BAZAAR: OPEN STREET SALES

During the late 1970s and early 1980s, the evolution of youthful street entrepreneurs who openly sold drugs (primarily heroin) at retail became an important element in drug distribution mechanisms in inner-city communities (Geberth 1978; Johnson et al. 1985; Mieczkowski 1986). In Detroit this attention centered around an organization called the "Young Boys, Incorporated" (YBI). The appearance and alleged citywide dominance of the heroin trade by these youths (the terms "young boys," "quarterkids," and "runners" were applied to them) appeared to be a new strategic approach to drug selling. The phenomenon of young operatives selling drugs en masse "on the street" was strikingly innovative. Geberth (1978) first reported this sales style in New York. In fact, these operations were so highly developed that certain areas of Detroit (and other cities) were labeled "drug bazaars" or "open drug markets." In these locales, drug sales were carried out overtly, in plain sight, and with apparent indifference to potential police action or other forms of legal interference. Such open action represented a dramatic departure from earlier "closed" selling strategies, which were characterized chiefly by secrecy, stealth, and the need to develop social entree and "connections" or "contacts" to secure drugs.

This new marketing method consisted of "runners" and their "crew boss," who operated a "runner system" for retailing drugs (Mieczkowski 1986). They worked in relatively large teams or units ("crews"), sold openly and often brazenly, and had an identifiable and effective organizational structure. Post-World War II drug trafficking had been based on a covert system that avoided gross public displays (Courtwright, Joseph, and Des Jarlais 1989). These older retail selling operations used a quiet, private strategy to distribute drugs, primarily heroin. Typically a house or apartment was used as a base. Heroin was obtained by house operators and was "broken down" from wholesale into small retail units of sale. From these base locations, operatives conducted their sales. Customers were referred by social networking; prior knowledge or personal introduction was requisite to making drug purchases.

In sharp contrast to the older procedure, the "crew system" of open-air street sales is based on the active "hawking" of drugs by a staff (the "crew") of out-of-pocket retailers servicing a drive-up or walk-up trade in open public spaces. This departure from the fixed-locale system represents a profound strategic marketing innovation based on rational business concerns, and is a critical functional evolutionary step in the drug distribution business.

The runner system exploits specific marketing advantages, with emphasis on high-volume street sales. It offers a potentially greater profit than the older style. The street crew style of sales offers at least three identifiable advantages:

First, the runner system allows the sellers to escape the confinement of a fixed selling locale, and makes police surveillance and control more difficult. A fixed locale is easier for police to observe, to isolate, and to seize; it represents a catchment of persons and contraband. In contrast, a diffused crew of teenagers, each carrying a little contraband, requires duplicate observational and legal processes. Furthermore, a "flight strategy" is a very effective deterrent to police invasions of the operation: in the event of a police raid or intrusion, crew members simply melt into the surrounding neighborhood. The crew system imposes a heavy burden of duplication on the police and diffuses the cache of contraband in such a way as to reduce the risks of large, simple seizures.

Second, when the trade is taken directly into the streets, a larger volume of sales can be achieved. Thus the economic return is more attractive.

A third advantage is the use of minors as street operatives. Minors strongly dominate the composition of the street crews. This arrangement reduces legal difficulties for syndicate executives, who themselves are adults. In addition, syndicate executives believe that minors work more cheaply and are easier to regulate and control than adult operatives. They are also easy to recruit.

Considering these advantages, we would expect widespread adoption of this sales strategy as time passes. The street "crew system" would become the standard for retail drug sales, at least in those urban communities which are comparable to Detroit and New York in size and composition. With the appearance and the relatively dramatic popularization of crack, one then could hypothesize that this technique would proliferate. It is not within the scope of this article to evaluate the national adoption of the technique, because a deep ethnographic data base does not yet exist for crack selling. Even so, we can evaluate Detroit, one of the origination sites of this system, and can speculate on the likelihood of multicity adoption.

STREET-LEVEL CRACK DISTRIBUTION

The findings from the Detroit Crack Ethnography Project do not support the hypothesis that the open-air street crew system is prominent in the crack marketplace. Several generalizations about the Detroit crack trade are supported by the DCEP data:

First, according to reports by crack sellers, open street sales are not a typical or popular method for selling crack. As an organizational format, this method is a distant third to crack houses and "beeper-men" (who deliver crack to their customers or rendezvous with customers at designated locations).

Second, according to reports by consumers, purchasing crack from street corner or curbside crack vendors is not popular with customers. They strongly prefer to purchase from fixed locales or from established vendors who "work off the beeper."

Finally, although the crew system per se has not survived intact as a marketing device, elements of the process have persisted and have

Table 22.1   Buying Preferences, DCEP Sample
(*N* = 100)

Crack House	35% (35)
Street Sellers	4% (4)
Beepermen	20% (20)
Combinations	41% (41)

Table 22.2   Buying Preferences, DUF Sample

Crack House	63.7% (135)
Street Sellers	10.4% (22)
Beepermen	11.8% (25)
Combinations	14.1% (30)

been incorporated into the crack trade. The most prominent of these are the dominant role of youthful managers and a management system that bears similarities to a crew system. It is applied, however, in the management of crack house locations and disparate "beeper" networks.

THE POPULARITY OF THE CRACK HOUSE

Table 22.1, based on DCEP data, shows customers' preferences for particular sales settings. The most frequent method of purchase is the crack house, which 35 of the respondents named as their principal setting. Twenty others relied upon a "touter" or "beeperman," who delivered the contraband to them. This delivery was reported variously as "home service" (i.e., delivery to the customer's residence) or as a delivery by rendezvous in an agreed-on public locale. Table 22.1 also shows that 41 respondents reported "combination" purchasing—that is, using both crack houses and beepermen. Only four respondents named the street as their exclusive source for crack.

Overt street sales of crack have not achieved the prominence and popularity that street sales of heroin reached in Detroit at the end of the 1970s (Mieczkowski 1986). In comparing Detroit DUF data to the DCEP data, we find these buying preferences reaffirmed. Table 22.2 documents a low regard for buying off the street. In a survey of 211 regular users of crack, fully 64 percent typically bought at a crack house. In contrast, only 10 percent reported making regular purchases from street dealers.

Overall, the data from these studies show that the system for crack sales is quite different from the heroin sales system. Interview transcripts from the DCEP data show that only 26 informants *ever* used street purchasing sites in their lifetime; as noted in Table 22.1, only four

individuals purchased consistently from street sellers. Although the percentage of regular street purchasers is somewhat higher in the DUF sample (Table 22.2), it still rates as the least popular of all methods, alone or in combination.

EXPLAINING THE PATTERNS

The reasons for this pattern are revealed in the characterizations of street sales by DCEP informants. They offer two perspectives on the disadvantageous nature of such transactions. The first is the view of the purchasers of crack; the second is that of the sellers.

Qualms of the Buyer

The two major issues of concern for crack buyers were quality control of the drug and their personal safety during the purchasing process. Informants reported that crack sold as "rocks" on the street was of very poor quality. This contempt for the quality of the merchandise led them to prefer a fixed locale. They believed that a seller in a fixed locale had to provide good quality because the transaction process was less furtive. If the customer smoked the crack at the locale, the quality would be revealed immediately. Also, the seller at the fixed locale was concerned with continuing business, so he had to pay attention to the quality of the merchandise.

In regard to danger, open street transactions were defined as the least physically secure. One was more likely to be "burned" on the street because the vendor, having no fixed locale, could not be held readily accountable for the quality of the merchandise. The furtive nature of the public sale was not conducive to examining and verifying even the most elementary aspects of the sale, such as adequate quantity, the "look" of the contraband, and similar concerns. Street sales, for example, virtually preclude a sample (a "taste") of the goods. Although sampling is not very common in crack houses, at least it is possible in certain types of crack houses (see Mieczkowski 1990b).

Buying crack on the street had another negative aspect: users reported that reliance on street crack was typical of people who had reached extreme stages of dependence. In effect, one was "reduced" to buying from the street as the craving for crack increased because using other sources required some measure of gratification delay and discipline. For optimum quality control, users ought to "rock up" their own cocaine, but it takes time and effort to "rock up 'caine." Thus, controlled users stigmatized street transactions as being associated with "fiending": acute, high-rate, compulsive crack use. Buying from the street was a sign that the user was growing imprudent and wasteful. In effect, "only a fool or a fiend" would buy from a street vendor.

Qualms of the Seller

In general, sellers also viewed open street sales negatively. The negative aspects were the ease of exploitation by both customers and employers, the relatively high degree of violence and exposure to potential violence, and the rate of financial return that resulted from the lack of high-volume "sales organization." The following three accounts by street dealers illustrate these points.

Interview 12: 27-year-old black male

Interviewee 12 began at the lowest level of sales as a street vendor recruited by a young gang of drug distributors (the "Pony Downs"). Eventually he reached a relatively high level in another gang, was injured seriously in a violent event, spent time in federal prison after conviction on felony drug trafficking charges, and then was severely addicted to crack cocaine. Interview 12 is cited here to demonstrate the concern with violence and the importance of a violent reputation in order to exert control; skill in violence is a necessary asset for a crack retailer. Note that the informant was recruited by the "Pony Crew" because he had a strong reputation for the ability to act violently; thereby he could deter or control predation by customers and his employers. Here he talks about his early days as a seller.

> I come against this, uh, this gang called Pony Crew, you know. And, uh, I had came back and they needed me around the neighborhood anyway cause I always liked to fight. I always liked to go in a disco and start a fight or end up with a fight and come out on top.
>
> (You had a violent reputation?) Yes. So next thing I know I was with em. I raised up with em so we went to gettin together. First they wouldn't show me no lotta dope, you know, it was like they was bringin me packs . . . (Nobody tried messing with you?) I wadn't worried about that, you know . . . I ain't got to worry bout em jumpin on me you know cause they knew where I was and what state of mind I was then, you know. Wadn't worried about nothin, you know. (If there) was somebody jumped on me they knew where I'd get back with 'em, you know. Either way, they know if they jumped on me they'd had to kill me, so my reputation was alright far as bein in there. . . .

Interview 13: 27-year-old black male

A seller's concerns did not focus only on potential violence from customers. In the following excerpt a seller recounts an event arising from competition on the street, which centered around a "turf issue." The visibility of street selling exposes one to observation and increases vulnerability; here the source of concern is retailers, who are in competition for customers. The group in competition to which the speaker refers is a group he once worked with.

> One night I'm up on Woodward in Highland Park doin my business, you know. They still doin they business, you know. They watch me pick up money and stuff. I was sellin off a beeper then and, uh, they decide w'elp I got enough money for em to rob tonight, you know, which I didn't have but a couple hundred dollars, you know . . .
>
> I had took my car when I broke away from 'em. I also cut 'em short on transportation too cause we was rentin cars. But I had a car also and that car that I had was for our other activities besides doin drugs. And so they felt in a lotta ways I left em hangin but they had asked me for a ride back over to the joint. But before we got there

the guy that I was in prison with asked me to drop him off somewhere, right? So I stopped and let him out the car. But his friend is still in the front seat and when he get out the other guy pulled a pistol on me, you know, sayin you know what time it is, right? But all along he had been tellin me and I wasn't goin for it. I looked at him, I said, "Man I want to talk to" (name), "you know this is the guy. . . ." I go to get out my car and he shot me in the back up under my shoulder blade with a .25. It punctured my lung, ricocheted off my rib cage, and it's in front of my spine. It surprises me cause personally I have killed, and I know I'm not tryin to brag or nothin like that, but I am a killer. If I shoot you I'mmo kill you, you know. I figured he just didn't want to kill me cause from what he was tellin me was just don't come back to Highland Park. It was just a warnin to run me out of Highland Park cause my legs was outside of the car, he put me back in the car, took my money, he coulda killed me, he coulda killed me but he didn't. I'm thankful for that.

Interview 61: 41-year-old black male

Interviewee 61 reports the "ripoff" of a solo street seller by robbers posing as customers. This particular scenario, robbery by violent predators, is one of the events feared most by the street dealer. This man was stabbed severely.

(Ever get stuck up?) Oh yeah, I got stabbed two or three times. Damn near killed me. One Saturday night we was on the corner and it was crowded that night. There was three different kinds of dope on the corner. Competitors. They say "Hey boy, come on man, walk us around here"—they want to get some. So I ain't thinkin and I ain't scared or nothin. I had been drinkin— I ain't drunk or nothin but I'm high. I walked around the alley and they say "Give it up" with a gun. I thought they was bullshittin and that's when he jugged it. I said "Man don't kill me, here take this bullshit, here take the dope, take the dope." He said "I should shoot your mutherfuckin ass" . . . I said "Man fuck that. Y'all can have the dope, man. I got to go to the hospital." So they took all the dope but not the money. They knew me. One's dead, shot in the head. Started fightin up there on the corner. Somebody came back and shot him in the head.

THE REEMERGENCE OF THE FIXED LOCALE: THE CRACK HOUSE AS A RESPONSE TO THE CREW SYSTEM

The disadvantages of open street sales and the resulting dislike are understandable when we review these accounts. The use of a crack house (combined to some extent with a beeper network) allows more control over the danger and uncertainty that surround the open street system. The differences between heroin and cocaine markets are important in conceptualizing these dangers and in understanding why the different systems emerged. The following marketing innovations are responses based on several key factors.

First, there is a qualitative difference between the heroin and the crack subcultures, both in selling and using. Elements of relative stability characterize the subculture of heroin users but are largely absent from the crack use world. Informants believe this difference is due to the relative novelty of crack. Long-term normative codes are absent in crack retailing because it achieved its widespread popularity so recently.

The older crew system included a relatively strict prohibition against drug intoxication "on the job," and reports indicated virtually no heroin use among street runners. This restriction, although cited by crack sellers, is less universally accepted and enforced. DCEP data show very large proportions of sellers using crack compulsively and destructively.

Second, the domination of crack sales organizations by youthful operatives stands in relatively sharp contrast to the historic control of heroin sales by adults. One can characterize this youth domination as an outcome of the crew system, in which youths were incorporated as low-level operatives and now have come to be dominant in the business. The YBI system in Detroit, for example, consisted of the recruitment of youths by adults to perform relatively low-level functions (Taylor 1990). Now the youths apparently have succeeded in dominating and commandeering the management of crack sales organizations. The DCEP data contain many cases of adults employed by teenage "bosses." Adults also state frequently that these younger boys are "crazy" or "scandalous" and that they enjoy gratuitous violence.

Third, there are differences in the methods of consuming the substances. Rates of crack use are high, both in Detroit and nationwide, among young black males (DUF 1991). Thus there are more crack users than heroin users, and volumes of crack sales are higher. One possible reason for the transition in sales systems is the addictive potential of crack. The method of crack use (smoking), as contrasted to heroin use (primarily injection), removes an important barrier to use. Therefore the relatively rapid proliferation of crack use has created entrepreneurial opportunities not seen in the heroin market.

Finally, the psychotropic and physiological effects of crack are different from those of heroin and translate into different behaviors. The stimulant and mania-inducing properties of crack cocaine, as contrasted with the analgesic and soporific properties of heroin, represent a major distinction which influences the evolution of marketing structures. The consequences are seen quite clearly in the evolution of different types of crack houses, each with its own peculiar ethos and methods of operation (Mieczkowski 1990b).

All of these elements contribute to the crack market's aura of instability and immaturity, and are compounded by a literal laissez-faire entrepreneurialism and a self-stimulating code of violence. Many of these underlying propensities are aggravated by the psychotropic effects of crack.

SUMMARY

It appears that the distinctive street-sales technique which emerged in Detroit and other cities during the heroin-dominated 1960s and 1970s has not survived intact in the era of crack cocaine. The crack trade appears to be dominated by a "crack house system" rather than by an expansion of the crew style of street selling. It remains to be seen whether the current constellation of sales methods will stabilize the crack trade, or whether further innovations will emerge from the entrepreneurial actions of crack dealers.

At this juncture it is difficult to generalize the Detroit-based findings of this study. Some elements of Ansley Hamid's (1990) work in New York parallel the Detroit experiences, especially the susceptibility of successful dealers to becoming disablingly addicted to crack. Yet several elements of Hamid's work are also distinctive, notably the domination of the trade by West Indians. The earliest descriptions of the YBI-style heroin gangs in New York by Geberth (1978) and Johnson (1985) showed very striking operational similarities to gangs in Detroit. Bourgois's material on New York Hispanic involvement in crack reflects the violent elements of Detroit, and reports the use of fixed locales (bodegas or botanicas) as retailing sites.

Until we have more ethnographic data on crack sales over a larger number of urban centers, however, the degree of generalizability of the Detroit experience will be a matter of speculation. Yet even with that restriction, Detroit's experiences with crack serve to inform us; they may prove quite useful to others who examine this behavior in different locales and in somewhat different social circumstances.

REFERENCES

Adler, Patricia A. (1985) *Wheeling and Dealing: An Ethnography of an Upper-Level Drug Dealing and Smuggling Community.* New York: Columbia University Press.

Albini, Joseph (1991) "The Distribution of Drugs: Models of Criminal Organization and Their Integration." In T. Mieczkowski (ed.), *Drugs, Crime, and Social Policy,* pp. 75–109. Boston: Allyn and Bacon.

Anderson, Elijah (1978) *A Place on the Corner.* Chicago: University of Chicago Press.

Bourgois, Philippe (1988) "Fear and Loathing in El Barrio: Ideology and Upward Mobility in the Underground Economy of the Inner City." Paper presented at the fortieth annual meeting of the American Society for Criminology, Chicago.

—— (1990) "In Search of Horatio Alger: Culture and Ideology in the Crack Economy." *Contemporary Drug Problems* 16(4): 619–50.

Courtwright, David, Herman Joseph, and Don Des Jarlais (1989) *Addicts Who Survived.* Knoxville: University of Tennessee Press.

Drug Use Forecast (DUF) (1991) *DUF Survey Data: 1988–1990.* Washington, DC: National Institute of Justice.

Fagan, Jeffrey and Ko-lin Chin (1989) "Initiation into Crack and Cocaine: A Tale of Two Epidemics." *Contemporary Drug Problems* 16(4): 579–618.

Geberth, Vernon (1978) "The Quarterkids." *Law and Order* 26(9): 42–56.

Hamid, Ansley (1990) "The Political Economy of Crack-Related Violence." *Contemporary Drug Problems* 17(1): 31–78.

Johnson, Bruce D., Paul J. Goldstein, Edward Preble, Thomas Miller, James Schmeidler, Barry Spunt, and Reuben Norman (1985) *Taking Care of Business: The Economics of Crime by Heroin Abusers.* Lexington, MA: Lexington Books.

Johnson, Bruce D., Ansley Hamid, and Harry Sanabria (1991) "Emerging Models of Crack Distribution." In T. Mieczkowski (ed.), *Drugs, Crime and Social Police,* pp. 56–70. Boston: Allyn and Bacon.

Johnson, Bruce D., Terry Williams, Kojo A. Dei, and Harry Sanabria (1990) "Drug Abuse in the Inner City: Impact on the Hard Drug Users and the Community." In J.Q. Wilson and M. Tonry (eds.), *Drugs and Crime,* pp. 9–68. Chicago: University of Chicago Press.

Mieczkowski, Tom (1986) "Geeking Up and Throwing Down: Heroin Street Life in Detroit." *Criminology* 24(4): 645–66.

—— (1989a) "Studying Life in the Crack Culture." *NIJ Reports* (Fall): 7–10.

—— (1989b) "The Detroit Crack Ethnography: Summary Report." Submitted to the Bureau of Justice Assistance, Washington, DC.

—— (1990a) "Crack Distribution in Detroit." *Contemporary Drug Problems* 17(1): 9–30.

—— (1990b) "The Operational Styles of Crack Houses in Detroit." NIDA Research Monograph 103. Washington, DC: National Institute on Drug Abuse.

Reinarman, Craig and Harry G. Levine (1989) "The Crack Attack: Politics and Media in America's Latest Drug Scare." In Joel Best (ed.), *Images of Issues: Typifying Contemporary Social Problems,* pp. 115–37. Hawthorne, NY: Aldine.

Reuter, Peter and John Haaga (1989) *The Organization of High-Level Drug Markets: An Exploratory Study.* Santa Monica: RAND.

Reuter, Peter, Robert Macoun, and Patrick Murphy (1990) *Money from Crime: A Study of the Economics of Drug Dealing in Washington, D.C.* Santa Monica: RAND.

Taylor, Carl (1990) *Dangerous Society.* Lansing: Michigan State University Press.

Williams, Terry (1989) *The Cocaine Kids.* New York: Addison-Wesley.

23

PERCEIVED RISKS AND CRIMINAL JUSTICE PRESSURES ON MIDDLE-CLASS COCAINE DEALERS

DAN WALDORF AND SHEIGLA MURPHY

The stereotypical perception of a drug dealer is not one of a middle-class person selling cocaine. Waldorf and Murphy dispel the myth that drug dealers are all from lower socioeconomic areas of the community and are deeply involved in the criminal underworld culture. The authors conducted interviews with 80 ex-sellers who sold cocaine steadily for a minimum of 1 year and have stopped dealing for 6 months or more. They reported that most former middle-class cocaine dealers felt that they could avoid detection and arrest by being able to control the network of purchasers they deal with and did not perceive any great threat from law enforcement agencies. Middle-class dealers were always more concerned about informants, angry customers, and being robbed than any other factor. Their own cocaine use posed a great risk, which was an important variable that caused them to quit drug dealing, more so than any pressures faced from criminal justice agencies.

INTRODUCTION

Despite the pronouncements of successive "drug wars" and extensive efforts by law enforcement agencies cocaine, in both powdered form and as crack, is readily available to large segments of users in most of the urban centers of the United States. Drug sales has become for many people not only a way to get drugs they might use but also a means to attain some of the All American dream—house, car and all the attractive consumer goods that are part and parcel of our notions of the "good life." Periodically politicians and criminal justice representatives promise interdiction of supplies by increased funding for narcotics law enforcement, special task forces, and various changes in laws that allow police more latitude to arrest and incarcerate drug sellers. Most recently, changes in laws have taken the forms of new legislation that allows the police to arrest sellers as continuing criminal enterprises (RICO) and

EDITORS' NOTE: From Waldorf, D. & Murphy, S., "Perceived risks and criminal justice pressures on middle-class cocaine dealers," in *Journal of Drug Issues, 25,* pp. 11–30. Copyright © 1995. Reprinted with permission.

allowed prosecutors to confiscate money and property of sellers derived from drug sales prior to actual trials. With the implementation of these two laws it would appear on the surface that risks for drug sellers are becoming greater and greater. And it was [with] this new legislation in mind that we undertook an exploratory, descriptive study of the perceived risks associated with cocaine sales and criminal justice pressures to quit sales.

METHODS AND SAMPLE

The sample consists of in-depth interviews with eighty ex-sellers from eight different levels of sales. To be eligible for the study a respondent had to have sold cocaine steadily for at least a year and had to have stopped selling for at least six months.

Respondents were located for the study by chain referral methods (Biernacki and Waldorf 1981; Watters and Biernacki 1989). This is a method used commonly by sociologists and ethnographers to locate hard to find groups and has been used extensively in the drug field (Lindesmith 1968; Becker 1953; Feldman 1968; Preble and Casey 1969; Biernacki 1986). The first of the location chains was initiated in 1974 and 1975. At that time the authors of the study were conducting a short term ethnography of the cocaine use and sales taking place among a friendship group of thirty-two persons (Waldorf et al. 1977). Other chains were developed during a second study conducted during 1986–1987 (Waldorf et al. 1991), and several other chains were developed during the present effort.

Characteristics of the Sample

There is no way to tell if the sample is representative of all cocaine sellers as there are no known populations of sellers with which to compare them. It should also be noted that the majority of ex-sellers were middle class and well educated.

There are, as well, other less violent drug sales scenes and many people who eschew violence and sell drugs in a friendly, non-threatening way. For example, within one network we discovered an accountant, an engineer,

a stock broker, an office manager, a clerk and a projectionist who all sold cocaine to small networks of users and sellers and did not resort to violence or threats of violence in any way. In fact, the portrayals of the lifestyles of all these sellers were very mundane, low key and without violent incident. Of the six persons from this group only one owned a gun and none of them had ever used them. The person who owned the gun said that he bought it for his protection when he began to sell parts of ounces, eight balls (eights of ounces) and quarters, and had some thoughts of being robbed. They were in short, middle-class citizens who held steady jobs, made mortgage payments, and were responsible parents. More often than not most of the sellers from our sample were the antithesis of common portrayals.

PERCEIVED RISKS OF DRUG SALES

In general, the study envisioned that perceived risks associated with drug sales could be categorized into eight groupings:

- General Fears of Arrest,
- Fears About Informants,
- Fears About Police Investigations,
- Fears About Confiscations of Property by Police or Other Authorities,
- Fears About Internal Revenue Service Audits,
- Fears About Robbery and Violence,
- Fears Associated with Customers or Suppliers, and
- Unanticipated Risks.

Each category was explored in the in-depth focused interviews with questions that asked how individuals viewed such risks, what were their specific experiences of risks, and what actions were taken to minimize risks. We begin our description with general fears of arrest.

General Fears of Arrest

In general, most respondents realized that there was a general possibility that they might be detected and apprehended by the police for drug sales, but it was not an abiding concern of most of the sellers. In fact, most believed that they could minimize this possibility of arrest by restricting their sales to a small group of friends, people who they work with and associates—people who they have known for some time and were known *not* to be police or persons who would reveal their activities to the police. In short, they believed that they need not fear arrest if they restricted their sales activities to persons they knew and trusted. Here is a typical example of a seller who had little concern about the risks involved in drug sales:

> I didn't think I was at any risk. My biggest risk was when I had to go over and pick up from my dealer. But I didn't think I was at any risk at all when I was selling.

Other persons were more cognizant of how to minimize the risks involved as one kilogram and pound dealer explained:

> Yeah, well I think the major thing is just keep your network closed, you know. If you've got a network, you only expand it to the people that you know and the people that they know. And if they want to buy for them, and if you could afford to give them a deal so they can go sell to their friends, you know . . . and in that way they can start their own small network. I think that was the easiest way. Like I had ten interactive networks going (that he sold to), you know, expanded families type thing.

One way to manage possible risks or to minimize the general possibility of arrest was to establish personal rules about who one would sell cocaine to. For example, an ounce dealer expressed some concerns about traffic in and out of his house to buy cocaine and rules about who he sold to:

> . . . I worried about like traffic, so most of the time I would deliver the product, you know. I kept decent hours, you know, if somebody wanted to come over late at night I refused, you know, I kept control of it pretty much so I wasn't worried in that sense. It's like the police, the only thing that could happen is like somebody introducing you to somebody and just being a little sloppy or a little greedy. If you deal with the same people, or let someone roll over . . . otherwise I think it's really hard for them to legitimately bust you, you know.

Traffic seemed to be an important consideration when sellers dealt out of their homes, most particularly, how neighbors would respond to a large volume of persons coming to and from a house or apartment. The strategies to deal with traffic were various. Some persons delivered the drugs themselves rather than have customers come to their home. Some were sure to locate their homes in areas that had high densities and large natural traffic, rather than live on a suburban cul de sac were traffic would be obvious. Other sellers required that buyers act like friends and other visitors and stay for a period of time, rather than rush in and out. One very outgoing and social pound dealer made it a point of inviting policemen that he knew socially to come to his house to socialize, feeling that if neighbors saw him socializing with the police that they would never suspect that he was a drug seller.

The rule about restricting sales only to persons known by the seller generally works for wholesalers, but is rather difficult to maintain for street and bar dealers. Street and bar dealers contact customers in public places and are more likely to take on new customers that they do not know; therefore, they are anxious to develop new customers and expand their business. Selling cocaine in small units, parts of grams and grams requires a larger number of customers to realize a reasonable profit. If a street or bar seller is ambitious and/or is using his own product too much then he may be less prudent about screening the persons he will sell to. And the less careful you are and more willing you are to develop new customers the more likely you are to sell to an undercover narcotics officer.

Fears About Informants

Cocaine sellers usually know that most arrests for drug sales come as the result of information provided to the police by informants. The use of informants by narcotics police has a long and enduring history for perhaps several reasons: drug sales with the exception of street sales is usually clandestine; drug users do not, as a rule, complain to the police about drug sellers; investigations of sellers requires a good deal of time and effort; and most undercover police are easily identified as such.

Fears about informants is a particular concern for heroin users who sell cocaine because of ways the police use heroin-addicted informants to gather information. This was illustrated very well by a heroin addict who sold ounces and parts of ounces of cocaine for six years and was arrested only after he went to treatment for his heroin addiction:

(R) Like when I did get busted I had a feeling in the beginning [that something was wrong] and then these guys kept calling . . .

(I) How did you meet that guy, the fisherman?

(R) [Through] the one guy that I was in treatment with in the hospital.

(I) He introduced you?

(R) When he came in the first time, he says I want to come and see you, right, I'm at the bar and they both [two DEA agents] come in and they both look scruffy as hell, they both looked like they just came in, you know, like they both had been fishing in Alaska. And he [the informant] introduces me to the guy and he [the undercover agent] says, "We're fishing in Alaska and I got a chance to make some big money in Alaska and we need some coke," and I said, "Well, I don't do it anymore," and he said, "Well can you get me a quarter gram or something so we can snort now?" So I go down to the end of the bar and get him a quarter gram and that in turn was one of the counts [charges] against me. There were three different counts, it was a pound sale that they got me the last time and then I got a quarter of an ounce the next time they came into the bar and I set it up to get him a quarter of an ounce but he brought this guy in and he leaves the scene. Marty, the guy that I knew, and the other guy came.

(I) So the guy in treatment he is an informant?

(R) He is a paid informant. Paid informant, that's all. He wasn't in trouble, it's just that he wanted money.

(I) Right. And so he introduced the D.E.A. to you?

(R) Right. Sure. I mean it's hard when you like somebody and you open up to somebody, I mean you're not thinking of everything you're saying is going to be used against you. I mean this guy had his kids over to the house and I met his family, you know.

Another woman who was selling parts of grams in a Latino community in Oakland expressed similar concerns about an informant. In this case it was her sister-in-law, who became angry with her brother who was selling heroin and informed on him to the police. This police activity against her brother was also viewed as a possible threat to herself.

Again street and bar dealers are also subject to this risk more than larger, more clandestine dealers because of their high visibility and their lack of caution.

Fears About Police Investigation

Fears about a possible police investigation were perceived as a regular risk for most of this group of sellers. This usually took one of two forms. The first form had to do with individuals' cocaine consumption and was most apparent when individuals were abusing the drug. The second form had to do with actual or possible police observations of individual sellers and the people they were associating with.

In the case of the first, fears associated with a seller's own cocaine consumption, there is a general paranoia that accompanies regular and/or heavy use which causes many dealers to imagine possible police surveillance and instigates certain cautions. One female dealer who sold parts of pounds for her lover described this form in combination with concerns about traffic:

We heard a lot of people say, "Well we know that the cops are watching this house for sure." But sometimes when somebody puts that in your mind that they [the police] can be upstairs . . . like I would walk through the halls and I would be tweaking out and feel like they were following me upstairs . . . stupid sick stuff. But it would make you wonder. You would look outside and you would see figures in other houses across the street. And I would think people were watching me through binoculars. I'm sitting there not moving because I'm tweaked out but that figure would just be there not moving too. Like I thought this lamp was a person that just never left and this person always watched us. It's sick. (Case #042).

Eventually this woman became so cautious about concerns with her neighbors, and the fears of being observed by the police that she developed a rather elaborate scheme where she did not allow anyone to come to their house for the drug and she went to considerable trouble to deposit the drug in an airport locker and then sold the key to the locker to the buyers.

Actual police observation was reported by a Latino smuggler who made regular forays to Mexico to bring back kilograms of cocaine which was eventually distributed by persons which he called the "Mexican Mafia."

(I) Do you think that you ever had your phone tapped or they were investigating you at all?

(R) Yeah. For sure they were investigating me at one point. The last part of the year that I was dealing I'm positive of that. I was stopped more than a few times and they were drug agents that were stopping me.

(I) Oh really, in your car?

(R) I think basically because of the association. Like the people that I knew were heavily into it and of course there was always arrests of people like that. You know when they get one person . . .

(I) When you got stopped what would they do to you?

(R) Just give me a bunch of shit, you know, search me and the vehicle I was in. Feed me a bunch of bullshit.

(I) And you never had any on you?

(R) No, they'd always tell me something like, "We're going to get you. We know what you're doing and you know so and so." And I said, "I don't know that person," And they say, "You know so and so." And I say, "No I don't." They'd say, "Well we seen you with him." I said, "Well, I don't know that person by that name so I don't know who you're talking

about." I said, "I have a lot of friends." So they always, you know, the fact that they were watching other people.

Fears About Confiscations of Property by Police or Other Authorities

In general, sellers did not express much concern that police would confiscate their personal property derived from drug sales should they be arrested. Low level sellers did not believe that the police would use that tactic with them because they had so little property that it would not matter to the police. High level sellers knew about the powers of the police to confiscate property and were either careful not to buy property or large conspicuous material items if they could not prove that the purchases came from legal income or did not spend drug money on items that could be confiscated. Only one seller reported that she felt considerable concerns about possible appropriation of her property. She was a long time marijuana and cocaine seller who regularly received kilograms and sold pounds and half-pounds to a small network of eight customers. She became concerned about her property when an old marijuana supplier who she had not done business with for four years was arrested and had all his property confiscated as well as all the property of his parents. Eventually, the parents of the accused regained their property, but the seller never did. Her way of handling the risk was to always work at a legal job and speculate on the stock market so that she had some way of verifying any purchases that she and her husband might make.

Fears About IRS Audits

In general, most sellers who did not work, did not file income tax returns and always dealt in cash. None had any problems with the IRS when they returned to work and filed returns. For persons who worked regularly while they sold cocaine it was seldom an issue, they just reported their legal income and forgot their illegal income. Some of the large sellers had legitimate businesses in which they invested cash and had ways to launder illicit drug money to make it appear that they earned the money legally.

Fears About Robbery and Violence

When sellers felt comfortable with the persons they were buying or selling cocaine from they had no particular fears about robberies or violence, but if they felt uncomfortable then it was a very real possibility. There was also some concern about customers revealing their activities to thieves who in turn might rob them. This was a particular concern for some women who dealt on their own, without the assistance of a man.

There were only six mentions of robbery, violence or threats of violence and we will illustrate three. The first is a robbery. A forty-six year old woman, who had worked as a prostitute for twenty years and sold cocaine for four years, reported how she was robbed and both she and her lover were badly beaten by three men.

> I didn't have any drugs in the house. I had some money but I didn't have any drugs and it was about four in the morning and a bunch of people had just left my house and these people [the thieves] must have been waiting outside . . . and there was a knock on the door and I went downstairs and I was really high and the guy knocked on the door and I asked, "Who is it?" He goes, "It's me, open the door," and I thought it was my son. But that was not my son, my son was in his room asleep. And I opened the door and three of them came in with stocking masks on their face and one of them said, "Let's blow them away now," and then the other one said, "no man, don't do that." And one beat me so bad that they had to shave my whole head. They knocked my old man out and took his jewelry and they got about $1,000 in the envelope that I had just laying around but the rest of my money they didn't get. But they got $1,000 and they pulled a gun on my son and they told him to freeze before they blew his brains out. And I played like I was passed out because I knew if the guy was to pull me outside he was gonna kill me. But when he did that to my son, I didn't care about me anymore. I go, "Please don't hurt my son" and he goes, "We know, we won't hurt your kid." So they left and after that I stopped dealing and I went back to prostitution. And then about a year later I started dealing again

and I dealt for about six months and then I stopped.

The second illustration is of a woman who dealt multiple kilograms with her lover until he was murdered. She was not exactly sure if he was killed because of his sales activity, but we were advised by an associate of the woman that the police believed her husband's death was drug-related.

(I) But basically you do believe that his death was drug-connected?

(R) Yeah, it was either someone jealous because they knew he had a lot of money or maybe they . . . it could have been that he was robbed and they could have taken a package that he had that I just didn't know about. Or they could have taken the money because when he got killed I didn't have the money and it wasn't found in his place. So, you know, I don't really know what happened to the money. I don't know. He had been telling me that he wasn't transacting anymore business and, you know, there was a time when I tested him because I didn't believe he was really out of the business. So I called him up and I said, "Hey, you know so and so, it will be simple cut and dry," and I'm talking like a $10,000 or $15,000 transaction and a lot of profit for us and he did not do it. I don't know whether he realized I was testing him and he said no or if he was really sincere and was out of the business. But it was either someone robbed him and he lost his life through that or, you know, he had couple little girl friends who didn't want to see him getting married. And they could have really been angry and feeling like they had been used and they could have set him up for the robbery, you know, and didn't realize he was gonna get killed too. I really don't know. He was the kind of person who kept a lot of protection around those he love . . .

Another good illustration of threats of violence was related by a thirty-eight year old black seller who sold pounds of powder and crack to several sellers in San Francisco housing projects. It also illustrates how a gang took over his crack business by use of violence against his network of customers.

(R) Well the thing that really caused me to get out of the business is the fact that . . . the whole process is being organized and organized crime is moving in. Okay they've got the L.A. Crypts that are up here now . . . Well I started getting some heat and I started getting people beat up on the street and threatened not to come back out there.

(I) So someone knew it was your man.

(R) No, they didn't know it was my man but I mean people were coming back to me saying, "Hey man, somebody robbed me and took all of my shit," or "Somebody told me not to come back out there unless I've got a gun." And the clientele really started to really decline so that's why I got out of it. And you would have to deal with people that, you know, if you turned your back they would steal the white off of your teeth . . . that's when I decided to get out of it and, you know, the people getting killed and robbed and stabbed and I didn't want that around. And I made a couple of enemies too because people would sit up and start getting high and they would want me to give them credit or give them some or something and it's like I'm sorry. Then, you know, they're like, "Well this mother fucker is selling coke up in that house," and before you know it they tell another person and the other person says, "Yeah well he's selling coke and he's got a bunch of freaky women up there," and they keep the story on until it snowballs to where it's I'm selling coke and selling women and selling hot cars and apartments and everything else.

There was a second report of gang activity to take over drug sales by a kilogram/pound dealer who reported that he was approached by a friend of a customer (who vouched for him) about putting together a ten kilogram deal. After he arranged to get the ten kilograms and the transaction was made the customer returned with two others, who were heavily armed, and demanded that he buy back the cocaine, not for the $100,000 for which he sold it for but $150,000. Furthermore, they demanded that he deal with them exclusively in the future and made threats that he could not disregard. His response to this threat was to pay them $150,000, take his family

on a long vacation and retire. He was in a position to do this because he had bought several income producing properties and did not need either the trouble or the money from cocaine sales.

Fears Associated With Customers or Suppliers

Customer Fears

Customers of sellers can be a problem for sellers. Very often customers will make unusual demands upon the seller that become irritating—calling at all hours of the day or night, talking about drugs on the telephone, being unable to pay for the drugs that they wish to buy and consume, and as we saw earlier by introducing narcotics police into the network who could pose some threat to the seller.

Freebasers, crack smokers and persons who cannot afford to buy the drugs they are using are considered to be particularly problematic. Sellers usually establish rules to deal with some of these issues, but there is a general rule that many dealers lived by, "Do not make your customers mad at you so they would not drop a dime on you." This usually translated into not turning troublesome customers down when they approached them for cocaine, but telling them that you did not have any cocaine for sale. As one man explained:

> No, not too many deals went bad because we would give more [when customers complained about quality or short weight] . . . we didn't want to have any complaints or hassles. We didn't want anybody to snitch on us, you know.

A white, male gram dealer expressed his problems with customers:

> Being woken up in the middle of the night by phone calls from customers who wanted to score. Having trouble collecting money from people who owed it to me. Having that lead to cutting people off and then having arguments with them about that and then being afraid of retaliation.

Freebasing customers could be particularly problematic as a twenty-nine year old female, black, ounce dealer elucidated:

(R) Well the people around me, the people who were buying from me, were getting a little bit weird . . . a lot of people started basing and I didn't want any part of that or feel like I was contributing to that.

(I) How about the friends you were hanging with?

(R) They started using a lot more too, you know . . . They wanted more and asking for more and wanting to base and stuff and it just started getting out of control and when it's out of control it's bad and you run more of a risk of getting busted when it's out of control . . .

(I) So did you just suddenly say you were going to quit?

(R) It started building up in me and then one day I just said this is it. I think I went and bought some more and paid off the last one and I said this is it. But I didn't tell my connection that or anybody else . . . Made one more cop and cleaned it out, paid the man and I said that's it for me.

Some sellers have continuing problems with customers and found them to be extremely troublesome to deal with:

(R) Well as I learned how difficult coke fiends are to deal with, when it come to money and they're reliability and how they will transgress on your life in order to get their powder, and you see people in really bad shape coming to you with nothing, you know, with fantastic stories of how they can pay you later cause something has happened to them, they just sold a screenplay or, you know, all this weird stuff they're telling you so you can front them just a few lines of blow, cause they're jones'n [addicted]. And seeing that too often and then having people who were real nice to me when I had blow and when I didn't have any or wouldn't sell to them, have them scream at me and threaten me, and I think well Jeez, I don't want this guy to be picked up by the police . . .

(I) Yeah. Did you have a lot of bad customers?

(R) . . . No, no, they aren't actually all bad customers, no. No, I sold many times without any problems at all.

(I) But there were some bad customers. What did bad customers do?

(R) They would call you at any hour of the night, okay, to see if you had something. They might show up at your door, any hour of the day or night, whether or not the lights were on then knock on the door to see if you just went to bed, maybe you just went to bed, "I was seeing if you just went to bed," it's 3:30 . . . So they're extremely egotistical and uhhh, and oh just coke fiends. Often they had no money at all, and would have all these stories they could pay you later and sometimes they are friends of yours and you figure you can front them the stuff, then each one of your friends always owes you $25 for the latest quarter and you're out a few $100 . . . It strains relationships too. Cause you're letting their drug problem, their inability to pay for their drugs, put you out.

Most sellers try to develop rules in dealing with customers; rules that will provide structure to the transactions much the same way shop keepers have rules about when businesses will be conducted, who will be given credit, etc.:

> . . . I only sold to the people that I knew and they knew they weren't suppose to bring anybody to my house, not even parked outside or anything like that . . . Yeah, and they knew that they shouldn't call me after ten and if they did my machine was usually on.

One gram dealer who had a number of troublesome customers had a particularly good tactic for dealing with them. He sold them very heavily diluted cocaine and they did not return. Upon getting his supply he would organize it into three baggies—his own personal stash which was not cut, a cut supply for good customers, and a very heavily cut supply for customers he wished to stop selling to. The tactic worked every time. Suppliers were plentiful and customers would not return to a seller that sold them low-quality drugs. His customers were persons who would not come back and demand their money back and the tactic seemed to work for non-violent customers, but by no means are all customers nonviolent.

Supplier or Connection Fears

Some large sellers were potentially violent, but most sellers were not fearful of connections unless they had problems paying for drugs that were being fronted. Again, freebasers and persons who injected the drug were particularly vulnerable to threats from connections when they could not make payments for fronted drugs. One ounce dealer, who was also an injector of both heroin and cocaine, told of one incident he had with an old friend who was connected to "the mob" and supplied him with large supplies from Miami.

(I)   You didn't feel any risks with the mob?

(R)   No, well one time I owed them money. I think I was into them for about four or five grand and I get a phone call and, uh, they are at the Miami airport . . . Perry who I would only see at his home in Fort Lauderdale, uh, my old roommate from years that past says me and Frank . . . are on our way out here. Now I owed him like, I don't know, four or five thousand and I don't have any money on me and I don't have any coke and they want me to rent them a car and they'll be here in five hours right so I get real nervous but at that time I was using a lot of drugs and strung out on heroin . . . So I'm loaded, so I got the car with a credit card and I met them at the airport and we drive like two miles outside the airport and Perry tells me to pull over and I go "Oh shit," you know, and I am thinking, "Get in the fuckin trunk," right and we get out and he slaps me, you know, not hard, you know, "Hey you fucked up and don't do it again, okay," and I say, "Okay." . . . And after that they proceeded to give me like three or four pounds of cocaine and I mean I don't know, I just beat them for like $4,000 and he slaps me in the face and you do it again I'm going to hurt you, that's what he said, "If you do it again I'm going to hurt you, now take care of business," you know.

A second case, a smuggler, told us that he regularly shaved off five or six ounces from kilograms that he smuggled across the Mexican border. He used these purloined supplies for his own rather outrageous freebasing sessions. Eventually the persons who financed his Mexican forays discovered the shortage and approached him about it. He denied the pilferage, but felt that he was in real danger of being shot or killed. He eventually convinced them that he did not do it, but felt that he only got away with it because he was related to one of the major financiers. Eventually when his own use began to get out of control the group simply cut him off from all money and supplies.

Unanticipated Risks

To our surprise we found only one person who did not use the product he sold. This man is an ex-convict (and was one of the models of John Irwin's "straight raised youth" in his book *The Felon*) who has seldom used drugs other than alcohol even though he always had access to them, even in prison. While in San Quentin prison he was the cell mate of an infamous addict (who has gone on to become a well known novelist and movie script writer) who always had heroin and other drugs smuggled into prison, but the respondent never used heroin or cocaine himself. He was literally allergic to cocaine so he never had any problems with his own consumption of the drug. He remarked that he was in a perfect position to sell cocaine since he never used it.

Many persons found that their personal use got completely out of control so that it either cut down on their profits or they had trouble meeting their debts to suppliers who were fronting them supplies. For example: one woman who sold for ten years and did very well when she only snorted the drug became a compulsive user and an unreliable seller when she began to freebase it:

(R)   . . . let met tell you what happened. When I started freebasing, I started losing money, you know, but the people that I was involved with and I had been involved with the same people for so many years they could not believe that I wasn't capable of doing it anymore. So they kept giving me these amounts of drugs, you know, half pounds, you know. And then they'd cut it down and I would end up with part of their money and not all of their money. Then they gave quarter pounds and would come back with hardly no money but, I mean, the average person would have been dead because they would have been killed probably. But by me knowing them since I was like nineteen or twenty years old and I had been dealing with them for so long and I'm sure I probably made them a lot of money too as well as for myself . . . they kept trying it but I kept slipping backwards. After awhile . . . after I started basing I tried to sell it for about a year after that and I couldn't. I just went so far in debt.

(I) Okay now let's talk about the circumstances that caused you to make a decision to stop selling? What caused you to come to that decision to stop selling cocaine?

(R) My own use. Freebasing. When I was using and selling, well, when I was snorting and selling cocaine, you know, I made money and I didn't have any pressure. Because like I said, over the years I had built my clientele to where they were all very close to me. Everything that I did worked out fine. . . . But in my head to quit selling cocaine completely was when I knew I couldn't handle it anymore and I couldn't make it pay . . . do you understand? It was costing me too much. Everything I had lost, just everything that I had. I tried to obtain more and the more I would try to climb the farther I would slip back. And it became a problem when I started freebasing cocaine because it gave such a craving that you can't stop. Whatever you have, whatever you're doing, whatever money you've got, you're gonna spend it. I haven't seen anyone do any different either.

Another twenty-four year old, student, ounce dealer, reported similar problems of being out of control, losing the confidence of his suppliers and physical problems:

(R) My own use was getting way out of hand and I was real skinny and way under nourished and I couldn't even stay awake anymore because I was . . . physically fatigued and I couldn't even keep my eyes open . . . so exhausting. I've been more tired from a night of bingeing than I could have ever been from back packing for two weeks and hiking one hundred miles a day. That's when my grades dropped in school because I was too tired for school, too tired to do anything. I would spend a lot of time in the shower or trying to suck some food down just to be somewhat nourished.

(I) So when you did decide to quit, what did you do? Did you leave the scene or did you get out for awhile?

(R) Well it also kind of happened the same time when my dealers weren't trusting me anymore. I was getting out of hand and not paying the debts fast enough and finally I think they got nervous. They were getting bigger and they had other customers so they didn't need me really as much anymore. And so they were kind of giving me the cold shoulder and cutting me out. And they were tired of me scaring them because I was scaring them to death. . . . I was getting too out of hand.

With only a few exceptions most persons who sold cocaine tended to use it very heavily and with heavy use came various unanticipated physical and psychological problems. There were of course certain variations among users, in the ways that they used the drug, but it was a very common occurrence for sellers to have myriad problems with their own use.

One man who was a very heavy drinker as well as a prodigious user of cocaine reported that he nearly died of peritonitis while he was using and selling cocaine.

One young, female pound dealer became anorexic with her heavy use and had to go to the hospital with a urinary infection:

(R) I was like anorexic. I probably ate ten times that year or maybe more like twelve times, like once a month. No really, I would not eat. I would maybe eat a bag of potato chips and that would keep me going for three days. Then I would sleep and I would wake up and I might have a piece of toast . . . would be sick all the time. I would be nauseous, faint . . . I had a constant cough that year and I would always be able to cough up black stuff.

(I)   So no serious illness?

(R)   No, the urinary infection was serious. I got to the point where I couldn't walk and I bled . . . I thought I was on my period for a full month and I knew something was going on. I put it off for another week and it was a urinary infection that went into my kidneys and it was in my back and I was bent over . . . and I was so thin and weak I just had to stay like that until they took me to the hospital. I was too weak to even pull myself up. I was just drained and I had no color. I had these big black things on my face and they would peel. Oh God. . . .

CRIMINAL JUSTICE PRESSURES

Our exploration of criminal justice pressures to stop selling cocaine were conceived as being of two types—direct and indirect criminal justice pressures. The first was pretty obvious—an arrest for some violation of the law, either drug related or not. Indirect pressures were a little more complicated—arrest of someone from a supplier's network, arrest of someone from a customer network, arrest of a drug selling partner, fears of being investigated by the police and fears of going to prison.

In general, the majority of ex-sellers we interviewed reported that they did not experience any criminal justice pressures to quit, neither direct nor indirect. Forty-eight persons or 60% of the eighty respondents reported that they did not experience any criminal [justice] pressures to quit. Of the remaining thirty-two who reported some pressures, sixteen (20%) said that they stopped selling because of some arrest, the majority of which was for sales and possession; and sixteen (20%) reported indirect pressures to quit.

Direct Pressures

The most direct criminal justice pressure, as might be expected, was arrest for sales and possession of cocaine and six persons were arrested for sales and possession. Another four were arrested for possession of cocaine with intent to sell, one for possession only and another two for possession of other illicit drugs.

One kilogram and pound dealer was arrested after being stopped for a traffic violation and the police officer discovered that there was an outstanding warrant for his arrest in Nevada on a conspiracy charge. The respondent hired a well known criminal lawyer and was finally released from the charge when the prosecutor's file was lost. Both the respondent and the interviewer, who has known this man for six years, believed that the attorneys had arranged to have the file stolen by someone in the prosecutor's office. As a result of the arrest the respondent lost his customers, they stayed away from him while he was awaiting trial, and never did resume sales when the charges were dropped.

Three respondents were arrested for other crimes: one for driving under the influence, another for assault, and a third turned himself into the police for check forgery under the pressure of his live-in girl friend.

Indirect Criminal Justice Pressures

Indirect pressures to stop sales could take several forms as we mentioned earlier. The most common form was arrest of someone from a supplier's network; ten people gave such responses. A good illustration of this was the case of a forty-year-old ounce dealer who sold cocaine on and off for six years. In general, she was very careful about restricting her sales to a small group of customers that she knew and trusted, but her major connection, who drank very heavily, was not as careful. Just prior to her voluntary retirement two incidents happened.

The first incident involved four thieves masquerading as police who broke into her connection's house, held him, his wife and two children at gun point and stole $30,000 and approximately a pound of cocaine. Three weeks later police contacted the wife of the connection and told her that they knew that her husband was selling cocaine and asked for their cooperation in identifying the gang. Both agreed to help the police, but could not identify anyone in a police line up. The connection continued to sell despite the protests of his wife.

The second incident occurred one month later. Whenever the connection went out of town she assumed delivery to several of his suppliers in the South Bay. On this occasion she was out of town herself, taking her children to a Girl Scout camping outing, and an associate took over the tasks of delivery. The associate was arrested with several ounces and the respondent believed that had she made the delivery the same thing would have happened to her. She decided that it was time to quit selling shortly thereafter.

A second illustration is the account of a forty-six year old pound dealer who regularly traded guns for pounds and kilograms of cocaine. The guns were not military weapons, but rifles and pistols that were eventually smuggled to Mexico by his connection who was said to be part of the "Mexican Mafia."

(I) What were the circumstances of your decision to quit selling?

(R) Well the circumstances is that this whole network seemed to have gotten busted and I haven't heard from any of them in about thirteen or fourteen months.

(I) So your connection dried up.

One stock-broker seller told an unusual story about his growing concerns about being arrested, a portentous dream he had that his whole supplier's network was arrested and his subsequent decision to retire from drug sales:

(R) I don't remember any significant changes until I started feeling that these people [his connections] are sloppy. And there is a lot of busts going down, and I suggested to them [his suppliers] several times that they cool the activities and let this storm [possible arrests] that is obviously coming blow over and we will see how the waters are at that point but everyone else was really into it and so I pretty much pulled the plug on it.

And I also had these dreams. I had this dream three times it was like the movie, "Little Big Man" where I had certain morality play . . . but the bottom line was that I kept dreaming that everybody got caught. And, uh, and the vividness where it took place and what happened and how it went down it was like playing a video tape. It was the same dream three times. Well I had stopped dealing [regularly], . . . somebody had asked me to pick something up for them and this was a very important guy . . .

(I) You are saying that you stopped dealing already?

(R) These were the one occasions where I would say, "Okay at my convenience I will do this for you." I think I did it once or twice. And one time there was this painting crew at the house next door [to the connection] and they were there painting . . . I think I made three trips in four days or five days and this fuckin crew of painters was there for five days and there were about nine guys. I mean you could have painted a Taj Mahal in five days. And I said, "I don't like this at all." I mean they had this van parked out there and nobody had paint on them and they were all standing around and doing their stuff and I don't know what was going on. I expected that they were filming everybody walking in and I don't know what the hell they were doing but it was weird and it scared me a little bit.

And another time I went back, there was a cement truck there. I went back in the morning and my dealer was not there he blew an appointment on me. And I went back later that day and there was the same cement truck, the license plate was the same. The cement pump truck had been there for eight hours and he would have to be pumping it by the shovel full to be there that long. And nobody rents a cement truck that long, I mean it just didn't make any sense . . . anyway there was two circumstances there that I felt very uncomfortable about.

I was able to cover my going in and out in front of these people because I had luckily met the nurse that lived upstairs and ran into her one day and ran into her roommate the other day and introduced myself and talked and stuff like that in front of these people. And I felt fortunate well there is a sly cover there and maybe they think I'm here to see the nurse on floor 2 instead of dealer on floor 3 . . . So a couple of weeks later I'm coming back from Tahoe on the behalf of these people that I decided to make an exception for and I phoned [the connection] and phoned and nobody is there. And, uh, then I picked up the newspaper and there is everybody in a drawing. They picked up so many people they are all being arraigned in a jury box 'cause there are so many of them and I said, "The heats gone done."

(I)   Your people?

(R)   My people and all of their North Bay connections and all of the lower people in the Haight. Everything blew over and around, and somehow I ducked and I missed it all.

(I)   So your dream came true, everybody got busted but you.

(R)   And they didn't finger me. By that time I'd been out [of it] for a few months and they didn't see me on a weekly basis or any kind of regular basis. I was the drop-in.

Arrest of Customers

Indirect pressures stemming from the arrest of customers was not reported as a reason to stop selling cocaine. When customers were arrested sellers usually denied them all access, refused to take phone calls, refused to sell to them and generally avoided them.

Arrest of Business Partners

Arrest of drug selling partners was viewed as indirect pressure and was reported by two cases. One was a woman who sold ounces and parts of ounces for four years and who at one point financed two lesbian friends to go to South America to smuggle cocaine into the United States. The two eventually became regular and reliable connections for the respondent. This woman's partner (another woman) was arrested for possession of one gram of cocaine was judged to be guilty and served six months in a county jail. During her imprisonment she suffered considerable psychological distress and her hair turned gray during the imprisonment. The respondent thought that the sentence was severe and noting the effect it had upon her partner she decided to stop using and stop selling herself.

Another woman who sold kilograms and pounds and was closely associated with Colombian smugglers stopped her activities when the DEA arrested fifteen persons in her supplier and customer network, one of whom was her brother-in-law. When she was not arrested herself, other Colombians in the supply network threatened her and others with violence to make sure that they did not reveal information to the DEA.

Police Investigations

Only two cases reported that they stopped selling cocaine because of an imminent police investigation. One woman first sold marijuana for three years and when the price of marijuana was increasing she and her lover decided to move on to cocaine. Initially, they traded marijuana for cocaine, but eventually they found a good supply for high quality cocaine and began to buy three or four kilograms a month. They sold it to a small group of eight trusted customers who would buy pounds, half pounds and ounces.

Early in their transition from marijuana to cocaine sales she was arrested by the police for possession of marijuana after a police informant directed narcotic officers to her home with the information that they sold cocaine. The police did not find the cocaine that was hidden there, but they did find several ounces of marijuana. She remained in jail over a weekend, plead guilty to the charge, and was sentenced to a diversion program which she described as a joke. "The counselor regularly asked me to score for him and tried to convince me to give up my marijuana use by utilizing meditation techniques."

Upon being released from jail she and her boyfriend (who she eventually married) moved to a new location and kept selling both drugs but at a reduced rate. Gradually they increased their business and began to make regular money, but she continued to work throughout the whole time that they sold cocaine. During the last year of their cocaine selling career they learned that an old marijuana supplier who lived in Arizona had been arrested by the police and was being prosecuted. Police confiscated the supplier's property and began to arrest other customers of the supplier. Although they had not done business with the old supplier in a number of years, they began to feel that they might be investigated themselves. To avoid possible investigation they moved a second time to a town in another county and were careful not to give out their new address and telephone number to any of their old associates. At the time her husband was using the drug very heavily and experiencing paranoid ideation about the investigation. Both became fearful that the police were imminent and decided to cut back on their sales activity. Six months later after she took on a new more responsible job, separated from her husband they both stopped selling completely.

Fears of Re-arrest or Imprisonment

Two respondents reported that they stopped selling because of fears of re-arrest or imprisonment. Sellers who had previous convictions for drug sales and possession and/or were ex-convicts who had a good deal to lose if they were re-arrested had considerable fears about going to or returning to prison. This naturally acted as an indirect pressure on some to stop selling. One ex-convict illustrated this type of indirect pressure very well. He had a long and checkered arrest history from an early age and had spent ten years in San Quentin and Folsom prisons and had no desire to return. Prior to going to prison he had sold various types of drugs—opiates, cocaine, barbiturates, amphetamines. After serving his last prison term he built a small but profitable contracting business and managed to change his life considerably. He was a regular user of marijuana and on occasion he used cocaine. When cocaine became more plentiful he undertook selling ounces and parts of ounces to a very small network of friends to finance his own and his wife's use. He knew his

connection very well, she was the best friend of his wife, so he had no fears on her part. He never allowed customers to bring any strangers into the house and on one instance when it happened he became livid with the culprit.

In general, he felt he had too much to lose if he were arrested and this acted as continuing indirect pressure on him to limit the scope of his sales activities. Eventually he decided to stop when his connection stopped selling and he did not seek any other sources, even though two others were available to him.

OTHER REASONS TO STOP SELLING DRUGS

There are of course other pressures exerted on sellers to stop—pressures from family, spouses/lovers, friends and children that can be just as effective, though not as obvious, as criminal justice pressures. There are as well other reasons why drug sellers stop selling. Some are simply inept as sellers, some can not meet the financial responsibilities of fronted supplies, some have severe physical and psychological problems with their own drug abuse. Some do not view drug sales as an enduring career and go into other legal occupations.

SUMMARY

The general possibility of arrest for sales was perceived by middle class ex-cocaine sellers as a risk that most could minimize by restricting their selling activities to small networks of persons that they knew and trusted. Such a possibility was not an abiding concern, but very clearly sellers put themselves at risk when they sold to strangers or people who were not personally known by the seller. Sellers were generally more concerned about disgruntled customers, informants and possible rip-offs and violence than they were of police investigations, possible confiscation of personal property by the police and IRS audits and investigations.

Heavy abuse of their product causes many persons to become generally paranoid about their drug sales and to maintain some caution about customer traffic and possible police surveillance. Abuse also causes many to experience unanticipated physical and psychological problems with the drug which are important factors in their reasons to stop sales.

Criminal justice pressures, either direct or indirect, were not a particularly recurrent factor in this sample's accounts of their decisions to stop selling cocaine. More than half reported that they felt no criminal justice pressures at all to stop sales. Of those who reported pressures there was near equal percentages of direct and indirect pressures. The most frequently mentioned indirect pressure to stop was an arrest of a member of a supply network.

ACKNOWLEDGMENT

The study from which this article emanates was supported by an NIJ grant #7–0363–9-CA-IJ, Bernard A. Gropper, Program Manager; Drugs, Alcohol and Crime Programs, Center for Crime Control Research.

REFERENCES

Becker, H. S.
  1953   Becoming a marijuana user. *American Journal of Sociology* 59:235–42.

Biernacki, P.
  1986   Pathways from Heroin Addiction. Philadelphia: Temple Univ. Press.

Biernacki, P. and D. Waldorf
  1981   Snowball sampling: Problems and techniques of chain referral sampling. *Sociological Methods and Research* 10:141–63.

Feldman, H. W.
  1968   Ideological supports to becoming . . . and remaining a heroin addict. *Journal of Health and Social Behavior* 9:131–9.

Irwin, J.
  1970   The Felon. Englewood Cliffs, N.J.: Prentice Hall.

Lindesmith, A.
  1968   Addiction and opiates. Chicago: Aldine.

Preble, E. and J. H. Casey, Jr.
  1969   Taking care of business—the heroin users' life on streets. *The International Journal of the Addictions* 4:1–24.

Waldorf, D., S. Murphy, C. Reinarman, and B. Joyce
  1977   Doing coke: An ethnography of cocaine users and sellers. Washington, D.C.: Drug Abuse Council, Inc.

Waldorf, D., C. Reinarman, and S. Murphy
  1991   Cocaine changes: The experience of using and quitting. Philadelphia: Temple Univ. Press.

24

WOMEN IN THE STREET-LEVEL DRUG ECONOMY: CONTINUITY OR CHANGE?

LISA MAHER AND KATHLEEN DALY

The issue Maher and Daly present is whether women played a greater entrepreneurial role, both in sales and in distribution, in the crack cocaine markets of the 1980s and early 1990s than in the heroin markets of the 1960s and 1970s. By utilizing an observational and interview format, the authors examined and analyzed women drug abusers in a high-crime, drug-infested area of New York. Their findings, contrary to other research reports, found that crack cocaine markets have not offered very much opportunity for women to monetarily prosper in the sales and distribution of crack. Their research suggests that the newer drug markets remain male dominated and have not seen any significant movement of women dealers' assent to the higher profit circles of the drug trade. Instead, women predominantly remain at low-level auxiliary positions of the business.

This article presents the results of an ethnographic study of women drug users conducted during 1989–1992 in a New York City neighborhood. We assess whether women's involvement in U.S. drug markets of the mid-1980s onward reflects change, continuity, or a combination of change and continuity from patterns in previous decades. We find that contrary to the conclusions of Baskin et al. (1993), Fagan (1994), Inciardi et al. (1993), Mieczkowski (1994), and C. Taylor (1993), crack cocaine markets have not necessarily provided "new opportunities" for women, nor should such markets be viewed as "equal opportunity employers" (Bourgois, 1989; Wilson, 1993). Our study suggests that recent drug markets continue to be monopolized by men and to offer few opportunities for stable income generation for women. While women's *presence* on the street and in low-level auxiliary roles may have increased, we find that their *participation* as substantive labor in the drug-selling marketplace has not.

RESEARCH SITE AND METHODS

Research Site

Bushwick, the principal study site, has been described as hosting "the most notorious drug bazaar in Brooklyn and one of the toughest in New York City" (New York Times, October 1, 1992:A1).

Between 1988 and 1992 drug distribution in Bushwick was intensely competitive; there were constant confrontations over "turf" as organizations strove to establish control over markets. Like many drug markets in New York City (see, e.g., Curtis and Sviridoff, 1994; Waterston, 1993), Bushwick was highly structured and ethnically segmented. The market, largely closed to outsiders, was dominated by Dominicans with networks organized by kin and pseudo-kin relations.

Fieldwork Methods

Preliminary fieldwork began in the fall of 1989 when the senior author established a field presence in several Brooklyn neighborhoods (Williamsburg, East Flatbush, and Bushwick). By the end of December 1991, interviews had been conducted with 211 active women crack users in Williamsburg, East Flatbush, and Bushwick. These were tape recorded and ranged from 20 minutes to 3 hours; they took place in a variety of settings, including private or semiprivate locations (e.g., apartments, shooting galleries, abandoned buildings, cars) and public locales (e.g., restaurants, parks, subways, and public toilets). From January to March 1992, a preliminary data sort was made of the interview and observational material. From that process, 45 women were identified for whom there were repeated observations and interview material. Contact with these women was intimate and extensive; the number of tape recorded interviews for each woman ranged from 3 to 15. Unless otherwise noted, the research findings reported here are based on this smaller group of 45 Bushwick women.

Profile of the Bushwick Women

The Bushwick women consisted of 20 Latinas (18 Puerto Ricans and 2 Dominicans), 16 African-Americans, and 9 European-Americans; their ages ranged from 19 to 41 years, with a mean of 28 years. At the time of the first interview, all the women used smokable cocaine (or crack), although only 31% used it exclusively; most (69%) had used heroin or powder cocaine prior to using crack. The women's average drug use history was 10.5 years (using the mean as the measure); heroin and powder cocaine initiates had a mean of about 12 years and the smokable cocaine initiates, about 6 years.

These 45 women represent the range of ages, racial-ethnic backgrounds, life experiences, and histories of crack-using women among the larger group of Brooklyn women interviewed.

Structure of New York City Crack Markets

In selling crack cocaine, drug business "owners" employ several "crew bosses," "lieutenants," or "managers," who work shifts to ensure an efficient organization of street-level distribution. Managers (as they were known in Brooklyn) act as conduits between owners and lower-level employees. They are responsible for organizing and delivering supplies and collecting revenues. Managers exercise considerable autonomy in the hiring, firing, and payment of workers; they are responsible for labor force discipline and the resolution of workplace grievances and disputes. Next down the hierarchy are the street-level sellers, who perform retailing tasks having little discretion. Sellers are located in a fixed space or "spot" and are assisted by those below them in the hierarchy: lower-level operatives acting as "runners," "look-outs," "steerers," "touts," "holders," and "enforcers." Runners "continuously supply the sellers," look-outs "warn of impending dangers," steerers and touts "advertise and solicit customers," holders "handle drugs or money but not both," and enforcers "maintain order and intervene in case of trouble" (Johnson et al., 1992:61–64).

In New York City in the early 1990s, it was estimated that 150,000 people were involved in selling or helping to sell crack cocaine on any given day (Williams, 1992:10). Crack sales and distribution became a major source of income for the city's drug users (Hamid, 1990, 1991; Johnson et al., 1994). How, then, did the Bushwick women fit into this drug market structure? We examine women's involvement in a range of drug business activities.

Selling and Distributing Drugs

During the entire three years of fieldwork, including the interviews with the larger group of over 200 women, we did not discover any woman who was a business owner, and just one worked as a manager. To the limited extent that they participated in drug selling, women were overwhelmingly concentrated at the lowest levels. They were almost always used as temporary workers when men were arrested or refused to work, or when it was "hot" because of police presence.

Of the 19 women (42%) who had some involvement, the most common role was that of informal steerer or tout. This meant that they recommended a particular brand of heroin to newcomers to the neighborhood in return for "change," usually a dollar or so. These newcomers were usually white men, who may have felt more comfortable approaching women with requests for such information. In turn, the women's perceptions of "white boyz" enabled them to use the situation to their advantage. Although they only used crack, Yolanda, a 38-year-old Latina, and Boy, a 26-year-old African-American woman, engaged in this practice of "tipping" heroin consumers.

They come up to me. Before they come and buy dope and anything, they ask me what dope is good. I ain't done no dope, but I'm a professional player. . . . They would come to me, they would pay me, they would come "What's good out here?" I would tell them, "Where's a dollar," and that's how I use to make my money. Everyday somebody would come, "Here's a dollar, here's two dollars." (Yolanda) [What other kinds of things?] Bumming up change. [There ain't many people down here with change.] Just the white guys. They give you more faster than your own kind. [You go cop for them?] No, just for change. You tell them what's good on [the] dope side. Tell them anything, I don't do dope, but I'll tell them anything. Yeah,

it's kicking live man. They buy it. Boom! I got my dollar, bye. (Boy)

Within the local drug economy, the availability of labor strongly determines women's participation in street-level distribution roles. Labor supply fluctuates with extramarket forces, such as product availability and police intervention. One consequence of police activity in Bushwick during the study period was a recurring, if temporary, shortage of male workers. Such labor market gaps promoted instability: The replacement of "trusted" sellers (i.e., Latinos) with "untrustworthy" drug users (i.e., women and non-Latinos) eroded the social and kinship ties that had previously served to reduce violence in drug-related disputes (see also Curtis and Sviridoff, 1994).

Early in the fieldwork period (during 1989 and early 1990), both men and women perceived that more women were being offered opportunities to work as street-level sellers than in the past. Such opportunities, it turned out, were often part of a calculated risk-minimization strategy on the part of owners and managers. As Princess, a 32-year-old African-American woman observed, some owners thought that women were less likely to be noticed, searched, or arrested by police:

Nine times out of ten when the po-leece roll up it's gonna [be] men. And they're not allowed to search a woman, but they have some that will. But if they don't do it, they'll call for a female officer. By the time she gets there, (laughs) if you know how to move around, you better get it off you, unless you jus' want to go to jail. [So you think it works out better for the owners to have women working for them?] Yeah, to use women all the time.

As the fieldwork progressed and the neighborhood became more intensively policed, this view became less tenable. Latisha, a 32-year-old African-American woman, reported that the police became more aggressive in searching women:

[You see some women dealing a little bit you know.] Yeah, but they starting to go. Now these cop around here starting to unzip girls' pants and go in their panties. It was, it's not like it was before. You could stick the drugs in your panties 'cause you're a female. Now that's garbage.

Thus, when initially faced with a shortage of regular male labor and large numbers of women seeking low-level selling positions, some managers appear to have adopted the opportunistic use of women to avoid detection and disruption of their businesses. How frequent this practice was is uncertain; we do know that it was short-lived (see also Curtis and Sviridoff, 1994:164).

In previous years (the late 1970s and early 1980s), several Bushwick women had sold drugs in their roles as wives or girlfriends of distributors, but this was no longer the case. During the three-year study period only 12 women (27%) were involved in selling and distributing roles. Of this group of 12, only 7 were able to secure low-level selling positions on an irregular basis. Connie, a 25-year-old Latina, was typical of this small group, and in the following quotation she describes her unstable position within the organization she worked for:

I'm currently working for White Top [crack]. They have a five bundle limit. It might take me an hour or two to sell that, or sometimes as quick as half an hour. I got to ask if I can work. They say yes or no.

Typically the managers said no to women's requests to work. Unlike many male street-level sellers who worked on a regular basis for this organization and were given "shifts" (generally lasting eight hours), Connie had to work off-hours (during daylight hours), which were often riskier and less financially rewarding. Temporary workers were usually given a "bundle limit" (one bundle contains 24 vials), which ensured that they could work only for short periods of time. As Cherrie, a 22-year-old Latina, said,

The last time I sold it was Blue Tops [crack]. That was a week ago. [What, they asked you or you asked them to work?] Oh, they ask me, I say I want to work. [How come they asked you?] I don't know. They didn't have nobody to work because it was too hot out there. They was too full of cops.

Similarly, although Princess was well-known to the owners and managers of White Top crack, had worked for them many times in the past year, and had "proved" herself by having never once "stepped off" with either drugs or money, she was only given sporadic employment. She reported,

Sometime you can't [sell]. Sometime you can. That's why it's good to save money also. So when you don't get work. [How come they wouldn't give you work on some days?] Because of some favor that someone might've done or y'know, jus' . . . [It's not like they're trying to punish you?] No, but they will do that y'know. Somebody go and tell them something, "Oh, this one's doin' this to the bags or this one's doin' this to the bottles." OK, well they check the bags and they don' see nothin' wrong, but they came to look at it so they're pissed off so they'll take it away from you, y'know.

Violence and Relationships

In addition to being vulnerable to arrest and street robbery, street-level sellers who use drugs constantly grapple with the urge to consume the product and to abscond with the drugs and/or the money. Retaliation by employers toward users who "mess up the money" (Johnson et al., 1985:174) was widely perceived to be swift and certain. Rachel, a 35-year-old European-American woman, said,

Those Dominicans, if you step off with one piece of it, you're gonna get hurt. They don't play. They are sick people.

The prospect of violent retaliation may deter women from selling drugs. Boy, a 26-year-old African-American woman, put it this way:

I don' like their [the managers'] attitude, like if you come up short, dey take it out on you . . . I don' sell no crack or dope for dese niggers. Because dey is crazy. Say for instance you short ten dollars, niggers come across you wit bats and shit. It's not worth it, you could lose your life. If dey say you are short, you could lose you life. Even if you were not short and dey say you is short, whatever dey say is gonna go, so you are fucked all the way around.

However, considerable uncertainty surrounds the likelihood that physical punishment will be meted out. This uncertainty can be seen in the comments by Princess, who had a long but sporadic history of street-level sales before and after the advent of crack:

It's not worth it. Number one, it's not enough. Come on, run away, and then *maybe* then these people want to heavily beat the shit out of you. And then they *may* hit you in the wrong place with the bat and *maybe* kill you (emphasis added).

Such disciplinary practices resemble a complex interplay between "patronage" and "mercy," which features in relations of dependence (Hay, 1975). The unpredictability of punishment may work as a more effective form of control than actual punishment itself. In Bushwick, the actuality of violent retaliation for sellers who "messed up" was further mediated by gender and ethnicity. In this Latino- (mainly Dominican) controlled market, the common perception was that men, and black men especially, were more likely than Latinas to be punished for "stepping off." Rachel described what happened after an African-American man had been badly beaten:

[What happened to him. I mean he stepped off with a package, right?] Yeah, but everybody has at one time or another. But it's also because he's a black and not a Puerto Rican, and he can't, you know, smooze his way back in like, you know, Mildred steps off with every other package, and so does, you know, Yolanda, they all do. But they're Spanish. And they're girls. So, you know, they can smooze their way back in. You know, a guy who's black and ugly, you know, so they don't want to hear about it.

Relationships in the drug economy are fueled by contradictory expectations. On the one hand, attributes such as trust and reliability are frequently espoused as important to drug-selling organizations. On the other hand, ethnographic informants often refer to the lack of trust and solidarity among organization members. This lack of trust is evident in the constant "scams" sellers and managers pull on each other and the everpresent threat of violence in owner-manager-seller relations.

Strategies of Protection and "Being Bad"

Women who work the streets to sell or buy drugs are subject to constant harassment and are regularly victimized. The Bushwick women employed several strategies to protect themselves. One of the most important was the adoption of a "badass" (Katz, 1988), "crazy," or "gangsta bitch" stance or attitude, of which having a "bad mouth" was an integral part. As Latisha was fond of saying, "My heart pumps no Kool Aid. I don't even drink the shit." Or as Boy put it,

Ac' petite, dey treat you petite. I mean you ac' soft, like when you dress dainty and shit ta come over here an' sit onna fuckin' corner. Onna corner an' smoke an you dressed to da teeth, you know, you soft. Right then and there you the center of the crowd, y'know what I'm sayin'? Now put a dainty one and put me, she looks soft. Dey look at me like "don't fuck wid dat bitch, she looks hard." Don' mess wit me cause I look hard y'know . . . Dey don't fuck wit me out here. Dey think I'm crazy.

Acting bad and "being bad" are not the same. Although many Bushwick women presented themselves as "bad" or "crazy," this projection was a street persona and a necessary survival strategy (see also Spalter-Roth, 1988). Despite the external manifestation of aggression, a posture and rhetoric of toughness, and the preemptive use of aggression (Campbell, 1993), women were widely perceived (by men and women alike) as less likely to have the attributes associated with successful managers and street-level sellers. These included the requisite "street cred" and a "rep" for having "heart" or "juice"— masculine qualities associated with toughness and the capacity for violence (Bourgois, 1989; Steffensmeier, 1983; Waterston, 1993). Women's abilities to "talk tough" or "act bad" were apparently not enough to inspire employer confidence. Prospective drug business employers wanted those capable of actually "being bad" (Bourgois, 1989:632). Because female drug users were perceived as unreliable, untrustworthy, and unable to deploy violence and terror effectively, would-be female sellers were at a disadvantage.

Selling Drug Paraphernalia

In Bushwick the sale of drug paraphernalia such as crack stems and pipes was controlled by the bodegas, or corner stores, whereas syringes or "works" were the province of the street. Men dominated both markets although women were sometimes employed as part-time "works" sellers. Men who regularly sold "sealed" (i.e., new) works had suppliers (typically men who worked in local hospitals) from whom they purchased units called "ten packs" (10 syringes). The benefits of selling syringes were twofold: The penalties were less severe than those for selling drugs, and the rate of return was higher compared to the street-level sale of heroin or crack.

The women who sold works were less likely than their male counterparts to have procured them "commercially." More often they "happened across" a supply of works through a family member or social contact who was a diabetic. Women were also more likely to sell works for others or to sell "used works." Rosa, a 31-year-old Latina, described in detail the dangerous practice of collecting used works strewn around the neighborhood. While she often stored them and later exchanged them for new works from the volunteer needle exchange (which was illegal at the time), Rosa would sometimes select the works she deemed in good condition, "clean" them with bleach and water, and resell them.

Although crack stems and pipes were available from neighborhood bodegas at minimal cost, some smokers chose not to carry stems. These users, almost exclusively men, were from outside the neighborhood. Their reluctance to carry drug paraphernalia provided the women with an additional source of income, usually in the form of a "hit," in exchange for the use of their stem. Sometimes these men were "dates," but more often they were "men on a mission" in the neighborhood or the "working men" who came to the area on Friday and Saturday nights to get high. As Boy put it,

I be there on the block an' I got my stem and my lighter. I see them cop and I be askin' "yo, you need a stem, you need a light?" People say "yeah man," so they give me a piece.

An additional benefit for those women who rented their stems was the build up of crack residues in the stems. Many users savored this resin, which they allowed to accumulate before periodically digging it out with "scrapers" fashioned from the metal ribs of discarded umbrellas.

Some women also sold condoms, another form of drug-related paraphernalia in Bushwick. Although condoms were sold at bodegas, usually for $1 each, many of the women obtained free condoms from outreach health workers. Sometimes they sold them at a reduced price (usually 25 cents) to other sex workers, "white boyz," and young men from the neighborhood. Ironically, these same women would then have to purchase condoms at the bodegas when they had "smoked up" all their condoms.

Running Shooting Galleries

A wide range of physical locations were used for drug consumption in Bushwick. Although these sites were referred to generically as "galleries" by drug users and others in the neighborhood, they differed from the traditional heroin shooting gallery in several respects. Bushwick's "galleries" were dominated by men because they had the economic resources or physical prowess to maintain control. Control was also achieved by exploiting women drug users with housing leases. Such women were particularly vulnerable, as the following quotation from Carol, a 40-year-old African-American woman, shows:

I had my own apartment, myself and my daughter. I started selling crack. From my house. [For who?] Some Jamaican. [How did you get hooked up with that?] Through my boyfriend. They wanted to sell from my apartment. They were supposed to pay me something like $150 a week rent, and then something off the profits. They used to, you know, fuck up the money, like not give me the money. Eventually I went through a whole lot of different dealers. Eventually I stopped payin' my rent because I wanted to get a transfer out of there to get away from everything 'cause soon as one group of crack dealers would get out, another group would come along. [So how long did that go on for?] About four years. Then I lost my apartment, and I sat out in the street.

The few women who were able to maintain successful galleries operated with or under the control of a man or group of men. Cherrie's short-lived effort to set up a gallery in an abandoned burned-out building on "Crack Row" is illustrative. Within two weeks of establishing the gallery (the principal patrons of which were women), Cherrie was forced out of business by the police. The two weeks were marked by constant harassment, confiscation of drugs and property, damage to an already fragile physical plant, physical assaults, and the repeated forced dispersal of gallery occupants. Within a month, two men had established a new gallery on the same site, which, more than a year later, was thriving.

Such differential policing toward male- and female-operated galleries is explicable in light of the larger picture of law enforcement in low-income urban communities, where the primary function is not so much to enforce the law but rather to regulate illegal activities (Whyte, 1943:138). Field observations suggest that the reason the police did not interfere as much with activities in the men's gallery was that they assumed that men were better able than women to control the gallery and to minimize problems of violence and disorder.

Other factors contributed to women's disadvantage in operating galleries, crack houses, and other consumption sites. Male drug users were better placed economically than the women in the sample, most of whom were homeless and without a means of legitimate economic support. When women did have an apartment or physical site, this made them a vulnerable target either for exploitation by male users or dealers (as in Carol's case) or for harassment by the police (as in Cherrie's). Even when a woman claimed to be in control of a physical location, field observations confirmed that she was not. Thus, in Bushwick, the presence of a man was a prerequisite to the successful operation of drug-consumption sites. The only choice for those women in a position to operate galleries or crack houses was between the "devils they knew" and those they did not.

Copping Drugs

Many Bushwick women supplemented their income by "copping" drugs for others. They almost always copped for men, typically white men. At times these men were dates, but often they were users who feared being caught and wanted someone else to take that risk. As Rachel explained,

> I charge them, just what they want to buy they have to pay me. If they want twenty dollars they have to give me twenty dollars worth on the top because I'm risking my free time. I could get busted copping. They have to pay me the same way, if not, they can go cop. Most of them can't because they don't know the people.

Those who cop drugs for others perform an important service for the drug market because as Biernacki (1979:539) suggests in connection with heroin, "they help to minimize the possibility of infiltration by undercover agents and decrease the chance of a dealer's arrest." In Bushwick the copping role attracted few men; it was regarded by both men and women as a low-status peripheral hustle. Most women saw the female-dominated nature of the job to be part of the parallel sex market in the neighborhood. Outsiders could readily approach women to buy drugs under the guise of buying sex. As Rosa recounted,

> You would [be] surprise. They'd be ahm, be people very important, white people like lawyer, doctors that comes and get off, you'd be surprised. Iss like I got two lawyer, they give me money to go, to go and cop. And they stay down over there parking. . . . [How do you meet them?] Well down the stroll one time they stop and say you know, "You look like a nice girl though, you know, you wanna make some money fast?" I say, how? So they say you know, "Look out for me." First time they give me like you know, twenty dollars, you know. They see I came back, next time they give me thirty. Like that you know. I have been copping for them like over six months already.

Sometimes this function was performed in conjunction with sex work, as Latisha's comment illustrates,

> He's a cop. He's takin' a chance. He is petrified. Will not get out his car . . . But he never gets less than nine bags [of powder cocaine]. [And he sends

you to get it?] And he wants a blow job, right, okay. You know what he's givin' you, a half a bag of blue (blue bag cocaine). That's for you goin' to cop, and for the blow job. That's [worth] two dollars and fifty . . . I can go to jail [for him]. I'm a piece of shit.

Women also felt that, given the reputation of the neighborhood as very "thirsty" (that is, as having a "thirst" or craving for crack), male outsiders were more likely to trust women, especially white women, to purchase drugs on their behalf. Often this trust was misplaced. The combination of naive, inexperienced "white boyz" and experienced "street smart" women produced opportunities for additional income by, for example, simply taking the "cop" money. This was a calculated risk and sometimes things went wrong. A safer practice was to inflate the purchase price of the drugs and to pocket the difference. Rosa explained this particular scam,

> He think it a ten dollar bag, but issa five dollar. But at least I don't be rippin' him off there completely. [But you're taking the risk for him.] Exactly. Sometime he give me a hunert dollars, so I making fifty, right? But sometime he don't get paid, he got no second money, eh. I cop then when I come back the car, he say, "Dear I cannot give you nothin' today," you know. But I still like I say, I gettin' something from him because he think it a ten dollar bag.

Similar scams involved the woman's returning to the client with neither drugs nor money, claiming that she had been ripped off or, less often, shortchanging the client by tapping the vials (removing some crack) or adulterating the drugs (cutting powder cocaine or heroin with other substances). These scams reveal the diversity of women's roles as copping agents and their ingenuity in making the most of limited opportunities.

Discussion

While in theory the built-in supervision and task differentiation of the business model, which characterized drug distribution in Bushwick, should

have provided opportunities to both men and women (Johnson et al., 1992), our findings suggest that sellers were overwhelmingly men. Thus, the "new opportunities" said to have emerged with the crack-propelled expansion of drug markets from the mid-1980s onward were not "empty slots" waiting to be filled by those with the requisite skill. Rather, they were slots requiring certain masculine qualities and capacities.

Continuity or Change?

The only consistently available option for women's income generation was sex work. However, the conditions of street-level sex work have been adversely affected by shifts in social and economic relations produced by widespread crack consumption in low-income neighborhoods like Bushwick. The market became flooded with novice sex workers, the going rates for sexual transactions decreased, and "deviant" sexual expectations by dates increased, as did the levels of violence and victimization (Maher and Curtis, 1992). Ironically, the sting in the tail of the recent crack-fueled expansion of street-level drug markets has been a substantial reduction in the earning capacities of street-level sex workers.

Of the four elements that have been used to explain women's restricted involvement in drug economies of the past, we see evidence of change in two: a diminishing of women's access to drug-selling roles through boyfriends or husbands, especially when drug markets are highly structured and kin based, and decreased economic returns for street-level sex work. Because few Bushwick women had stable households or cared for children, we cannot comment on changes (if any) in discretionary time. Underworld institutionalized sexism was the most powerful element shaping the Bushwick women's experiences in the drug economy; it inhibited their access to drug business work roles and effectively foreclosed their ability to participate as higher-level distributors. For that most crucial element, we find no change from previous decades.

Those making a general claim about "women's emancipation" in the current drug economy ignore the obdurateness of a gender-stratified labor market and associated beliefs and practices that maintain it. Those making the more restricted claim that male-dominated street networks and market processes have weakened, thus allowing entry points for women, need to offer proof for that claim. We would expect to see variation in women's roles, and we would not say that Bushwick represents the general case. However, assertions of women's changing and improved position in the drug economy have not been well proved. Nor are they grounded in theories of how work, including illegal work, is conditioned by relations of gender, race-ethnicity, and sexuality (see, e.g., Daly, 1993; Game and Pringle, 1983; Kanter, 1977; Messerschmidt, 1993; Simpson and Elis, 1995).

Our findings suggest that the advent of crack cocaine and the concomitant expansion of the drug economy cannot be viewed as emancipatory for women drug users. To the extent that "new opportunities" in drug distribution and sales were realized in Bushwick and the wider Brooklyn sample, they were realized by men. Women were confined to an increasingly harsh economic periphery. Not only did the promised opportunities fail to materialize, but the expanding crack market served to deteriorate the conditions of street-level sex work, a labor market which has historically provided a relatively stable source of income for women drug users.

REFERENCES

Baskin, Deborah, Ira Sommers, and Jeffrey Fagan
1993   The political economy of violent female street crime. Fordham Urban Law Journal 20:401–407.

Biernacki, Patrick
1979   Junkie work, "hustles," and social status among heroin addicts. Journal of Drug Issues 9:535–549.

Bourgois, Philippe
1989   In search of Horatio Alger: Culture and ideology in the crack economy. Contemporary Drug Problems 16: 619–649.

Campbell, Anne
1993   Out of Control: Men, Women, and Aggression. London: Pandora.

Curtis, Richard and Michelle Sviridoff
1994   The social organization of street-level drug markets and its impact on the displacement effect. In Robert P. McNamara (ed.), Crime Displacement: The Other Side of Prevention. East Rockaway, N.Y.: Cummings and Hathaway.

Daly, Kathleen
1993   Class-race-gender: Sloganeering in search of meaning. Social Justice 20:56–71.

Fagan, Jeffrey
1994   Women and drugs revisited: Female participation in the cocaine economy. Journal of Drug Issues 24:179–225.

Game, Ann and Rosemary Pringle
1983   Gender at Work. Sydney: George Allen and Unwin.

Hagedorn, John M.
1994   Homeboys, dope fiends, legits, and new jacks. Criminology 32: 197–219.

Hamid, Ansley
1990   The political economy of crack-related violence. Contemporary Drug Problems 17:31–78.

1991   From ganja to crack: Caribbean participation in the underground economy in Brooklyn, 1976–1986. Part 2, Establishment of the cocaine (and crack) economy. International Journal of the Addictions 26:729–738.

Inciardi, James A., Dorothy Lockwood, and Anne E. Pottieger
1993   Women and Crack Cocaine. New York: Macmillan.

Johnson, Bruce D., Ansley Hamid, and Harry Sanabria
1992   Emerging models of crack distribution. In Thomas M. Mieczkowski (ed.), Drugs and Crime: A Reader. Boston: Allyn & Bacon.

Johnson, Bruce D., Mangai Natarajan, Eloise Dunlap, and Elsayed Elmoghazy
1994   Crack abusers and noncrack abusers: Profiles of drug use, drug sales, and

nondrug criminality. Journal of Drug Issues 24:117–141.

Kanter, Rosabeth Moss
1977   Men and Women of the Corporation. New York: Basic Books.

1981   Women and the structure of organizations: Explorations in theory and behavior. In Oscar Grusky and George A. Miller (eds.), The Sociology of Organizations: Basic Studies. 2d ed. New York: The Free Press.

Katz, Jack
1988   Seductions of Crime: Moral and Sensual Attractions of Doing Evil. New York: Basic Books.

Messerschmidt, James D.
1993   Masculinities and Crime. Lanham, Md.: Rowman and Littlefield.

Mieczkowski, Thomas
1994   The experiences of women who sell crack: Some descriptive data from the Detroit crack ethnography project. Journal of Drug Issues 24: 227–248.

Murphy, Sheigla, Dan Waldorf, and Craig Reinarman
1991   Drifting into dealing: Becoming a cocaine seller. Qualitative Sociology 13:321–343.

Simpson, Sally S. and Lori Elis
1995   Doing gender: Sorting out the caste and crime conundrum. Criminology 33:47–81.

Spalter-Roth, Roberta M.
1988   The sexual political economy of street vending in Washington, D.C. In Gracia Clark (ed.), Traders Versus the State: Anthropological Approaches to Unofficial Economies. Boulder, Colo.: Westview Press.

Steffensmeier, Darrell
1983   Organization properties and sex-segregation in the underworld: Building a sociological theory of sex differences in crime. Social Forces 61:1010–1032.

Taylor, Carl S.
1993   Girls, Gangs, Women and Drugs. East Lansing: Michigan State University Press.

Waterston, Alisse
1993   Street Addicts in the Political Economy. Philadelphia, Pa.: Temple University Press.

Whyte, William Foote
1943   Street Corner Society. Chicago: University of Chicago Press.

Williams, Terry
1992   Crackhouse: Notes from the End of the Line. New York: Addison-Wesley.

Wilson, Nancy Koser
1993   Stealing and dealing: The drug war and gendered criminal opportunity. In Concetta C. Culliver (ed.), Female Criminality: The State of the Art. New York: Garland Publishing.

Part VII

GENDER AND CRIME

Until the 1970s, interest in women's involvement in crime was thought to be not very serious in nature. Due to gender roles, women offenders were most often involved in prostitution, child abuse, shoplifting, and the like. However, in the era of the 1970s through the 1990s, women's involvement in more serious crimes was thought to have increased. But according to official crime data, women are still arrested for the same minor crimes that they have committed historically. Although women's crime rate is somewhat higher than in the past, their poverty rate, victimization, and economic dependency is more likely an explanation for their increased offense rate.

In order to provide some insight into women's involvement in what is considered more masculine criminal behavior, the selections in this part of the anthology analyze female crimes that are atypical of the majority of offenses that women usually commit.

In an attempt to explore women's involvement in homicide, Brownstein, Spunt, Crimmins, and Langley examine the causes of murder by women who killed in drug market situations. They analyze the ways in which changes in drug markets have influenced women's involvement in the use of lethal violence. The authors found that women who commit homicide against men do so in order to protect or enhance their own economic interests in the context of selling drugs, but they often murder out of fear of a man or on behalf of a man.

In her analysis of female street robbers, Jody Miller examines the role that gender plays in the commission of robbery. In a comparison with male street robbers, Miller found that women do not appear to commit robberies differently than men. However, those gender differences that do exist reflect the practical decisions made on the basis of physical perceptions that men are thought to be stronger than women. Although women's motivations to participate in violent street robbery are similar to those of male offenders, gender remains an important factor in shaping the method they choose to enact the crime, which is very different from that of their male counterparts.

An example of a female entrepreneur in the drug dealing business is the focus of the case study authored by Dunlap, Johnson, and Manwar. By utilizing a single woman crack dealer, the researchers explore the systematic career of an inner-city minority woman's successful operation. Rachel, the authors claim, is a deviant among deviants because she is a woman operating an illegal enterprise in a traditional male occupation, and further, she is educated and comes from a middle-class family background, which is atypical of drug dealers in her operational environment. Her clientele consist of older, employed hidden crack users as opposed to the more traditional stereotypical young, inner-city males, and she follows business practices common among middle-class drug sellers. The authors note that Rachel is herself a drug user who is addicted to crack yet has avoided the negative consequences of addiction as well as arrest. Somehow she is able to balance the worlds of the straight and drug subcultures, middle-class and inner-city dealers of narcotics, and casual crack users and addicted drug abusers. Her case illustrates how particular drug markets periodically change in this nation, how the use of drugs is prevalent in urban minority areas, how differing socioeconomic groups are involved in drug use, and how female dealers function in an illegal activity typically dominated by males.

The last article on gender utilizes studies of methamphetamine among female drug users and sellers in three cities: San Francisco, San Diego, and Honolulu. Authors Morgan and Joe examine patterns of experiences, beliefs, and drug environments of women in these locales. They report that one of the most important factors that emerged from their research was the large number of female methamphetamine dealers and distributors who had substantial experience in this criminal enterprise. The authors discovered many misconceptions in prior studies of women dealers. Such misplaced notions that women have less control over their use of drugs as compared with men, that female drug users suffer more guilt and lower self-esteem than men, and that women dealers are in subordinate positions to males in this illicit business world were found to be untrue. Morgan and Joe conclude from their study that most of the women involved in sales of methamphetamine considered their experience in a positive light, which resulted in their economic independence, high self-esteem, and professionalism, an increase in their ability to function on their own, and a source of ethics in an illegal social world. Finally, dealing provided them with a perceived need to maintain control over their social and intimate relationships, responsibilities of their daily lives, and their use of drugs.

25

WOMEN WHO KILL
IN DRUG MARKET SITUATIONS

HENRY H. BROWNSTEIN, BARRY J. SPUNT,
SUSAN M. CRIMMINS, AND SANDRA C. LANGLEY

Prior research indicates that a woman's involvement in the act of murder is often more complex an act than merely self-defense against an abusive man. Brownstein, Spunt, Crimmins, and Langley examined incidents of homicide by female drug dealers and explore the ways in which women's changing role in drug market opportunities has affected their participation in lethal violence. The authors analyzed 19 women's homicide cases to see whether the effect of gender prevents women who occupy high positions in drug sale operations from committing murder over economic conflicts in drug market situations. The authors report that women will utilize violent means (homicide) in situations that involve augmenting or protecting moneymaking in the drug market. Furthermore, the authors conclude that there is evidence that women in the drug trade will murder or act as coparticipants if they encounter a threat to their business associate or to their significant other. The authors suggest that females involved in the narcotics market kill for purposes unrelated to their gender.

The literature comparing homicides by women and by men has focused primarily on cases involving domestic situations. It has been suggested that women are more likely than men to have killed a relative, especially a spouse or partner (Hazlett and Tomlinson 1987; Jurik and Winn 1990; Riedel, Zahn, and Mock 1985; Wilbanks 1984; Wolfgang 1958), and more likely to have killed in a domestic dispute (Jacobs and Rosen 1983; Swigert and Farrell 1976; Wilbanks 1984). This statement may be statistically accurate, but it does not negate the fact that women also kill for reasons unrelated to concerns of the family or home.

In this paper we examine women who have killed in circumstances related to a market economy rather than in a domestic setting. Through the experiences of women who have killed in an economic rather than a domestic context, we explore whether the significance of gender is altered by the specific circumstances of the economy of the situation. Specifically, we focus on women who have killed in drug market situations.

Drug markets in the late 1980s provided women, at least in theory, with the opportunity for economic liberation in a social setting where problems of socioeconomic relationship were commonly resolved with violence. In this paper we do not explore whether the crack market actually provided women with widespread economic opportunities. Rather, we examine situations in which women used lethal violence in relation to the drug market, and ask whether or not they killed in relation to an economic interest related to the market. Alternatively, even in a violent market situation, did these women kill for gender-based reasons?

METHOD AND SAMPLE

This analysis is part of a larger study of women sentenced in New York State for homicide. We selected respondents from among 280 women identified by the State Department of Correctional Services as residing in a New York State prison when the study began in April 1992, from an additional 71 women admitted during the following seven months, and from 69 women who were on active parole supervision in New York City for murder or manslaughter during winter and spring 1993, whom the State Division of Parole identified as meeting the sample criteria for a parole subgroup. Of these women, 300 were located by interviewers; 217 of the 300 agreed to be interviewed and 83 refused, for a response rate of 72 percent. Two interviews were incomplete, so data are available for 215 respondents. Interviews were conducted during 1992 and 1993 and took place at eight prisons (with 88% of the respondents) and nine parole sites.

Of the 215 women interviewed, most had been involved with drugs before the homicide event. When asked, 95 percent said they had tried one drug or more, including alcohol; 90 percent had tried alcohol, 76 percent marijuana, 54 percent cocaine, 29 percent crack, and 27 percent heroin. Of the 215 women, 39 percent told the interviewer that they had been involved in drug sales or some other drug distribution activity.

As was true for drugs, the women in the sample were not strangers to violence. About one-third grew up in communities that they remembered as violent. As for their own use of violence, 65 percent said they had participated in violent activity (e.g., hitting others, robbery), and 64 percent said they had seriously harmed someone else (e.g., beating up, threatening with a weapon) when they were growing up. In discussing their own violent victimization, 58 percent said they had been the victims of serious physical harm as a child, and 79 percent said they had been the victims of such harm as an adult. In addition, 49 percent said they had been the victims of sexual harm (i.e., inappropriate touching or forced sexual penetration) as a child; 43 percent said they had experienced such harm as an adult.

Sixty percent of the 215 women told the interviewer that they were personally responsible for the homicide for which they were convicted and sentenced to prison. An additional 12 percent said they were accomplices, 2 percent said they were conspirators, and 10 percent said they were simply bystanders. The remainder said they were not present or not involved in the killing, could not remember, or did not want to discuss the event in detail.

Most of the killings took place in a residence (64%), usually that of the victim (24%) or a place shared by the victim and the respondent (26%). About one-third of the women used a knife or other cutting instrument to kill the victim; 28 percent used a gun. The victim was the respondent's spouse or partner in only 20 percent of the incidents; more often, the victim was a stranger (23%); and most often, an acquaintance (33%). Sixteen percent of the victims were the children of respondents, and 9 percent were other relatives. In 68 percent of the incidents the respondent said the killing was precipitated by the victim's actions or words, including threats, insults, and arguments.

We determined the drug-relatedness of homicide incidents by the women's responses to one of two questions. First, an incident was classified as systemic if the respondent said "yes" to the question "Do you think the homicide was in any way related to the drug business? [If yes] How was it related?" Second, an incident was classified as systemic if the respondent said that it was related to the victim's involvement in the drug business; she was asked, "Would you say the homicide was in any way related to the victim's drug or alcohol involvement? [If yes] How was it related?" Using these criteria, we ultimately classified 19 incidents as related to drugs through participation of the woman or her victim in the drug business.

Motives of Women Who Kill

In any given year in the United States, the number of women who commit homicide is smaller than the number of men, and the number of women who kill for economic reasons is a small percentage of all female homicide offenders (Federal Bureau of Investigation 1994). In this section we discuss those women with economic motives for killing in order to study women who kill in situations related to social or economic interests.

Every homicide event is a complex social interaction between "an offender, victim, and possibly an audience" in what Luckenbill, almost two decades ago, called a "situated transaction" (1977:176). As the event occurs over time and through the course of progressive interaction, individual motives are subject to change. Thus a single homicide event can have a variety of associated motives.

All of the 19 incidents of women who killed in drug market circumstances (defined earlier as "systemic") took place, by definition, in an economic context. Yet, they did not all necessarily have purely economic motivations. Of the 19, 13 (68%) were coded as having at least one economic motive; even these sometimes had additional motives that were not necessarily economic.

Eight of the 13 killings were related to an attempt to get drugs or money for drugs. In Incident 099, for example, the respondent was involved in the killing of a taxicab driver to get money to repay drug money that had been stolen. In Incident 265, the respondent was collecting money from the spot where she sold drugs, and she shot and killed someone who tried to rob her. According to the respondent in that case,

> I went to collect some money from one of my [drug-selling] spots. I drove up, parked my car, and went to get the money. This guy from the neighborhood, a heroin user in fact, one of my customers probably, comes up to my car, pointed a gun at me, and demanded the money. I kept my own gun with me, and before he had a chance to shoot me, I shot him first. I was really scared. I thought he was going to kill me.

As described by the respondent, this case is a representative example of an incident that began with an economic motive and ended as a killing motivated by fear.

Of the remaining five of these 13 incidents, one was a case of someone being paid to kill, one involved getting money for something other than drugs, and three involved angry retaliation for "messing" with drugs or drug money. In one of the latter incidents, Number 433, the motivation was as close as possible to being purely economic. The respondent was the lieutenant in a drug-dealing operation, and she killed a worker who had run off with her boss's money.

Of the six drug market incidents that were coded as not having economic motivations, the respondents either claimed noneconomic reasons for their participation in the killing (four respondents) or stated that they had nothing to do with it (two respondents). One of the latter said she was too high to remember what happened; the other said she was just a bystander. The bystander, Respondent 078, was the girlfriend of a dealer and was present during a robbery of another dealer in a dispute over drug territory. The woman who was too high to remember what happened, Respondent 014, was convicted of killing a drug dealer she was robbing while he was picking up his money.

The respondent in Case 014 might not have been able to remember what happened, but the police produced three witnesses who said they did. The woman was 26 years old in 1986, when she was arrested for killing a man over crack and money. The police witnesses claimed they saw her get out of a car and shoot a man to get his money and his crack. They told her, she said to the interviewer, that the man was in the drug business and was in her building picking up his money.

The woman in Case 014 had been selling crack for about a year when the killing occurred. For $50 she bought crack worth about $100 on the street from a dealer who had several sellers working for him. In this case, she told corrections officials, the victim owed her money, and she was trying to scare him. Although she reports that she does not remember the killing, she knows, as she said to the interviewer, that "when using crack, I was always into fighting over anything and with anybody, especially men."

Among the four other cases involving women who killed in the context of the drug business but said they killed for other than economic reasons, two said they killed in self-defense, one was trying to scare the victim, and one was protecting herself from sexual attack. In the case involving a sexual attack (086), the respondent was the girlfriend of a major dealer who attacked her in part because on one occasion she refused to carry a package of drugs for him. The motivation for the killing clearly was not economic, but it was just as clearly connected with the drug business.

The boyfriend of the respondent in Incident 086 was much older than she and was well known in the neighborhood as a major heroin dealer. She worked for him as a carrier, delivering heroin to another part of the city. One night after midnight he asked her to go to pick up a package. She was afraid to go to that part of the city alone at that hour, and refused. He slapped her. A friend of his who was in the room held her down while he beat her. Then he ripped off her clothes and sodomized her. When he finished, he hit her again and sent her to bed.

While she lay in bed waiting for him, she kept thinking, "I'm gonna kill him." When he came to bed, she got up and went to get a gun that he always left under the dresser. She came back to bed and shot him twice. She took a handful of money and went to a shooting gallery where she "shot a lot." Then she turned herself in to the police.

WOMEN AND CRACK MARKET KILLING

In theory, the opportunities to gain an economic interest in the crack market should have been available to enterprising young women, who typically found opportunities only at the lower levels of the more firmly established drug markets (compare Brownstein et al. 1994; Fagan 1994; Inciardi et al. 1993; Sommers and Baskin 1992, 1993). As noted earlier, we do not attempt here to assess to what extent women took this opportunity and became active in the crack market. Rather, we focus on women who became involved in drug markets during this period, both as traffickers and as users, and we consider to what extent they engaged in lethal violence in

pursuit of their social and economic interests. Given the central role of violence as a means of social control in the crack markets, we would expect that women who acquired a stake in the early crack markets would have engaged in market-related violence.

Of the 19 killings related to the drug market, 10 (53%) involved crack as the primary drug. All of these took place between 1986 and 1991. One involved a dispute over territory; one, the collection of a drug debt; four, the robbery of a drug dealer; one, the "messing up" of drug money; and three, other reasons.

We have hypothesized that women with their own economic interest in the crack market would use lethal violence to protect that interest. In six of the 10 incidents involving crack, the respondent herself was a participant in the drug trade. In one of these (250) she denied involvement in the killing, and in another (399) she was the codefendant. In the remaining four cases the respondent said she was the killer. One of these was Incident 014, described earlier, in which the woman killed another dealer for his money and crack. The circumstances of the other three are described below.

In Incident 223, in 1990 a woman in the drug business killed someone who was working for her. Originally she had her own business selling drugs on the street, and felt safe in her belief that being on the street made it easy to get away if the police came. One day, when she was short of money and was looking for a way to make some, a bigger dealer offered her an opportunity to work for him selling crack from an apartment. She disliked inside selling because it made her feel trapped, and she feared the beatings workers faced when they were accused (rightly or wrongly) of "messing up the money." Still, the bigger dealer was insistent and she needed the money.

At first, things went well. The respondent was able to control her own smoking while on the job and she showed herself to be a responsible worker, taking care of herself and her son and not "messing up the money." The bigger dealer saw this and set her up in her own business, selling his drugs and hers. She started to make $6,000 daily, but the money was "possessing" her, and she spent days locked in the apartment from which she sold the crack. She told the bigger dealer, "I got to go home and take care of my personal needs." He agreed, but she had to leave another woman locked in the apartment to take care of business.

After a day of shopping, being with her son, and sleeping, the respondent became worried about her money and the woman she had left in the apartment. At 10 that night she returned and found that the other woman had given the bigger dealer his share of the earnings but did not have the respondent's $6,000.

On the way to the apartment, regular customers stopped the respondent to complain that bottles were short. To keep her customers satisfied, she paid $200 to the people who had complained. She also sent someone up to the apartment to cop two bottles, to see if in fact they were short. They were, so she went home and got her gun.

In her anger, the respondent returned to the apartment and demanded her money. Because the apartment was locked from the outside and only the bigger dealer had the key, she could not get in and the woman inside could not get out. They argued. She called to the woman, "You don't have none of my money?"

The woman responded, "I don't have nothing. I smoked it. I bought dope. I bought works. I bought cigarettes."

The respondent was enraged. As she left, the woman asked her if she would bring back a pack of cigarettes. That only made her angrier. She went to a liquor store and came back with a few dollar bottles of Bacardi. The apartment was lit with candles. She threw the Bacardi in and watched the apartment burst into flames while the other woman was locked inside. Before she walked across the street to watch, she said, "Smoke that."

In Incident 433 the respondent was a lieutenant in a crack-dealing business. She killed a worker who had absconded with company funds. In 1989 a man who was working for her ran off with $8,500, and it was her job to notify the boss. He told her, "You're the lieutenant and you're responsible for it. Either you find him and get the money back or kill him, or I'll have to kill you."

When the respondent found the man, she did not ask any questions. She did not ask for the money; she simply shot him in the eye and killed him. She was not arrested until a year later.

Incident 405 involved a woman who was the girlfriend of a crack dealer and had worked in his drug business. She killed a man who tried to rob her. On an "average day" in 1992, she decided to go out to cop some crack. In the past she had taken her drugs from her boyfriend, but that arrangement had led to too many fights. She still worked for him occasionally, but now she mostly hustled on her own. On this occasion she robbed a house of a television set and a video recorder, and sold the television to buy some crack. After she smoked the crack she ran into a man who knew that her boyfriend was a dealer. He assumed she was carrying for her boyfriend, and demanded that she give him money or drugs. When she said she had none, he called her a liar and began fighting with her. He took a knife from his pocket, it fell to the floor, and she grabbed it and stabbed him in the chest. Respondent 405 was involved infrequently in dealing crack for her boyfriend; she killed the man who attacked her because he thought she had an economic interest in the boyfriend's business. Respondents 223 and 433 are examples of women who killed in defense of their own economic interest in the crack trade. Both of these women, however, were someone else's employees rather than owners of their own business. It could be argued that these two women killed not only to defend their own interests but also in the economic interest of or from fear of their male employers.

Of the two women who were in the drug business and were involved in killings related to the crack trade, but denied being the killers, one (399) was present when a male associate did the killing. The other (250) claimed in the interview that a man killed her drug partner, and stated in the EOC report that she killed her partner over a fight with her boyfriend. These cases are described below.

Incident 399 involves a woman who participated in 1991 in the robbery of a drug dealer, during which the dealer was killed. The dealer was a friend to whom she had loaned $1,500 to set up his own crack house. The friend had worked previously for Dominican dealers, who wanted to kill him because he owed them money. With her loan he was supposed to buy drugs from her son and her nephew, who were in business as distributors, and to use his profits to pay back the Dominicans and the loan from the respondent.

The respondent discovered that her friend was buying most of his drugs directly from the Dominicans and cutting out her son and nephew. Also, he was not paying back the loan. She and her nephew and son decided to rob him: she would get her $1,500 and they would keep the rest. Then he would be unable to pay the Dominicans, and they would likely kill him.

On the night of the planned robbery, the respondent went in to the dealer's apartment first to buy $10 worth of crack. She thought too many people were present, but her nephew and son came in anyway. An argument ensued, and the dealer pulled a machete and cut the nephew badly. The nephew and the son took out their guns, and ended up killing the dealer and two other people in the room.

In case 250, a 32-year-old woman describes the 1989 killing of another woman who had "a good package of crack." The respondent was on hand when it happened, but her role in the killing is uncertain. She had started using drugs (inhalants) at age 14, and by the time of the killing she believed she had been dependent on crack for six years. During the years before her arrest for the homicide, she had used and sold drugs with her boyfriend. Because he was always high, she "was always taking care of business." He smoked everything he had and then tried to take her drugs. He also tried to take her money. They fought often.

Respondent 250 first became involved in selling drugs when she was 21. She was very successful; customers came constantly. Sometimes people from other areas came to her territory and tried to sell their drugs there. Whenever she saw them, an argument started. So, as she said, "a lot of times I might pull out the gun [I carried] to scare them off." At other times she might get into fights with people whose drug stash she had stolen, or with someone whose money she had "messed up" after they had given her drugs on consignment.

The woman who was killed was the respondent's drug-dealing partner. The victim was stabbed 17 times and shot between the eyes. According to the respondent, they had a fight over drug money in a residence hotel room. Fighting was not unusual, but they were best friends, and the respondent knew that her partner was really looking out for their mutual interests. The respondent used to get high, smoking up their product, and her partner did not.

On this occasion they stopped fighting and the respondent went downstairs, where someone approached her about buying crack. He told her he had heard that her partner had a good package upstairs. She took the money and brought it back upstairs to the room where her partner was. She claims she knocked on the door and no one answered. It was unlocked, and she pushed it open. Her partner never left it unlocked. The respondent looked around but did not see her partner. When she looked into the bedroom, she saw a body and ran into the hall, screaming. She returned to the room with her boyfriend, and they called the police.

About a month later, the police arrested the respondent for the killing. She told the interviewer that she thought her partner was robbed and killed by the man who sent her upstairs to get the drugs for him to buy. She told corrections officials that she was guilty of the killing, but that she did it "in an attempt to intervene in an argument between the victim and [her own] boyfriend." The boyfriend denied being present at the time.

In the remaining four crack market cases, the respondents reported that they were not participants in the drug business, but in two of these cases they said they were the killers. In one case (340) the respondent killed a female crack dealer who repeatedly harassed her and threatened her family after she called the police about the crack traffic in the neighborhood. She stabbed the dealer to death during a fight on the street.

In the other case (295) a woman killed her cousin, who stole the crack she was holding for a male dealer. The dealer had promised the respondent a cut of the profits for watching his drugs. According to the EOC, this was her first encounter with the criminal justice system but not the first time she had killed someone. In 1988 the courts decided she was justified in stabbing her boyfriend to death after he attacked her during a drunken incident, and she was not prosecuted. She also claims she stabbed her stepfather once, but that was not reported.

Respondent 295 told the interviewer that she started using alcohol and marijuana when she was a teenager, but was still using only powder cocaine and crack at the time of the killing. She never was involved in dealing drugs or any other crime, except for the month when she tried prostitution. She did that to earn money for drugs. At the time of the killing, she was holding crack in her house for somebody else. As she told the interviewer,

> I had possessions of money and drugs in my house that belonged to somebody else. My cousin smoked at the time, and she stole it—the drugs and the money. The person that it belonged to came for the money, which I didn't have. He threatened my life and put a gun to my head. I told him I'd try to get the money back. My cousin came back about four days later and she wanted to go to sleep, and I told her no. She acted like she ain't did nothing wrong. So I threw her clothes out the house and she tried to force her way back in, and that's when I stabbed her.

The woman originally was holding the drugs because the dealer promised to give her a cut if she did so. Thus one could argue that fear of reprisal for losing the drugs might have been her main motive, but protecting her own economic interest was also a consideration.

The two other cases that were related to the crack trade involved women who said they were not drug dealers and did not commit the killings. In Incident 078, a woman was present in 1988 when her boyfriend, a crack dealer, shot and killed two men and wounded another in a dispute over drug business territory. Incident 401 involved a woman who was trying to sell a gun to a drug dealer. She denies responsibility for the killing, but that conclusion is uncertain even according to her account of the circumstances.

Respondent 401 was convicted and sentenced for shooting and killing a drug dealer during the course of her dealings in a business that was tangential to the drug trade: she was trying to sell a gun to a crack dealer. One day in 1991, she was smoking crack and drinking alcohol when she came into possession of a gun that she believed she could sell. She took it to a woman she knew, whose boyfriend was a crack dealer. She negotiated with the dealer until he

agreed to buy the gun for $75. Then he decided he wanted his money back. He held the money and the gun, playing with the gun for about an hour before returning it.

Meanwhile a young boy who sold drugs for the dealer entered the building, and the respondent asked him if he wanted to buy the gun. He brought some friends to look at it. Suddenly the crack dealer asked to see it again. He took it, then slammed it into the respondent's hand. By her account, at the same time he turned the gun toward the boys and pushed the clip into it. The gun went off, and one boy was shot and killed. Then she remembers being beaten unconscious and left to take the blame for what she assumes was the punishment of a drug worker for "messing up the money."

DRUG MARKET KILLINGS AND ECONOMIC INTERESTS

Of the 19 women who killed in drug market situations, 10 killed in cases primarily involving the crack business, and only six of those were participants in the business. When all illicit drugs are considered, however, 11 of the 19 women were personally involved in the drug business. Beside the six who trafficked in crack, two were dealers of other drugs (045 and 265), two delivered drugs for someone else (086 and 441), and one guarded cocaine for a dealer (099). In addition, two women worked at an occupation that depended on the drug business for its profits; these women (369 and 440) were in the business of robbing cocaine or crack dealers. Thus 13 of the 19 women had an economic interest in the drug business.

Only one of the 13 women with an economic interest in the drug market killed for reasons that were clearly not economic. That was Respondent 086, who killed her dealer boyfriend. He raped her after she refused to carry drugs for him; the killing was in response to the abuse. In one other case (405), a woman killed someone who was trying to rob her, thinking she was running drugs or money for her boyfriend. According to her interview, she was more interested in defending herself than in protecting money or drugs—hers or her boyfriend's—so the significance of the economic interest is unclear.

The remaining 11 women clearly killed or participated in a killing to protect or enhance an economic interest. In four of these cases the economic interest was their own. Two reported that they were in business for themselves and did the killing on their own. Respondent 014, described earlier, told corrections officials that she was robbing a man of his money and crack to try to scare him into paying her the money he owed for her drugs. Respondent 265, also discussed earlier, was defending her own economic interest as a drug dealer when she shot and killed a man who tried to rob her while she was collecting money at her drug spot. A third respondent, 223, was employed by a male dealer but killed in her own interest. This woman set fire to a locked apartment from which another woman was selling drugs for her. The fourth respondent in this group, 045, was present when a male associate killed someone in her interest.

In Incident 045, the woman was a dealer who, in 1983, ran low on drugs and money while partying and getting high on cocaine, heroin, and marijuana. She remembered that a customer owed her $6,000 for cocaine she had sold him earlier in the week. She called him and asked:

> "You got my shit? You got my money and stuff?" So he starts giving me a song and dance routine and finally I'm like, "Look, forget about the bullshit, okay. It's over a week now. You said you was going to pay me in a week and I haven't heard anything." So he says, "Okay, come over. I'll have everything for you." I said, "Fine." Then when I come over to the house his girlfriend tells me that he left to go to Puerto Rico.

This made the respondent angry, especially because she recently had had to stop her people from killing his mother over a previous debt. She felt that "he played me for a fool." Her lover, who had accompanied her to get the money, was even angrier. He shot the woman who answered the door, the girlfriend of the man who owed the money.

The respondent claimed she was "shocked" when the killing took place. She said, "You know, being in the drug business you hear about people getting killed, and it's no biggie. But

actually being there and witnessing it for yourself is totally different." The EOC, however, tells another story. According to corrections records, this woman was involved in at least one other killing earlier in the same year. She was present when a male acquaintance shot a Colombian man twice in the head during a robbery. The murdered man apparently was a partner of the woman in a cocaine-trafficking business. According to the EOC, "The defendant felt that she had done nothing wrong in that she was merely a business-woman, admittedly dealing in cocaine for profit and enjoyment."

Four additional women who had an economic interest in the drug market killed or participated in a killing related to their mutual interest with others in that market. One respondent, 250, was linked to the killing of her drug-dealing partner, with whom she had argued earlier over her personal use of their product and whom she told the interviewer she found dead when she went to her to pick up drugs for a male customer. Respondent 399 was present when her nephew and son shot and killed three people during the robbery of a drug dealer who she claimed owed her money. She had loaned him that money to buy crack from a distribution business run by her son and nephew; as noted earlier, the dealer had used the money to buy his drugs from their competitors.

In both of the above cases, the woman was a dealer whose shared interest in the drug market was at stake. In the other two cases involving women who killed to protect or augment a shared interest, the women were robbers rather than drug dealers. In both of these cases the woman was the killer.

The respondent in Case 369 was visited in 1983 by two friends, who offered her $500 to drive a car for them during a robbery. She accepted the job, saying, "Shit, I do that shit with my eyes closed." They left to rob a shooting gallery where heroin and cocaine were sold and used.

Their plan was to gain entrance to the apartment as if they were potential customers. When the first member of the team was denied admittance, the respondent was given a gun to hold in case she needed to scare anyone. She went upstairs with the woman who had not been admitted alone. When the people inside saw the respondent, they opened the door and admitted her and her partner.

For a while they stood around getting anxious, while other people sat around getting high. Her partner went off with someone and got high too. At that point the respondent got nervous and said to her, "Come on. Let's do this shit and get the fuck out of here." Then she pulled the gun and yelled, "Everybody against the wall. This is a fucking stickup." A man in the room laughed at her and told her to stop playing around with the gun, which was already cocked, and to try some cocaine. When he saw she was serious, he grabbed the money in the cash box. He started to stand up, and she saw the handle of his gun. She got scared and started to back up toward the wall. Her arm hit the wall, and the gun went off. The man fell dead. Her friends came up, and they took the money and ran.

The respondent in Case 440 went out with her brother late one night in 1990 to rob a "coke spot." When they arrived, she was high on alcohol and "hyped" by the situation. "I'm not a woman now," she said. "I'm a man."

When they got to the coke spot, they announced the robbery, and words were exchanged. As respondent 440 recalls,

> I had noticed one of the guys that had been standing behind the scale went for his pocket, and I was always told, "Never allow anybody to move after the specific orders were given." By me being under the influence, I did what I did because of the simple fact [that] I didn't know what you were going in your pocket for 'cause I knew you was told "Don't nobody move, nobody get hurt." I told you not to get anything, just something in reach where I could see, but I didn't know what you had inside of your pocket. So when he went to go, I pistol-whipped him. When I pistol-whipped him, the bullet hit the next guy. Actually, the one I took his life, it wasn't called for. The bullet wasn't meant for him. The bullet wasn't meant for either one of them. It was to show [that] when orders are given, don't do nothing but what you are supposed to do.

As in the previous case, this respondent was a robber who shared her economic interest in the drug market with others with whom she preyed on drug dealers.

The three remaining women who were involved in killings related to drug market interests did not kill primarily in their own economic interest. Rather, they participated in killings in someone else's interest. Respondent 433, the lieutenant in a crack-dealing business, in 1989 killed a worker who had absconded with company funds. When she notified her boss, he told her it was her responsibility to get the money back or kill the absconder. She killed him. A second case (099) involved a woman who was guarding drugs for a dealer; the third (441) a woman who was delivering drugs for a dealer. These two cases are described below.

According to the EOC, Respondent 099 killed a man to get money in 1983, at the dawn of the crack era in New York. She needed the money for a cocaine dealer who had threatened her with the loss of a leg unless she paid him. She and a male accomplice robbed and killed a cab driver. The EOC does not make clear who pulled the trigger.

In the story she told to the interviewer, Respondent 099 was only a bystander. She had been using a variety of drugs since she was a teenager, and sold drugs a few times when she was 19. She worked out of her home for her boyfriend, selling drugs to people she knew. He used to "sell weight" between New York City and Washington, DC. On the day of the killing, after using "a lot" of cocaine and crack, she was spending the evening with her boyfriend.

Besides her boyfriend, with whom she was living at the time of the killing, Respondent 099 had a husband. For some reason she did not explain, she had been forced into an arranged marriage to this man, which she now blames for all her troubles. One evening a few days before the killing, she was alone in the apartment when her husband, from whom she was estranged, came by and begged her to drive to his uncle's house with him. She finally agreed, and they walked to his car, which was parked in the dark under a tree. Two men jumped out and began yelling at her but not at him. When she heard them "speaking the same damn Jamaican language" as her husband, she guessed they were his friends. The two men forced them back to her apartment, where they stole whatever drugs and money they could find. Then they and her husband left her alone. She wondered what she would tell her boyfriend, to whom the drugs and money belonged.

Respondent 099 acquired a pistol, and looked for her husband for two days. When she finally found him at his girlfriend's house, she was ready to kill him. He grabbed a small girl and held her in front of him. The woman would not shoot the child. He talked to her until she agreed to go with him to find the men who had taken her boyfriend's money and drugs. On the way, they apparently shot a cab driver to get the money to give back to her boyfriend. In this case her motive was fear rather than money.

Respondent 441 was delivering angel dust for a dealer when she became entangled in an argument that resulted in a fatal stabbing. She claims she was only a bystander.

One day in 1991, the respondent took an order for $250 worth of angel dust to be delivered to an apartment in the building where she lived. Against the dealer's better judgment, she chose to make the delivery herself. She was surprised when the door was opened by an older woman whom she recognized as a churchgoer. The older woman invited her in but stalled when she was asked for the $250.

A man who was obviously high came out of a back room and began to argue with the respondent. He did not want to pay and refused to take the drugs. She told him, "This is a house call. First of all, I didn't even charge you for a house call, being that I know the project. Second of all, you have to pay for it even if you don't want to take it."

The older woman and the man began to argue. He hit her. The respondent intervened, asking to get her money so she could leave. Then she claims she fell and was knocked unconscious. When she awoke, the older woman was dead on the floor, and the man was standing over her, mumbling. He saw the respondent, and she claims he threw her from the window. She fell six stories and broke her ankle and hip.

CONCLUSION

Our analysis shows that some of the women participating in a drug market economy killed to protect or extend their own interests in that economy. Of 10 killings committed in the especially violent crack market of the late 1980s and early 1990s, six involved women who were in the trade; five of those clearly killed for economic reasons. Among 13 cases of women who had a personal interest in the drug market economy involving any illicit drug, we found four in which the woman killed in her own economic interest, four in which she killed in a shared interest with a partner, and three in which she killed in someone else's interest.

Our findings support the conclusion that in the economic context of drug markets, women who kill do so for purposes unrelated to their gender. We found evidence that in such circumstances, women kill to protect or enhance their own economic interests. Yet, we also found evidence to support the conclusion that even in a setting which gives women who kill the opportunity and the rationale to act in their own self-interest, they often act in terms of their relationship to a man, killing on behalf of a man or out of fear of a man. Thus in these cases, the effect of gender is not necessarily altered by the circumstances of the drug market economy.

REFERENCES

Adler, F. (1975) *Sisters in Crime: The Rise of the New Female Criminal.* New York: McGraw Hill.

Bannister, S.A. (1991) "The Criminalization of Women Fighting Back against Male Abuse: Imprisoned Battered Women as Political Prisoners." *Humanity and Society* 15:400–16.

Belenko, S. (1990) "The Impact of Drug Offenders on the Criminal Justice System." In R.A. Weisheit (ed.), *Drugs, Crime and the Criminal Justice System*, pp. 27–28. Cincinnati: Anderson.

Block, K.J. (1990) "Age-Related Correlates of Criminal Homicides Committed by Women: A Study of Baltimore." *Journal of Crime and Justice* 13:42–65.

Browne, A. (1987) *When Battered Women Kill.* New York: Free Press.

Browne, A. and K.R. Williams (1989) "Exploring the Effect of Resource Availability and the Likelihood of Female-Perpetrated Homicides." *Law and Society Review* 23:75–94.

Brownstein, H.H., H.R.S. Baxi, P.J. Goldstein, and P.J. Ryan (1992) "The Relationship of Drugs,

Drug Trafficking, and Drug Traffickers to Homicide." *Journal of Crime and Justice* 15:25–44.

Brownstein, H.H. and P.J. Goldstein (1990) "A Typology of Drug-Related Homicides." In R.A. Weisheit (ed.), *Drugs, Crime and the Criminal Justice System*, pp. 171–92. Cincinnati: Anderson.

Brownstein, H.H., B.J. Spunt, S. Crimmins, P.J. Goldstein, and S. Langley (1994) "Changing Patterns of Lethal Violence by Women: A Research Note." *Women and Criminal Justice* 5:99–118.

Canestrini, K. (1987) *1986 Female Homicide Commitments*. Albany: New York State Department of Correctional Services.

Chesney-Lind, M. (1993) "Girls, Gangs and Violence: Anatomy of a Backlash." *Humanity and Society* 17:321–44.

Daly, K. and M. Chesney-Lind (1988) "Feminism and Criminology." *Justice Quarterly* 5:497–538.

Daniel, A.E. and P.W. Harris (1982) "Female Homicide Offenders Referred for Pre-Trial Psychiatric Examination: A Descriptive Study." *Bulletin of the American Academy of Psychiatry and Law* 10:261–69.

Ewing, C.P. (1987) *Battered Women Who Kill*. Lexington: Lexington Books.

Fagan, J. (1994) "Women and Drugs Revisited: Female Participation in the Cocaine Economy." *Journal of Drug Issues* 24:179–225.

Fagan, J. and K. Chin (1990) "Violence as Regulation and Social Control in the Distribution of Crack." In M. De La Rosa, E.Y. Lambert, and B. Gropper (eds.), *Drugs and Violence: Causes, Correlates, and Consequences*, pp. 8–43. Washington, DC: National Institute on Drug Abuse.

Falco, M. (1989) *Winning the Drug War: A National Strategy*. New York: Priority Press.

Federal Bureau of Investigation (1994) *Crime in the United States 1993*. Washington, DC: U.S. Department of Justice.

Goetting, A. (1987) "Homicidal Wives: A Profile." *Journal of Family Issues* 8:332–41.

(1988) "Patterns of Homicide among Women." *Journal of Interpersonal Violence* 3:332–41.

Goldstein, P.J., H.H. Brownstein, and P.J. Ryan (1992) "Drug-Related Homicide in New York: 1984 and 1988." *Crime and Delinquency* 38:459–76.

Goldstein, P.J., H.H. Brownstein, P.J. Ryan, and P.A. Bellucci (1989) "Crack and Homicide in New York City: A Conceptually Based Event Analysis." *Contemporary Drug Problems* 13:651–87.

Hazlett, M.H. and T.C. Tomlinson (1987) "Females Involved in Homicides: Victims and Offenders in Two U.S. Southern States." Presented at the annual meetings of the American Society of Criminology, Montreal.

Inciardi, J.A. (1989) "Beyond Cocaine: Basuco, Crack, and Other Coca Products." *Contemporary Drug Problems* 14:461–92.

Inciardi, J.A., D. Lockwood, and A.E. Pottieger (1993) *Women and Crack Cocaine*. New York: Macmillan.

Jacobs, S. and R.A. Rosen (1983) *Homicide in New York State, 1981*. Albany: New York State Division of Criminal Justice Services.

Johnson, B.D., A. Hamid, and H. Sanabria (1992) "Emerging Models of Crack Distribution." In T. Mieczkowski (ed.), *Drugs, Crime, and Social Policy: Research, Issues, and Concerns*, pp. 56–78. Boston: Allyn and Bacon.

Jones, A. (1980) *Women Who Kill*. New York: Holt, Rinehart and Winston.

Jurik, N.C. and R. Winn (1990) "Gender and Homicide: A Comparison of Men and Women Who Kill." *Violence and Victims* 5:227–42.

Luckenbill, D. (1977) "Criminal Homicide as a Situated Transaction." *Social Problems* 25:176–86.

Mann, C.R. (1984) *Female Crime and Delinquency*. University: University of Alabama Press.

(1986) "Getting Even? Women Who Kill in Domestic Encounters." Presented at the annual meetings of the American Society of Criminology, Atlanta.

Massing, M. (1989) "Crack's Destructive Sprint across America." *New York Times Magazine*, October 1, pp. 38, 40–41, 58, 60, 62.

Mercy, J.A. and L.E. Saltzman (1989) "Fatal Violence among Spouses in the United States, 1976–85." *American Journal of Public Health* 79:595–99.

Mieczkowski, T. (1990) "Crack Distribution in Detroit," *Contemporary Drug Problems*, 17:9–29.

Miller, E.M. (1986) *Street Women*. Philadelphia: Temple University Press.

Office of the Attorney General (1989) Drug Trafficking: *A Report to the President of the United States*. Washington, DC: U.S. Department of Justice.

Rasche, C.E. (1990) "Early Models for Contemporary Thought on Domestic Violence and Women Who Kill Their Mates: A Review of the Literature from 1895 to 1970." *Women and Criminal Justice* 1:31–53.

Reuter, P., R. MacCoun, and P. Murphy (1990) *Money from Crime: A Study of the Economics of Drug Dealing in Washington, D.C.* Santa Monica: RAND.

Riedel, M., M. Zahn, and L. Mock (1985) *The Nature and Patterns of American Homicide*. U.S. Department of Justice, National Institute of Justice.

Simon, R.J. (1975) *Women and Crime*. Lexington: Lexington Books.

Simon, Rita J. and J. Landis (1991) *The Crimes Women Commit, The Punishments They Receive*. Lexington: Lexington Books.

Smart, C. (1978) "The New Female Criminal: Myth and Reality." *British Journal of Criminology* 19:50–59.

Sommers, I. and D.R. Baskin (1992) "Sex, Race, Age, and Violent Offending." *Violence and Victims* 7:191–201.

(1993) "The Situational Context of Violent Female Offending." *Journal of Research in Crime and Delinquency* 30:136–62.

Sparrow, G. (1970) *Women Who Murder*. New York: Tower.

Spunt, B.J., P.J. Goldstein, C. Tarshish, M. Fendrich, and H.H. Brownstein (1993) "The Utility of Correctional Data for Understanding the Drugs-Homicide Connection: A Research Note," *Criminal Justice Review* 13:46–60.

Spunt, B.J., P.J. Ryan, P.J. Goldstein, and H.H. Brownstein (1989) "Current Research on the Drugs-Homicide Relationship." Presented at the annual meetings of the Academy of Criminal Justice Sciences, Washington, DC.

Swigert, V. Lynn and R.A. Farrell (1976) *Murder, Inequality, and the Law*. Lexington: Lexington Books.

Totman, J. (1978) *The Murderess: A Psychological Study of Criminal Homicide*. San Francisco: R&E Research Associates.

Walker, L.E. (1989) *Terrifying Love—Why Battered Women Kill and How Society Responds*. New York: Harper and Row.

Ward, D., M. Jackson, and R. Ward (1979) "Crimes of Violence by Women." In F. Adler and R. Simon (eds.), *The Criminology of Deviant Women*, pp. 14–38. Boston: Houghton Mifflin.

Weisheit, R.A. (1984) "Female Homicide Offenders: Trends over Time in an Institutionalized Population." *Justice Quarterly* 1:471–89.

Wilbanks, W. (1984) *Murder in Miami: An Analysis of Homicide Patterns and Trends in Dade County (Miami) Florida, 1917–1983*. Lanham: University Press of America.

Wolfgang, M.E. (1958) *Patterns in Criminal Homicide*. Philadelphia: University of Pennsylvania Press.

(1967) "Criminal Homicide and the subculture of Violence," In M.E. Wolfgang (ed.), *Studies in Homicide*, pp. 3–12. New York: Harper and Row.

26

Up It Up: Gender and the Accomplishment of Street Robbery

Jody Miller

In this article Miller analyzes the role that gender plays in women's participation in violence. She based her study on interviews with 37 active street robbers, 23 men and 14 women. She compared women's and men's rationale for their involvement in street robbery as well as the role that gender plays in the tactical commission of this violent activity. Miller finds that both sexes offer similar reasons for robbing people in public areas, but the methods they utilize to carry out the crime differ greatly. Much of this difference can be explained in the contents of a gender-stratified environment, one in which men are perceived to be physically stronger than women, and this affects the practical choices women make in carrying out robbery.

With the exception of forcible rape, robbery is perhaps the most gender differentiated serious crime in the United States. According to the Federal Bureau of Investigation's Uniform Crime Report for 1995, women accounted for 9.3% of robbery arrestees, while they were 9.5%, 17.7%, and 11.1% of arrestees for murder/manslaughter, aggravated assault, and burglary, respectively (Federal Bureau of Investigation, 1996). And while recently there has been considerable attention among feminist scholars to the question of why males are more violent than females, there have been few attempts to examine women's participation in these "male" crimes. Though their numbers are small, women who engage in violent street crime have something significant to teach us about women's place in the landscape of the urban street world.

Examining violence as masculine accomplishment can help account for women's lack of involvement in these crimes, just as this approach offers explanation for women's involvement in crime in ways scripted by femininity (e.g., prostitution). However, it leaves unexplained women's participation in violent

EDITOR'S NOTE: From Miller, J., "Up it up: Gender and the accomplishment of street robbery," in *Criminology, 36,* pp. 37-65. Copyright © 1998. Reprinted with permission from the American Society of Criminology.

street crime, except as an anomaly. Perhaps this is because femininity in this approach is conceived narrowly—specifically "within the parameters of the white middle class (i.e., domesticity, dependence, selflessness, and motherhood)" (Simpson and Elis, 1995:51). Given urban African-American women's historical patterns of economic self-sufficiency and independence, this passive feminine ideal is unlikely to have considerable influence and is "much more relevant (and restrictive) for white females" (Simpson and Elis, 1995:71).

The strength of the current study is its comparative analysis of women's *and* men's accounts of the accomplishment of one type of violent crime—street robbery. In comparing both the question of *why* women and men report engaging in robbery, and *how* gender organizes the commission of robbery, this research provides insight into the ways in which gender shapes women's involvement in what is perhaps the typification of "masculine" street crime. As such, it speaks to broader debates about women's place in the contemporary urban street world.

METHODOLOGY

The study is based on semistructured in-depth interviews with 37 active street robbers. The sample includes 14 women and a comparative sample of 23 men, matched approximately by age and age at first robbery. The respondents range in age from 16 to 46; the majority are in their late teens to mid-twenties. All of the men are African-American; 12 of the women are African-American and 2 are white.

Criteria for inclusion in the sample included the following: the individual had committed a robbery in the recent past, defined him- or herself as currently active, and was regarded as active by other offenders.

Respondents were recruited from impoverished urban neighborhoods in St. Louis. St. Louis typifies the midwestern city devastated by structural changes brought about by deindustrialization. With tremendous economic and racial segregation, population loss, and resulting social isolation, loss of community resources, and concentrated urban poverty

among African-Americans, the neighborhoods the respondents were drawn from are characteristic of "underclass" conditions (Sampson and Wilson, 1995; Wilson, 1996). These conditions no doubt shape respondents' offending through the interactive effects of structural barriers and resulting cultural adaptations (see Sampson and Wilson, 1995). Thus, they should remain in the foreground in examining the accomplishment of robbery.

MOTIVATIONS TO COMMIT ROBBERY

In this study, active robbers' articulation of the reasons they commit robbery is more a case of gender similarities than differences. What they get out of robbery, why they choose robbery instead of some other crime, why particular targets are appealing—the themes of these discussions are overlapping in women's and men's accounts. For both, the primary motivation is to get money or material goods. As Libbie Jones notes, "You can get good things from a robbery." For some, the need for money comes with a strong sense of urgency, such as when the individual is robbing to support a drug addiction—a situation more prevalent among older respondents than younger ones. But for the majority of women and men in this sample, robberies are committed to get status-conferring goods such as gold jewelry, spending money, and/or for excitement.

If anything, imperatives to gain money and material goods through robbery appear to be stronger for males than females, so that young men explain that they sometimes commit robberies because they feel some economic pressure, whereas young women typically do not. Masculine street identity is tied to the ability to have and spend money, and included in this is the appearance of economic self-sufficiency. Research has documented women's support networks in urban communities, including among criminally involved women (see Maher, 1997; Stack, 1974). This may help explain why the imperative for young men is stronger than for young women: Community norms may give women wider latitude for obtaining material goods and economic support from a variety of sources, including other females, family members, and boyfriends; whereas the pressure of society's view of men as breadwinners differentially affects men's emotional experience of relying on others economically. This may explain why several young men specifically describe that they do not like relying on their parents in order to meet their consumer needs. As Mike J. notes, "My mother, she gives me money sometimes but I can't get the stuff like I want, clothes and stuff . . . so I try to get it by robbery." Though both males and females articulate economic motives for robbery, young men, more than young women, describe feeling compelled to commit robberies because they feel "broke."

Asked to explain why they commit robberies instead of other crimes with similar economic rewards, both women and men say that they choose robberies, as Cooper explains, because "it's the easiest." Libbie Jones reports that robbery provides her with the things she wants in one quick and easy step:

> I like robbery. I like robbery 'cause I don't have to buy nothing. You have a herringbone, I'm gonna take your herringbone and then I have me a herringbone. I don't have to worry about going to the store, getting me some money. If you got some little earrings on I'm gonna get 'em.

The ease with which respondents view the act of robbery is also reflected in their choice of victims—most frequently other street-involved individuals, who are perceived as unlikely to be able to go to the police, given their own criminal involvement. In addition, these targets are perceived as likely to have a lot of money, as well as jewelry and other desirable items. Less frequently, respondents report targeting individuals who are perceived as particularly easy marks, such as older citizens. However, most robberies, whether committed by females or males, occur in the larger contexts of street life, and their victims reflect this—most are also involved in street contexts, either as adolescents or young adults who hang out on the streets and go to clubs, or as individuals involved (as dealers and/or users) in the street-level drug economy. Because of this, it is not uncommon for robbers to know or at least know of their victims (for more on target selection, see Wright and Decker, 1997:Ch. 3).

In addition to the economic incentives that draw the respondents toward robbery, many also derive a psychological or emotional thrill from committing robberies. Little Bill says, "when my first robbery started, my second, the third one, it got more fun . . . if I keep on doing it I think that I will really get addicted to it." Likewise, Ne-Ne's comment illustrates the complex dynamics shaping many respondents' decisions to commit robberies, particularly the younger ones: "I don't know if it's the money, the power or just the feeling that I know that I can just go up and just take somebody's stuff. It's just a whole bunch of mixture type thing." Others describe a similar mixture of economic and emotional rewards. Buby notes, "you get like a rush, it be fun at the time."

When individuals on the street are perceived as "high-catting" or showing off, they are viewed by both male and female robbers as deserving targets. Ne-Ne describes the following dialogue between herself and a young woman she robbed: "[The girl] said 'if you take my money then I'm gonna get in trouble because this is my man's money.' He told you to keep it, not showboat. You talking 'nigger I got $800 in my pocket,' pulling it out. Yeah, you wanted us to know." Likewise, describing a woman he robbed at a gas station, Treason Taylor says, "really I didn't like the way she came out. She was like pulling out all her money like she think she hot shit." A few respondents even specifically target people they don't like, or people who have insulted or hurt them in the past.

For both women and men, then, motivations to commit robbery are primarily economic—to get money, jewelry, and other status-conferring goods, but they also include elements of thrill seeking, attempting to overcome boredom, and revenge. Most striking is the continuity across women's and men's accounts of their motives for committing robbery, which vary only by the greater pressure reported by some young men to have their own money to obtain material goods.

WOMEN'S ENACTMENTS OF STREET ROBBERY

The women in the sample describe three predominant ways in which they commit robberies: targeting female victims in physically confrontational robberies, targeting male victims by appearing sexually available, and participating with males during street robberies of men. Ten women (71%) describe targeting female victims, usually on the streets but occasionally at dance clubs or in cars. Seven (50%) describe setting up men through promises of sexual favors, including two women who do so in the context of prostitution. Seven (50%) describe working with male friends, relatives, or boyfriends in street robberies; three (21%) report this as their exclusive form of robbery.

Robbing Females

The most common form of robbery reported by women in the study is robbing other females in a physically confrontational manner. Ten of the 14 female respondents report committing these types of offenses. Of those who do not, three only commit robberies by assisting men, whose targets are other males, and one only robs men in the context of prostitution. Typically, women's robberies of other females occur on the streets, though a few young women also report robbing females in the bathrooms or parking lots of clubs, and one robs women in cars. These robberies are sometimes committed alone, but usually in conjunction with one or several additional women, but not in conjunction with men. In fact, Ne-Ne says even when she's out with male friends and sees a female target, they don't get involved: "They'll say 'well you go on and do her.'"

Most robberies of females either involve no weapon or they involve a knife. Four women report having used a gun to rob women, only one of whom does so on a regular basis. Women are the victims of choice because they are perceived as less likely to be armed themselves and less likely to resist or fight back. CMW explains, "See women, they won't really do nothing. They say, 'oh, oh, ok, here take this.' A dude, he might try to put up a fight." Yolanda Smith reports that she only robs women because "they more easier for me to handle." Likewise, Libbie Jones says, "I wouldn't do no men by myself," but she says women victims "ain't gonna do nothing because they be so scared." The use of weapons in these assaults is often not

deemed necessary. Quick explains that she sometimes uses a knife, "but sometimes I don't need anything. Most of the time it be girls, you know, just snatching they chains or jewelry. You don't need nothing for that." Quick has also used a gun to rob another female. She and a friend were driving around when they spotted a young woman walking down the street with an expensive purse they liked. "We jumped out of the car. My friend put a gun up to her head and we just took all of her stuff." However, this approach was atypical.

On occasion, female victims belie the stereotype of them and fight back. Both Janet Outlaw and Ne-Ne describe stabbing young women who resisted them. Janet Outlaw describes one such encounter:

> This was at a little basketball game. Coming from the basketball game. It was over and we were checking her out and everything and she was walking to her car. I was, shit fuck that, let's get her motherfucking purse. Said let's get that purse. So I walked up to her and I pulled out the knife. I said "up that purse." And she looked at me. I said "shit, do you think I'm playing? Up that purse." She was like "shit, you ain't getting my purse. Do what you got to do." I was like "shit, you must be thinking I'm playing." So I took the knife, stabbed her a couple of times on the shoulder, stabbed her on the arm and snatched the purse. Cut her arm and snatched the purse. She just ran, "help, help." We were gone.

Ne-Ne describes a similar incident that occurred after an altercation between two groups of young women. When one young woman continued to badmouth her, she followed the girl to her car, pulled out a knife, "headed to her side and showed the bitch [the knife]." The girl responded, "I ain't giving you shit," and Ne-Ne said, "please don't make me stick you." Then, "She went to turn around and I just stuck it in her side . . . She was holding her side, just bleeding. And so when she fell on the ground one of my partners just started taking her stuff off of her. We left her right there."

As with pulling guns on women, stabbing female victims is a rare occurrence. Nonetheless, women's robbery of other women routinely involves physical confrontation such as hitting, shoving, or beating up the victim. Describing a recent robbery, Nicole Simpson says, "I have bricks in my purse and I went up to her and hit her in the head and took her money." Kim Brown says that she will "just whop you and take a purse but not really put a gun to anybody's face." Libbie Jones says she has her victims throw their possessions on the ground, "then you push 'em, kick 'em or whatever, you pick it up and you just burn out." Likewise, CMW describes a recent robbery:

I was like with three other girls and we was like all walking around . . . walking around the block trying to find something to do on a Saturday night with really nothing to do and so we started coming up the street, we didn't have no weapons on us at the time. All we did was just start jumping on her and beating her up and took her purse.

According to Janet Outlaw, "We push 'em and tell them to up their shit, pushing 'em in the head. Couple of times we had to knock the girls down and took the stuff off of them." She explains the reason this type of physical force is necessary: "It's just a woman-to-woman thing and we just like, just don't, just letting them know like it is, we let them know we ain't playing." As discussed below, this approach is vastly different from women's approaches when they rob men, or when they commit robberies with males. It appears to be, as Janet Outlaw says, "a woman-to-woman thing."

As noted above, sometimes female-on-female robberies occur in or around night clubs, in addition to on the streets. Libbie Jones explains, "you just chill in the club, just dance or whatever, just peep out people that got what you want. And then they come out of the club and you just get them." Likewise, Janet Outlaw says, "we get a couple of drinks, be on the blow, party, come sit down. Then be like, damn, check that bitch out with all this shit on." Libbie Jones came to her interview wearing a ring she had gotten in a robbery at a club the night before, telling the interviewer, "I like this on my hand, it looks lovely." She describes the incident as follows:

This girl was in the bathroom. I seen the rings on her hands. Everybody was in there talking and putting their makeup on, doing their hair. So I went and got my godsister. She came back with her drink. She spilled it on her and she was like, "oh, my fault, my fault." She was wiping it off her. I pulled out my knife and said "give it up." The girl was taking the rings off her hand so when we got the rings we bounced up out of the club.

Though most of the women who rob females are teenagers or young adults and rob other young women, two women in the sample—Lisa Wood and Kim Brown—also describe targeting middle-aged or older citizens. It is notable that both are older (in their late 30s) and that both describe robbing in order to support drug habits, which make them more desperate. As with the younger women who choose to rob other young women because they believe them unlikely to resist, both of these women choose older targets because they won't fight back. Lisa Wood says sometimes they accomplish these robberies of non-street-involved citizens by getting victims to drop their guard when they are coming out of stores. She describes approaching the person, "say 'hi, how you doing,' or 'do you need any help?' A lot of times they will say yeah. They might have groceries to take to they car and get it like that." She says once they drop their guard she will "snatch they purse and take off running."

To summarize, notable elements of women's robberies of other women are that they most frequently occur within street-oriented settings, do not include male accomplices, and typically involve physical force such as hitting, shoving and kicking, rather than the use of a weapon. When weapons are used, they are most likely to be knives. In these contexts, women choose to rob other females rather than males because they believe females are less likely to fight back; they typically do not use weapons such as guns because they perceive female targets as unlikely to be armed.

SETTING UP MALES BY APPEARING SEXUALLY AVAILABLE

Women's robberies of men nearly always involve guns. They also do not involve physical contact. Janet Outlaw, who describes a great deal of physical contact in her robberies of other women (see above), describes her robberies of men in much different terms: "If we waste time touching men there is a possibility that they can get the gun off of us, while we wasting time touching them they could do anything. So we just keep the gun straight on them. No touching, no moving, just straight gun at you." The circumstances surrounding the enactment of female-on-male robberies differ as well. The key, in each case, is that women pretend to be sexually interested in their male victims, whose guard drops, providing a safe opportunity for the crime to occur. Two women—Jayzo and Nicole Simpson—rob men in the context of prostitution. The other five typically choose a victim at a club or on the streets, flirt and appear sexually interested, then suggest they go to a hotel, where the robbery takes place. These robberies may involve male or female accomplices, but they are just as likely to be conducted alone.

Nicole Simpson prostitutes to support her drug habit, but sometimes she "just don't be feeling like doing it," and will rob her trick rather than complete the sexual transaction. Sometimes she does this alone, and other times has a female accomplice. She chooses tricks she feels will make safe victims. She explains, "like I meet a lot of white guys and they be so paranoid they just want to get away." When Nicole Simpson is working alone, she waits until the man is in a vulnerable position before pulling out her knife. As she explains, "if you are sucking a man's dick and you pull a knife on them, they not gonna too much argue with you." When she works with a female partner, Nicole Simpson has the woman wait at a designated place, then takes the trick "to the spot where I know she at." She begins to perform oral sex, then her partner jumps in the car and pulls a knife. She explains, "once she get in the car I'll watch her back, they know we together. I don't even let them think that she is by herself. If they know it's two of us maybe they won't try it. Because if they think she by herself they might say fuck this, it ain't nothing but one person." Jayzo's techniques parallel those of Nicole Simpson, though she uses a gun instead of a knife and sometimes takes prospective tricks to hotels in addition to car dating.

Young women who target men outside the context of prostitution play upon the men's beliefs about women in order to accomplish these robberies—including the assumptions that women won't be armed, won't attempt to rob them, and can be taken advantage of sexually. Quick explains, "they don't suspect that a girl gonna try to get 'em. You know what I'm saying? So it's kind of easier 'cause they like, she looks innocent, she ain't gonna do this, but that's how I get 'em. They put they guard down to a woman." She says when she sets up men, she parties with them first, but makes sure she doesn't consume as much as them. "Most of the time, when girls get high they think they can take advantage of us so they always, let's go to a hotel or my crib or something." Janet Outlaw says, "they easy to get, we know what they after—sex." Likewise, CMW and a girlfriend often flirt with their victims: "We get in the car then ride with them. They thinking we little freaks . . . whores or something." These men's assumptions that they can take advantage of women lead them to place themselves at risk for robbery. CMW continues: "So they try to take us to the motel or whatever, we going for it. Then it's like they getting out of the car and then all my friend has to do is just put the gun up to his head, give me your keys. He really can't do nothing, his gun is probably in the car. All you do is drive on with the car."

Several young women report targeting men at clubs, particularly dope dealers or other men who appear to have a lot of money. Describing one such victim, Janet Outlaw says she was drawn to him because of his "jewelry, the way he was dressed, little snakeskin boots and all . . . I was like, yeah, there is some money." She recounts the incident as follows:

> I walked up to him, got to conversating with him. He was like, "what's up with you after the club?" I said "I'm down with you, whatever you want to do." I said "we can go to a hotel or something." He was like "for real?" I was like, "yeah, for real." He was like "shit, cool then." So after the club we went to the hotel. I had the gun in my purse. I followed him, I was in my own car, he was in his car. So I put the gun in my purse and went up to the hotel, he was all ready. He was posted, he was a lot drunk. He was like, "you smoke weed?" I was like, "yeah shit, what's up." So we got to smoking a little bud, he got to taking off his little shit,

laying it on a little table. He was like, "shit, what's up, ain't you gonna get undressed?" I was like "shit, yeah, hold up" and I went in my purse and I pulled out the gun. He was like "damn, what's up with you, gal?" I was like, "shit, I want your jewelry and all the money you got." He was like, "shit, bitch you crazy. I ain't giving you my shit." I said, "do you think I'm playing, nigger? You don't think I'll shoot your motherfucking ass?" He was like, "shit, you crazy, fuck that, you ain't gonna shoot me." So then I had fired the thing but I didn't fire it at him, shot the gun. He was like "fuck no." I snatched his shit. He didn't have on no clothes. I snatched the shit and ran out the door. Hopped in my car.

Though she did this particular robbery alone, Janet Outlaw says she often has male accomplices, who follow her to the hotel or meet her there. While she's in the room, "my boys be standing out in the hallway," then she lets them in when she's ready to rob the man. Having male backup is useful because men often resist being robbed by females, believing that women don't have the heart to go through with what's necessary if the victim resists. Janet Outlaw describes one such incident. Having flirted with a man and agreed to meet him, she got in his car, then pulled her gun on him:

> I said "give me your stuff." He wasn't gonna give it to me. This was at nighttime. My boys was on the other side of the car but he didn't know it. He said "I ain't gonna give you shit." I was like, "you gonna give me your stuff." He was like "I'll take that gun off of your ass." I was like, "shit, you ain't gonna take this gun." My boy just pulled up and said, "give her your shit." I got the shit.

In the majority of these robberies, the victim knows that the woman has set him up—she actively participates in the robbery. Ne-Ne also describes setting up men and then pretending to be a victim herself. Her friends even get physical with her to make it appear that she's not involved. She explains:

> I'll scam you out and get to know you a little bit first, go out and eat and let you tell me where we going, what time and everything. I'll go in the restroom and go beep them [accomplices] just to let them know what time we leaving from wherever we at so they can come out and do their little robbery type thing, push me or whatever. I ain't gonna leave with them 'cause then he'll know so I still chill with him for a little while.

Only Ne-Ne reports having ever engaged in a robbery the opposite of this—that is, one in which her male partners flirted with a girl and she came up and robbed her. She explains:

> I got some [male friends] that will instigate it. If I see some girl and I'm in the car with a whole bunch of dudes, they be like "look at that bitch she have on a leather coat." "Yeah, I want that." They'll say "well why don't you go get it?" Then you got somebody in the back seat [saying] "she's scared, she's scared." Then you got somebody just like "she ain't scared, up on the piece" or whatever and then you got some of them that will say well, "we gonna do this together." It could be like two dudes they might get out like "what's up baby," try to holler at her, get a mack on and they don't see the car. We watching and as soon as they pulling out they little pen to write they number, then I'll get out of the car and just up on them and tell them, the dudes be looking like, damn, what's going on? But they ain't gonna help 'cause they my partners or whatever.

Street Robberies With Male Robbers

As the previous two sections illustrate, women's accomplishment of robbery varies according to the gender of their victims. As a rule, women and men do not rob females together, but do sometimes work together to set up and rob males. In addition, half of the women interviewed describe committing street robberies—almost always against males—with male accomplices. In these robberies, women's involvement either involves equal participation in the crime or assisting males but defining their role as secondary. Three women in the sample—Buby, Tish, and Lisa Jones—describe working with males on the streets as their only form of robbery, and each sees her participation as secondary. The rest engage in a combination of robbery types, including those described in the previous two sections, and do not distinguish their roles from the roles of male participants in these street robberies.

Lisa Jones and Tish each assist their boyfriends in the commission of robberies; Buby goes along with her brother and cousins. Lisa Jones says "most of the time we'll just be driving around and he'll say 'let's go to this neighborhood and rob somebody.'" Usually she stays in the car while he approaches the victim, but she is armed and will get out and assist when necessary. Describing one such incident, she says, "One time there was two guys and one guy was in the car and the other guy was out of the car and I seen that one guy getting out of the car I guess to help his friend. That's when I got out and I held the gun and I told him to stay where he was." Likewise Buby frequently goes on robberies with her brother and cousins but usually chooses to stay in the car "because I be thinking that I'm gonna get caught so I rather stay in the back." She has never done a robbery on her own and explains, "I know what to do but I don't know if I could do it on my own. I don't know if I could because I'm used to doing them with my brother and my cousins." Though her role is not an active one, she gets a cut of the profits from these robberies.

Tish and Lisa Jones are the only white respondents in the study. Each robs with an African-American boyfriend, and—though they commit armed robberies—both reject the view of themselves as criminals. Lisa Jones, for instance, downplays her role in robberies, as the following dialogue illustrates:

Interviewer: How many armed robberies have you done in your life?

Lisa Jones: I go with my boyfriend and I've held the gun, I've never actually shot it.

Interviewer: But you participate in his robberies?

Lisa Jones: Yeah.

Interviewer: How many would you say in your whole life?

Lisa Jones: About fifteen.

Interviewer: What about in the last month?

Lisa Jones: Maybe five or six.

Interviewer: What other crimes have you done in your life, or participated with others?

Lisa Jones: No, I'm not a criminal.

It is striking that this young woman routinely engages in robberies in which she wields a weapon, yet she defines herself as "not a criminal." Later in the interview, she explains that she would stop participating in armed robberies "if I was to stop seeing him." She and Tish are the only respondents who minimize the implications of their involvement in armed robbery, and it is probably not coincidental that they are young white women—their race and gender allow them to view themselves in this way.

Both also describe their boyfriends as the decision makers in the robberies—deciding when, where, and whom to rob. This is evident in Tish's interview, as her boyfriend, who is present in the room, frequently interjects to answer the interviewer's questions. The following dialogue is revealing:

Interviewer: How do you approach the person?

Tish: Just go up to them.

Interviewer: You walk up to them, you drive up to them?

Boyfriend: Most of the time it's me and my partner that do it. Our gals, they got the guns and stuff but we doing most of the evaluating. We might hit somebody in the head with a gun, go up to them and say whatever. Come up off your shit or something to get the money. The girls, they doing the dirty work really, that's the part they like doing, they'll hold the gun and if something goes wrong they'll shoot. We approach them. I ain't gonna send my gal up to no dude to tell him she want to rob him, you know. She might walk up to him with me and she might hit him a couple of times but basically I'm going up to them.

These respondents reveal the far end of the continuum of women's involvement in robbery, clearly taking subordinate roles in the crime and defining themselves as less culpable as a result. Tish's boyfriend also reveals his perception of women as secondary actors in the accomplishment of robbery. For the most part, other women who participate in street robberies with male accomplices describe themselves as equal participants. Older women who rob citizens to support their drug habits at times do so with male accomplices. For instance, Lisa Woods sometimes commits her robberies with a male and female accomplice and targets people "like when they get they checks. Catch them coming out of the store, maybe trip 'em, go in they pocket and take they money and take off running." Among the younger women, robberies with male accomplices involve guns and typically come about when a group of people are driving around and spot a potential victim. Janet Outlaw describes a car jacking that occurred as she and some friends were driving around:

> Stop at a red light, we was looking around, didn't see no police, we was right behind them [the victims] . . . So one of my boys got out and I got out. Then the other boy got up in the driver's seat that was with them. My boy went on one side and I went on the other side and said "nigger get out of the car before we shoot you." Then the dudes got out. It was like, shit, what's up, we down with you all. No you ain't down with us, take they jewelry and shit off. It was like, damn, why you all tripping? Then my boy cocked the little gun and said take it off now or I'm gonna start spraying you all ass. So they took off the little jewelry, I hopped in, put it in drive and pulled on off.

Likewise, Ne-Ne prefers committing street robberies with males rather than females. She explains:

> I can't be bothered with too many girls. That's why I try to be with dudes or whatever. They gonna be down. If you get out of the car and if you rob a dude or jack somebody and you with some dudes then you know if they see he tryin' to resist, they gonna give me some help. Whereas a girl, you might get somebody that's scared and might drive off. That's the way it is.

It is not surprising, then, that Ne-Ne is the only woman interviewed to report having ever committed this type of street robbery of a male victim on her own. Her actions parallel those of male-on-male robbers described above. Ne-Ne explicitly indicates that his robbery was

possible because the victim did not know she was a woman. Describing herself physically, she says, "I'm big, you know." In addition, her dress and manner masked her gender. "I had a baseball cap in my car and I seen him . . . I just turned around the corner, came back down the street, he was out by his-self and I got out of the car, had the cap pulled down over my face and I just went to the back and upped him. Put the gun up to his head." Being large, wearing a ballcap, and enacting the robbery in a masculine style (e.g., putting a gun to his head) allowed her to disguise the fact that she was a woman and thus decrease the victim's likelihood of resisting. She says, "He don't know right now to this day if it was a girl or a dude."

DISCUSSION

What is most notable about the current research is the incongruity between motivations and accomplishment of robbery. While a comparison of women's and men's motivations to commit robbery reveals gender similarities, when women and men actually commit robbery their enactments of the crime are strikingly different. These differences highlight the clear gender hierarchy that exists on the streets. While some women are able to carve out a niche for themselves in this setting, and even establish partnerships with males, they are participating in a male-dominated environment, and their actions reflect an understanding of this.

To accomplish robberies successfully, women must take into account the gendered nature of their environment. One way they do so is by targeting other females. Both male and female robbers hold the view that females are easy to rob, because they are less likely than males to be armed and because they are perceived as weak and easily intimidated. Janet Outlaw describes women's robbery of other women as "just a woman to woman thing." This is supported by Ne-Ne's description that her male friends do not participate with her in robberies of females, and it is supported by men's accounts of robbing women. While women routinely rob other women, men are less likely to do so, perhaps because these robberies do not result in the demonstration of masculinity.

In sum, the women in this sample do not appear to "do robbery" differently than men in order to meet different needs or accomplish different goals. Instead, the differences that emerge reflect practical choices made in the context of a gender-stratified environment—one in which, on the whole, men are perceived as strong and women are perceived as weak.

Motivationally, then, it appears that women's participation in street violence can result from the same structural and cultural underpinnings that shape some of men's participation in these crimes, and that they receive rewards beyond protection for doing so. Yet gender remains a salient factor shaping their actions, as well as the actions of men.

REFERENCES

Federal Bureau of Investigation
1996   Crime in the United States, 1995. Washington, D.C.: U. S. Government Printing Office.

Maher, Lisa
1997   Sexed Work: Gender, Race and Resistance in a Brooklyn Drug Market. Oxford: Clarendon Press.

Sampson, Robert J. and William Julius Wilson
1995   Toward a theory of race, crime, and urban inequality. In John Hagan and Ruth D. Peterson (eds.), Crime and Inequality. Stanford, Calif.: Stanford University Press.

Simpson, Sally and Lori Elis
1995   Doing gender: Sorting out the caste and crime conundrum. Criminology 33(1):47–81.

Sommers, Ira and Deborah R. Baskin
1993   The situational context of violent female offending. Journal of Research on Crime and Delinquency 30(2): 136–162.

Wilson, William Julius
1996   When Work Disappears: The World of the New Urban Poor. New York: Alfred A. Knopf.

Wright, Richard T. and Scott Decker
1997   Armed Robbers in Action: Stickups and Street Culture. Boston: Northeastern University Press.

27

A Successful Female Crack Dealer: A Case Study of a Deviant Career

Eloise Dunlap, Bruce D. Johnson, and Ali Manwar

The exploration of a middle-class, minority crack dealer is presented as a case study by Dunlap, Johnson, and Manwar. Rachel, as the authors call her, is a deviant among deviants in that she operates in a male-dominated business whose clientele is composed of hidden crack users as opposed to street addicts, uses operating techniques more associated with middle-class drug dealers than is more typically found in her inner-city residence, and is a crack addict herself, who has managed to avoid arrest and detection in her drug dealing operation. In the world of crack cocaine dealing, Rachel surely represents an anomaly. She is a college graduate who abandoned the rewards of a legitimate life for drug sales.

This paper presents a case study of Rachel (a pseudonym), a petite, attractive, out-going African-American woman with a moderate Afro and fashionable appearance who looks much younger than her 40 years. A successful Harlem crack dealer, Rachel is a deviant among deviants, for several reasons. First, she is female in a traditionally male occupation. Second, her educational level, middle-class background, and professionalism are atypical of drug dealers. Third, her drug-related activities are distinctly different from those of both male crack dealers and other female dealers studied

(cf. Dunlap and Johnson 1992a, 1992b; Maher and Curtis 1992): her thriving business serves not the stereotypical young, male addict but the older, employed, "hidden" user; and even though her activities occur in the depths of the inner city, she conducts her business according to practices common among middle-class dealers. Finally, Rachel is herself an addict who uses sizable amounts of crack several times a day, yet she has avoided both arrest and the usual consequences of personal crack use. Neatly balancing the various roles and competing social expectations placed on her, she stands at the intersection of the straight and drug subcultures, between middle-class and inner-city drug dealers, between drug abusers and more casual users.

Careful analysis of deviant cases often provides important insights, both theoretical and practical, about little understood phenomena (Becker 1963). Particularly helpful are informants who, like Rachel, are articulate enough to discuss their deviant behavior and explain the pressures and rewards that mold it. Rachel's case illustrates how the market for illegal drugs has shifted in the United States from marijuana to cocaine to crack; how drugs permeate inner-city communities, reaching older, better educated, middle-class professional people as well as the young, the uneducated, and the poor; how women fit into a sphere of activity heavily dominated by men; and how drug dealers justify their illegal activities and cope with moral opprobrium.

METHODS

During ethnographic field work between 1989 and 1992, the senior author (Dunlap) developed close relationships with many crack dealers and their families. Her introduction to Rachel, whose cooperation was to prove crucial to the research, was arranged by a dealer for whom several others worked. With a college degree in psychology, Rachel quickly grasped the intent of the study and the importance of the issues to be examined. She was outgoing and friendly, and mutual rapport and trust soon emerged. As well as securing the ethnographer's personal safety (Williams et al. 1992), she also assured fellow dealers of the legitimacy of the research, which encouraged them to participate fully. Her story, elicited over a one-year period in ethnographic interviews and direct observations at her apartment (also her place of business), generated more than 700 pages of transcripts from tapes and field notes. The quoted material below is drawn from these transcripts.

In the world of illegal drug sales, the opposition of government and the absence of formal training means that individuals must discover by themselves how to deal with the complex contingencies involved in selling drugs (Faupel

1981; Waldorf et al. 1991). They must learn how to obtain supplies of high quality drugs to sell; create retail sales units; recruit buyers; avoid arrest, incarceration, and violence from competitors or customers; and handle and account for large amounts of cash, while evading both formal and informal sanctions. Individual dealers must develop informal rules or norms of conduct by which to conduct business (Zinberg 1984), determining what kinds of suppliers and customers to seek out; where and when to conduct sales; how to avoid the police and competitors; how to moderate personal crack use; how to spend cash income; and how to obtain shelter, food, and clothing. Perhaps their most difficult challenge, however, is to limit their own drug consumption so that they sell enough to "re-up," or purchase more wholesale units of drugs (Hamid 1992; Johnson et al. 1992).

DEVELOPMENT OF A DRUG-DEALING CAREER

Rachel's Early Involvement With Drugs

As a child, Rachel moved from Mississippi to Harlem, where at 15 she met her future husband, a heroin and marijuana dealer with several uncles in the illegal drug business. She became pregnant at 16, gave birth to a daughter, and married the baby's father a year later. As a teenage mother, Rachel remained uninvolved in drug use or sales, yet through her husband's activities she was introduced to drug dealing and a network of users and dealers. Her husband also provided a model of behaviors that contributed to his success in drug dealing: legitimate employment in addition to illegal activities; separation of drug involvement and family life; and expenditure of extra time and money on family rather than on street life.

After her daughter's birth, Rachel enrolled in an educational program targeting the poor, holding a program-related job as she worked toward her high school diploma. By 20 she had earned a high school equivalency degree. Then, when she was 21 and had been happily married for 5 years, her husband died suddenly of a kidney infection, leaving her to care alone for her

5-year-old daughter as well as her alcoholic mother. Rachel had negotiated adolescence without using drugs, had completed high school, held a legitimate job, lived in a household that was affluent by Harlem standards, and had well-developed household management and child-rearing skills. In short, by 21 she had acquired resources (housing and household furnishings, limited savings, close relationships with her husband's relatives) and skills (household maintenance, budgeting and accounting, child care, interpersonal negotiation) that would be valuable in her future career.

As a single black parent and informal custodian of an alcoholic mother, Rachel's new position placed conflicting demands upon her. In her bereavement, she began going to parties and smoking marijuana.

> With the marijuana, you know, it . . . started out as a social situation, you know . . . I would say that for myself, I was probably a late person, in trying any of . . . any of these drugs at all. You could go to a party, you know, when you smoking marijuana, everybody smoking it, you know, its, really—uh—nothing that you want to get over, you don't have worry about going into a special room or anything like that, you know. A lot of clubs and places, you know, and—um—a lot of people who you knew at the time, any people, say people . . . with money or—you know—like movers and shakers . . . it was all right, it was acceptable, you know?

Rachel's earnings at her poverty program job were insufficient for the lifestyle she wanted for herself and her family. Since she was already familiar with the drug business, she began selling marijuana, obtained from her husband's family.

> And then really, it's like—it's a good—you get a little chunk of money. A little boost in your checking account, and as you boost your checking account one time and you know you not planning to do it no more, then something will come up and maybe you want something, you know. . . . You don't be planning to do it, like be out there, you know. . . . I made good money, okay. . . . I'm always working, so I always had a good job, okay. Let's say I figure it like this. I predict what I

needed, and if I needed $1,000 then that's what I was gonna get. If I needed $500 then that's what I was gonna get. See, I'd make whatever I needed, you know. . . .

Rachel's success with her marijuana customers, most of whom were co-workers or other working-class drug users, increased her economic reliance on drug profits. At the same time, however, she was establishing herself as the provider for her family, a role that increased her commitment to conventional norms.

> You see, it (drug dealing) was about different things then, you know. 'Cause I wanted to travel, and then—you know—the kid, and we gave her everything . . . and then [I] gotta look out for my mother and stuff, so it was about—it was all about business then, you know . . . I wanna give her [daughter] everything in the world. When she's into college, no struggling, no sleeping in no dorm with two and three people. You get your own room, your own telephone, your own refrigerator, your own everything, you know.

From Marijuana to Cocaine

Thus far, Rachel had used and sold only marijuana, avoiding heroin. However, the drug scene—customers, dealing activities, patterns and places of consumption—was changing. By the mid-1970s, cocaine use had begun to increase among nonheroin drug users in New York (Johnson et al. 1985; 1990). Rachel, still in her 20s, began snorting cocaine at social gatherings, although marijuana remained her dominant drug of consumption.

> I was still working and stuff, you know, and I was still out there doing my thing (selling marijuana), and I used to deal with—you know—a lot of people of different ethnic origins, okay, and I can remember—I used to go camping a lot. And we take the van, you know, go upstate, and hustle on back down here to Manhattan, to the Bronx, get a load of herb, you know, take it up to the camp, you know . . . and then I noticed, I think it was probably like . . . one of the last times I went camping . . . we came down in the van and we had $200 and we (spent it) for cocaine! Instead of, you know, herb . . . Well, um . . . I started—all right, I

started snorting cocaine, okay? But I didn't stop using marijuana on account of that, you know. Uh, I used marijuana and cocaine, sniffing cocaine, okay? Uh, a lot together, you know, and still basically under the same social situations. I . . . had a close friend that I worked with, and he got into the habit, like, um, bringing it to the job, okay? And you know, we started sniffing a little bit. But, um . . . I would still primarily (snort cocaine) in a social situation, you know.

By the early 1980s, Rachel, like many other inner-city marijuana users, had begun to consume cocaine more regularly. She continued to sell and use marijuana, although good quality marijuana was becoming hard to find.

> The marijuana . . . it just got to a point where it was hard to get. I mean, it's not hard to get marijuana, herb is—you know—you can still get herb anywhere, but it's just not the same quality, you know. And if you been smoking herb a long time . . . it's nothing compared to what herb used to be. . . . The more cocaine started being utilized in the form of crack, the less good herb seem to be available, okay . . . when I was getting it, you know, for, um, other reasons, other than personal use, you know, there's people that I would get it from, you know . . . (and) the quality just started (going down). 'Til you . . . just start gradually moving away from it, you know . . .

While she used cocaine in the same social situations as she had smoked marijuana, the setting for cocaine use was slightly different.

> Um, with cocaine, okay, you still . . . supposedly with "in" people, but you know—you kinda—it started moving to the back room a little bit, so naturally you can't—because a lot of places you couldn't go in the place, smoke coke openly, you know, but you could go the ladies room, and you know, all the ladies—quote ladies—was in there doing it, you know. Or if you went to a party at somebody's house, you know, there was always that special place where you could do . . . a line, you know, snorting a line of cocaine, or you could spoon, you know . . . years ago people used to wear those spoons around their neck, you know, little gold chain thing . . . you would announce it, you know, in a way . . . it was fashionable!

At 24, Rachel enrolled in college, again under a social welfare program, and eventually earned a bachelor's degree in psychology. She was accepted into and began a graduate program, but dropped out to work as a rehabilitation psychotherapist.

Career and Drug Use Changes

Rachel continued to supplement her legal income with marijuana dealing, always keeping her drug-related activities separate from her family life. Prior to 1985, she had been motivated primarily by the desire to provide her mother and daughter with a comfortable lifestyle, and her personal drug consumption was limited by her commitment to them. But after her daughter married and moved away and her mother died, her involvement with drugs began to change; she sold drugs more and more to support her own habit. Eventually she left her legal job, although she retained the appearance and attitude of a legitimate professional woman.

> See, I'm, not . . . not so much into it the way I was before, when all I thought about was the dollar. I just feel like being comfortable, you know—um—support my own particular habit, you know, keep myself cared for. There was a time when it was a lot different, it was just all about the dollar bill, you know . . . when my daughter was going to college I was dealing, you know, because I needed, I wanted the money, okay? That was my primary thing. After my mother died and . . . my daughter went away to college . . . it was easier to get high and stuff.

As supplies of good quality marijuana became increasingly difficult to acquire, Rachel drifted into crack selling. The switch was accomplished through the same male co-worker who had previously provided her with cocaine.

> This particular friend . . . was also, as it turned out, when I first started dealing with crack and then started, you know, again with the selling a little bit, with the same person, you know, we just escalated (from) one thing to the other. . . .

Selling crack as she had marijuana, in an open, convivial, "teapad" atmosphere, attracted attention, and Rachel soon recognized that even though she was making "crazy money" she would have to change her style of doing business. News of her high-quality crack traveled so fast that overnight she had lines of people seeking to buy it. Not only did such obvious drug dealing put her in jeopardy from her nondrug-using neighbors, whose friendship she valued, and from the authorities, it also made her more prone to robbery by crack users and other dealers.

> I mean it was like lines of niggers, you know what I'm saying. Oh, God. I remember one time . . . I had the line up from the door. It was like, oh, God, I can't do this . . . I mean, it was like, it was really ugly in the beginning, you know. . . . Money was so fast, people so crazy, it was ugly.

Rachel quickly ended such sales, although occasionally she is still pressured into dealing with strangers who have heard of the quality of her crack.

> . . . you know, I'm the type of person, I'm friendly with people, you know, so some time it be like, "yo, I heard . . ."—especially when . . . I got that good quality from Long Island, then it be—it's like a nickel, you know, see for $5, somebody gonna go, child—I got this bad [i.e., good] stuff . . . I sell it to them. But that—that's not really what I . . . [have for] my clientele per se.

Since the advent of crack, Rachel's neighborhood has become permeated with crack users and dealers, even though most residents are still working people with families. Shootings, robberies, assaults, and rape have become commonplace. Indeed, rape is so common that most of its victims no longer consider it a reportable crime.

> It's a lot more violence associated with this drug. With cocaine in general, okay. But with crack in particular, you know. I never knew anybody on this—on this block that got killed for cocaine, but I've known, in the last year, three people getting killed for this crack business. They'll pull a gun up to you in a minute. In a minute. The young guys . . . I'm talking about organized drug dealers, okay, the young ones that are really into it and

working for real big movers and shakers in the drug business, okay. They have no respect for no one, male or female, okay? And they as soon whip it out on you as anything . . . seem like the taking of a life is not . . . important . . .

Relationships among crack users are different from marijuana users' social interactions, and Rachel had to readjust her selling strategy to accommodate the particular effects of crack on users. Whereas marijuana consumers enjoy a sociable environment with wine, music, and conversation, many of Rachel's crack customers wanted to be left alone, requiring only a safe, quiet place where they could consume their drug discreetly.

> . . . particularly for older customers, they don't like that feeling of being—you know—like moles under the ground or something . . . if you ever been in a (crack house), it's a awful, awful sight, you know what I'm saying.

Recognizing a need among middle class crack users for a safe haven, Rachel reorganized her business, developing techniques similar to those of the middle class/professional users and dealers studied by Waldorf et al. (1991) in the San Francisco Bay area. Some of her customers were former marijuana users who, like herself, had grown older and disliked the street crack users who were everywhere evident in the neighborhood.

With the demise of her obligations to her daughter, her mother, and her legal job, Rachel became more involved in drug consumption. In addition, the stress of dealing crack, combined with its easy availability and pleasurable high, encouraged her increased consumption.

> You see, I'm not the gun-toting mama type, you know what I'm saying. So . . . when it was like that, when it was like heavy at that time, it was like—I don't know, I would start to smoking too much my own damn self and then it was just—you know—when it's free . . .

Rachel in the Ethnographic Present

Four years after entering the crack market, Rachel is one of very few successful female crack dealers in the inner city. A freelancer, she operates as a "house connection"—a dealer who conducts drug sales from her apartment (Johnson, Dunlap, Manwar, and Hamid 1992). Most other freelancers are involved in networks of dealers who help one another to avoid robbery, competition, and arrest (Johnson, Dunlap, and Hamid 1992), but Rachel is not a part of any network. Nor does she fit comfortably into the established hierarchy of drug dealers, since she is neither a street dealer or lower level distributor like most other female crack dealers (Johnson, Hamid, and Sanabria 1991; Maher and Curtis 1992; Dunlap et al. 1992b) nor an upper level distributor.

Rachel's apartment, one of the few places where buyers can both buy and consume crack, is a sixth floor walk-up with bars on the windows and several locks on the door. Unlike the typical crack house, it is immaculate, much like the Harlem home of a typical lower middle-class nondrug user. The living room contains a worn but clean sofa, a coffee table, two end tables, a bookcase full of old books, a portable black and white TV, an ineffective table fan, and an old-fashioned entertainment set (a radio/record player/tape deck combination with a space for albums). A record player, long broken, and a stack of old albums provide evidence of Rachel's former marijuana-dealing days. Pictures of her mother, daughter, grandchildren, and son-in-law decorate the room, and she is eager to show visitors her family photo albums. The bedroom contains a single bed, a wardrobe, and a dresser with a mirror. The kitchen is furnished with an old table and chairs, refrigerator, stove, and a large barrel in one corner.

Rachel had more household possessions and more money when she was selling marijuana to supplement her income. During the early stages of her crack career, she lost possessions such as tape decks and color TVs by admitting "certain types" of customers into her apartment. Now, with her daughter educated and married and her mother deceased, she lives modestly, maintaining her business primarily to pay her rent and bills and to get high.

> So right now it ain't really all about (money) . . . it's like keeping a steady pace, keeping myself on a steady flow and just basically having what

I want, you know, like that. And it works out better, and it's a lot easier 'cause it's a lot less headache. . . .

Despite her satisfaction with her current lifestyle, Rachel has occasional feelings of remorse about being "a bad person," a judgment she believes her customers reflect.

> No matter who's buying it, it's like they always wanna look at the person that's selling it . . . therefore that makes me—you know—the dealer is the bad guy . . . I mean . . . it's the fact that you're doing something shady and you're doing something wrong, you know. Especially, like I said, the way most of the time I dealt with it, it was a lot of professional people, people that I worked with and stuff like that. So it's like I'm the shady character, you know—'cause I'm the dealer. So somehow or another, that makes me the shady character.

To counter this view, Rachel and her suppliers, whom she chooses according to the same criteria of adherence to conventional norms as she does her customers, attempt to bolster one another's self-images as respectable professionals.

> They (suppliers) are also human beings who wanna be looked at in another light . . . They wanna be looked at respectable and decent, you know. So you go out to dinner with me, I'm also going with you. I'm making you look good, you make me, and we all pretending that we normal people out here just like everybody else, hob-knobbing downtown and where ever you wanna go, you know. So you feed into each other's egos, because you don't really want to look at yourself as a bad person.

Her relationships with her (male) dealing associates take two different forms. Some treat her as a businessperson, according her the same respect they would give any business owner. But others treat her like a street dealer, making her vulnerable to the violence ordinarily found in drug dealing. Over the years she has developed the expertise required to offset the potential for violence with the ability to handle difficult situations, managing to tread a thin line between being a friendly, accommodating businesswoman and attracting unwanted sexual attention.

> What I dislike . . . men get the wrong idea about you, you know. Even when you straight up . . . they always gotta look at you somewhere lesser than what they are. If they being a businessman, why can't I be being a businessperson too? . . . They always feel like somehow or another—you know—there's supposed to be that little extra fringe benefit in there, you know . . . and especially if you try to treat 'em like a human being, you know, with respect. Oh, then—you know— you think—they start playing on you then, you know. Because if you try, then they wanna, "oh, would you like to go out to dinner." Then the next thing is, "cause you bought me a meal, I'm supposed to go to bed with you." And then that makes you a black bitch and all that kind of stuff . . . and stuck up, you know. Okay, I'm friendly, I'm a damn friendly personality, okay, and I like everybody, but then—you know—they read it all out of proportion . . .

STRATEGIES FOR A SUCCESSFUL DRUG-DEALING CAREER

In conducting her successful drug business, Rachel deviates considerably from the practices of most other drug dealers.

Catering to Working/Middle Class Users

Rachel's clients represent a hidden element in the world of illegal drugs about whom little is known (cf. Hamid 1992). Many have spouses and children who are unaware of their crack use. Their average age is 30, although many are older, with adult children. Some are newly middle class, the first in their families to have college degrees or to own businesses. Despite their secure, longtime positions, comfortable housing, and material wealth, however, they find themselves in empty households and holding unfulfilling jobs.

> I must say, of the people I know like myself . . . they older people, I have to say that, yeah. They're all older people . . . some of 'em got families and things like that, you know . . . everybody,

everybody works. . . . I know some, to be frank, I know some doctors, you know. Two doctors, yeah, and, um, a couple of social workers, you know . . . They got a sense of black identity about them also, okay? And they do believe in something of a higher being, you know. I mean, they know about faith and stuff like that . . . And they get high differently, you know, than the younger crowd . . . they don't get as paranoid, you know. And they don't want to talk about negative things . . . We be getting high but we be keeping talk about religion. What's happening to the black people in general, including ourselves. We can talk about these things, okay. . . .

Avoiding the Street Market

A key element in Rachel's success is her avoidance of both customers and dealers involved in the street crack market. Normally she declines to sell crack to street users or to anyone who threatens her or others. While she knows street dealers, she stays out of their territory and does not compete for their customers. In this way she avoids violence and maintains the safe, comfortable environment her customers prefer.

> Well, (I) try not to . . . get into no antagonistic situations, you know . . . You try to give a little respect to the neighbors around you, you know, so that they don't get too mad . . . they will tolerate you, you know. You try not to . . . go direct into no one's turf, and stuff like that. In other words, you have to maintain a low profile, you know. That's the main thing . . . And then . . . you know, like—I don't hit nobody, I don't rob nobody, I don't deliberately mess with people's heads, and I expect the same with them, okay . . . I think it's all in the way you treat people . . .

Maintaining Good Relations With Neighbors

Most crack users eventually lose their apartments (Dunlap 1992), but Rachel has lived in the same building for some 20 years and has established herself as a good neighbor to other long-time tenants, most of whom are not involved with drugs. Her neighborliness is motivated both by genuine affection for her neighbors and by an awareness that they could destroy her business. She retains their friendship (and avoids detection by the police) by restricting her business to a limited number of quiet, middle class customers.

> This is a quiet building and stuff like that. . . . I got neighbors that I've known, like, a long, long time, and (if I attracted the wrong clientele) I'd be outta here quick fast. 'Cause they got a law now, they kick you out . . . when you selling and they know.

Providing a Setting Appropriate to Clients' Needs

Aware of her customers' desire for a quiet, discreet atmosphere in which they can smoke crack without having to deal with the dangers of street life, Rachel strives to create a climate of tranquility in her apartment. She also makes much of the fact that she will not sell anything she would not smoke herself, which assures her customers that her crack is of high quality.

> I have it set up in a sense that now it's nice, easy, a social kind of situation. Person comes . . . they bring a certain amount of money, it's payday, you know, and . . . most of the time, they don't want to go out no more. They don't want to be seen, there's a lot of professional people, and they want to feel safe, you know. [I] sell it right to them, point blank period dot, and then smoke it with 'em too. You know what I'm saying? . . . And it's easy, it's subtle, it's quiet. Ain't a whole lot of noise, a whole lot of traffic . . . my main focus is to keep 'em comfortable.

Managing the Effects of Crack on Customers

Familiar with the negative physical and psychological reactions that crack consumption can cause, Rachel works hard to keep her customers from suffering "tension" (also a street name for crack). Her dimly lit bedroom provides those who prefer not to interact with others a place to be alone and those who are paranoid a place to feel safe.

> You don't know how this shit hits some people . . . If you could see some of the people

that I have to tolerate, people come here because they know I have patience to deal with them, with the looking under the bed, and looking in the closet. They feel safe here . . .

Managing Customers' Finances

Many of Rachel's customers need help managing their drug habits. She knows when they get their paychecks, and also that if they arrive at her apartment with a full paycheck they may not have paid their bills or set anything aside for necessities. If she is aware that a customer needs money, she may cash his paycheck but then purchase a money order for him to take home.

> Whatever they come with, 9 times outta 10, sometime what I do is I'll take some off the top and— you know—"what did you say you needed?'" . . . And I will go to the store and get it, get 'em whatever it is they needed . . . or even give 'em a money order or something. Especially the people that's living with families.

Controlling Unruly Customers

Rachel is quite capable of holding her own when customers attempt to "try her out" by violating her house rules. Prior to one interview, the ethnographer found Rachel scrubbing blood off the floor of the hall in front of her apartment. She reported that a customer had become violent—an unusual occurrence—and she had settled the matter. Since this was the second time this customer had acted this way, she barred him from returning. She keeps knives in her apartment for emergencies but no longer owns a gun, having devised other ways to discourage rowdy or threatening customers.

> Well, yeah, I got stuff like (knives), you know. I got quite a few of them . . . I just don't have weapons, I don't have a gun anymore, okay, 'cause it's hard for me to keep from not shooting somebody if I really did want to . . . and uh, I just went through last year a thing that—and it's like . . . I tried to kill that man. You don't come here and then bogart me around the house [i.e., act in a rough or bossy manner], okay, you can't do that, you know.

Choosing Suppliers Carefully

Rachel chooses stable suppliers who, like herself and her customers, lead superficially "normal" lives and have self-images of attachment to conventional norms and behaviors.

> . . . my best person, supplier, is a family man, you know, they have a home, everything. You know, nice people . . . Drive a nice car, nice kids, nice wife, the whole bit. That's the best kind of people to deal with, because they can't go but so far. You know, raunchy people, man, they'll be ready to take you off, you know, and then you gotta come out a whole 'nother bag, you know. Which I can do that too, but you see, why set yourself up for that?

Avoiding Unwanted Sexual Attention

Rachel knows that both clients and suppliers are drawn to her as an attractive woman, and she exploits this for business purposes. At the same time, she is aware that no special consideration is given female drug dealers and that the consequences of failing to anticipate and defuse potentially dangerous situations may be sexual harassment or physical harm.

> Being a female—you know, its good . . . 'cause they always think they gonna get something out the deal other than business, you know. I dress good, look good, you know. . . . (I) let them think whatever they wanna think, you know. It ain't about stringing them on, because . . . when push come to shove, you know what talk more than anything else is the money. . . . Most of the time—you know—you just, you gotta handle it. You either gotta do one of two things . . . 9 times out of 10, you know, you can kinda talk your way out of it . . . use that reverse psychology in Psychology 101 on 'em . . . (But) sometime there are—you know—you just have to get downright, like, you just gotta act like a bitch. I mean you gotta get downright nasty, you know. See—but that's always a problem in that, you know—because then you don't want nobody coming back on you, you know. But . . . it depends on the kind of person you are, you know. Because it's like I don't sell my body.

Avoiding Arrest

By dealing discreetly, catering to a middle-class clientele, and being a responsible neighbor and tenant, Rachel has successfully avoided the police and has no arrest record. The threat of arrest is

> . . . important enough to me to do it (deal) the way I do it, you know. In other words, I'm not looking for the big profits no more, okay . . . it's very, very, very essential that I do not go to jail!

Another of her tactics for avoiding the police is to dress like a working woman when buying cocaine. On one occasion, when she encountered police who had staked out her supplier's apartment, her appearance plus some creative lies helped her to evade arrest.

> It's in the wintertime, right? Put my boots on, had nice leather boots, long coat, you should have seen me . . . I put my makeup on and stuff, like I'm going to work, okay? Earrings and everything, okay? So I get to where I got to go, right. . . . Soon as I went in there [i.e., into the area in which weight cocaine is sold], I felt it. I said, it's too quiet around here, and I turned around to make my move and come on out and they [undercover police] stopped me. Five of em, okay? But I was looking the part, so they say, "Where you going?" I said, "I'm getting ready to go to work. I heard my girlfriend, she gets in a lot of trouble on the job, and I ain't know whether she had got fired and I came to see about her." . . . He said, "You know about this place," and I said "I've heard terrible, terrible things about this place here, but I was so worried about my girlfriend that I had to take a chance and come anyway." So they said "Well, where?" I say "Apartment one, right over there" . . . They said, "Get the hell outta here."

Controlling Personal Consumption

For the most part, Rachel has avoided the pitfalls of excessive crack use. She manages to eat properly by requiring her customers to bring food, which she then cooks, serves, and shares with them, and she is careful to set aside money to pay the rent. She also avoids consuming too much of her own supply of crack. Unlike the average consumer, she does not "go on missions"—nonstop 3- to 5-day crack smoking binges without food or sleep. When one of her customers goes on a mission, consuming a substantial part of her supply of crack, she re-ups with money from her next well-heeled customer. She considers herself an intelligent crack user.

> You know, that's all a part of using your head. When you take care of your body, okay, and you put some knowledge up here in your head, if you are doing these things, it's because you want to survive after you do it. . . . See, a lot of people . . . just want to get high, and don't care nothing else other than that. Getting high, okay . . . for me [is] like some people want to take a martini after work, okay? Then you work and you do what you have to do. For me it's delayed gratification, is what I call it, okay?

LESSONS FROM A SUCCESSFUL DRUG-DEALING CAREER

Rachel is in many ways unique, a product of her singular personality, life history, and experiences, yet her very uniqueness sheds light on a number of aspects of illegal drug sales and use in the United States.

First, Rachel's career illuminates trends in the history of illegal drug use in New York and the United States. The shifts in her involvement, from one drug to another, were engendered not so much by her personal decisions as by macrolevel social and logistical forces (Johnson and Manwar 1991; Dunlap and Johnson 1992a, 1992b). Her introduction to cocaine in the mid-1970s, for example, was typical of many inner-city drug users of that era, who frequented "after-hours clubs" to snort cocaine when they could afford it (Williams 1978). More broadly, Rachel's business illustrates general shifts in the drug market in New York, from marijuana to cocaine to crack, showing how supply, demand, and consumption feed into one another. Her customers' drug choices and consumption patterns and her marketing practices reflect the availability of certain drugs. Availability in turn influences demand, which drives the market.

Second, Rachel's business illuminates an unknown side of the drug economy: the world of the older, better educated, employed, middle-class, violence-averse drug user (Hamid 1992; Waldorf et al. 1991). In the United States, the stereotypical crack user is young, violence-prone, poorly educated, and unemployed. In addition, entertainers and sports figures are widely viewed as being involved with drugs; actor Richard Pryor provides an example. But Mayor Marion Barry's arrest for crack use came as a surprise to many because his position as mayor of Washington, D.C. was inconsistent with such stereotypes. Middle-class drug users may consume drugs discreetly to protect their jobs and reputations, but they are no less a part of the illegal drug industry.

Third, Rachel illustrates the effects of being female in a profession dominated by males, a subject about which little is known despite recent increases in convictions of female drug offenders. Females are often stigmatized in the world of illegal drugs, but Rachel has managed to turn being female into an asset by cultivating a friendly, accommodating, caring, noncompetitive image. She exploits the fact that males find her attractive, but is able to defuse unwanted sexual attention with professionalism.

Finally, Rachel's attempts to be "normal" suggest the tension inherent in moving between the legitimate world and the world of illegal drugs. Her feelings of remorse, reflected in her need to see herself as other than "a bad person," are assuaged by her ability to justify her actions. She prides herself on providing high quality crack for her customers and a safe, congenial place to smoke it, and on her ability to avoid the police. But most important, she sees herself as a legitimate businessperson with a commitment to conventional norms and chooses associates—suppliers, other dealers, and customers—who share this self-image.

CONCLUSION

Rachel's success in her profession can be attributed to a unique combination of historical contingencies, personal qualities, and career choices. On the one hand, she is a product of a particular background and of the macrolevel social forces existing in a particular temporal and geographical context. Yet within this context, her unique personal characteristics—her intelligence, sense of professionalism, commitment to conventional values, skill as a businessperson, and self-discipline—have contributed to her position as a deviant among deviants, as have her surprising personal choices. Formerly a conventional black mother and nondrug user, she became involved with drugs relatively late, after her husband died. She completed high school and college and was clearly capable of graduate work, yet abandoned the promise of a legitimate career for marijuana sales. And not until she was in her mid-30s did she shift from marijuana to crack. Together, these factors have contributed to the ingenious ways in which Rachel controls her environment, to the specific strategies she employs, and, ultimately, to her career success.

REFERENCES

Becker, Howard S. 1963. *Outsiders: Studies in the Sociology of Deviance.* Glencoe, IL: Free Press.

Dunlap, Eloise and Bruce D. Johnson. 1992a. "Structural and Economic Changes: An Examination of Female Crack Dealers in New York City and Their Family Life." Paper presented at the annual meeting of the American Society of Criminology, November, 1992. New Orleans, LA.

Dunlap, Eloise and Bruce Johnson. 1992b. "Who They Are and What They Do: Female Dealers in New York City." Paper presented at the annual meeting of the American Society of Criminology, November, 1992. New Orleans, LA.

Faupel, Charles E. 1981. "Drug Treatment and Criminality: Methodological and Theoretical Considerations." Pp. 183–206 in *The Drugs-Crime Connection,* edited by James A. Inciardi. Beverly Hills, CA: Sage.

Hamid, Ansley. 1992. "Drugs and Patterns of Opportunity in the Inner City." Pp. 209–239 in *Drugs, Crime and Social Isolation,* edited by Adele Harrell and George Peterson. Washington, DC: Urban Institute Press.

Johnson, Bruce. 1991. "Crack in New York City." *Addiction and Recovery* XX:24–27.

Johnson, Bruce D., Eloise Dunlap, Ali Manwar, and Ansely Hamid. 1992. "Varieties of Freelance Crack Selling." Paper presented at the annual meeting of the American Society of Criminology, November. New Orleans, LA.

Johnson, Bruce D., Eloise Dunlap, and Ansley Hamid. 1992. "Changes in New York's Crack Distribution Scene." Pp. 360–364 in *Drugs and Society to the Year 2000,* edited by Peter Vamos and Paul Corriveau. Montreal: Portage Program for Drug Dependencies.

Johnson, Bruce D., Paul J. Goldstein, Edward Preble, James Schmeidler, Douglas S. Lipton, Barry Spunt, and Thomas Miller. 1985. *Taking Care of Business: The Economics of Crime by Heroin Abusers.* Lexington, MA: Lexington Books.

Johnson, Bruce, Ansley Hamid, and Harry Sanabria. 1991. "Emerging Models of Crack Distribution." Pp. 56–78 in *Drugs and Crime: A Reader,* edited by Tom Mieczkowski. Boston: Allyn-Bacon.

Johnson, Bruce and Ali Manwar. 1991. "Towards a Paradigm of Drug Eras: Previous Drug Eras Help to Model the Crack Epidemic in New York City During the 1990s." Presentation at the American Society of Criminology, November, 1991. San Francisco, CA.

Johnson, Bruce, Terry Williams, Kojo Dei, and Harry Sanabria. 1990. "Drug Abuse in the Inner City: Impact on Hard-Drug Users and the Community." Pp. 9–66 in *Drugs and Crime,* edited by Michael Tonry and James Wilson. Chicago: University of Chicago Press.

Maher, L. and R. Curtis. 1992. "Women on the Edge of Crime: Crack Cocaine and the Changing Contexts of Street-Level Sex Work in New York City." *Crime, Law and Social Change* 18: 221–258.

Waldorf, Dan, Craig Reinarman, and Sheigla Murphy. 1991. *Cocaine Changes: The Experience of Using and Quitting.* Delphia, PA: Temple University Press.

Williams, Terry. 1978. "The Cocaine Culture in After Hours Clubs." PhD dissertation. New York: Sociology Department, City University of New York.

Williams, Terry, Eloise Dunlap, Bruce D. Johnson, and Ansley Hamid. 1992. "Personal Safety in Dangerous Places." *Journal of Contemporary Ethnography* 21:343–347.

Zinberg, Norman E. 1984. *Drug Set and Setting: The Basis for Controlled Intoxicant Use.* New Haven, CT: Yale University Press.

28

CITIZENS AND OUTLAWS: THE PRIVATE LIVES AND PUBLIC LIFESTYLES OF WOMEN IN THE ILLICIT DRUG ECONOMY

PATRICIA MORGAN

KAREN ANN JOE

Three cities, San Francisco, San Diego, and Honolulu, were utilized for this study of methamphetamine users. A culturally and ethnically diverse population of methamphetamine users was interviewed, combined with survey questionnaires to study the patterns and results of drug utilization. The one finding of major importance was the unexpected high proportion of women involved in methamphetamine distribution and dealing. The authors note that most of the female study subjects considered their involvement in the drug business to be a positive experience, which affected their financial independence, self-esteem, professional pride, confidence to function in a male-dominated business, and ethical behavior in an illegal trade. Women reported the major problems in dealing methamphetamine as avoiding arrest, violence, being betrayed by buyers, and becoming too dependent on methamphetamine in order to feel normal and continue to operate their dealing business without fear on a social level.

The historical compulsion to limit the contexts of women's drug use to gendered relationships continues to cling tenaciously to the research literature. For decades, patriarchal world views placed women's drug use as a symptom of an underlying moral weakness or sexual deviance. Consequently, research restricted them to passive or victimized roles in social worlds dominated by men. Rationales for use depended on race, class, or type of substance. Women drug users were either good citizens suffering from "addiction" or "bad" ones whose drug use reflected their roles as deviants. This conceptual framework limited the context of women's drug use almost totally to their relationship with men. It was rarely considered that a woman's drug use could be shaped by choices outside the gendered relationship.

Beginning in the 1980s however, growing evidence from research began to reveal broader contexts framing women's use of illicit drugs (Ettorre 1992; Fagan 1994). Ethnographic studies found that women had rationales for use outside the world of men, that women often control their context of use, and that they play a larger role in the illicit drug economy than previously thought (Rosenbaum 1981; Mieczkowski 1992; Maher 1992). However, because most research continues to focus on disenfranchised marginalized women, and on drug use in the context of subordinate power relationships and victimized lifestyles, it is difficult to uncover the full range of the values and motivations which guide behavior. Yet, the only way to move outside traditional patriarchal paradigms is to identify these unexplored and hidden worlds of women's drug use. Consequently, the purpose of this paper is to undertake this task by utilizing findings from a comparative study of methamphetamine use. Findings revealed complex systems of normative behavior framing overlapping contexts of drug use careers, personal lifestyles, and involvement in the illicit drug economy among women in the study. Moreover, it provides evidence that women have been especially safe in this hidden world of drug use for decades, overshadowed by the glare of traditional patriarchal interpretations.

CONSTRUCTING THE IMAGE

Until relatively recently, women were largely ignored in the drug and alcohol research field. The few studies which were conducted before the 1980s limited their focus to women on either end of the lifestyle continuum. One body of literature grew out of Alcoholics Anonymous and the "addiction model." These were mostly clinical studies of women alcoholics who were said to suffer more severe forms of pathology than men because their addiction inherently conflicted with their natural role as wives and mothers (Hirsh 1962; Curlee 1967). The other body of literature were criminologic studies of women occupying outlaw social worlds. These focused almost exclusively on women prostitutes suffering from heroin addiction. They were studied as members of "deviant subcultures," who were thought to automatically violate basic gender norms, by neglecting traditional roles as wives and mothers (Woodhouse 1992).

After 1990, several studies began showing women unexpectedly inhabiting diverse social worlds of illicit drug users equally complex as their male counterparts: Dunlap and Johnson; Fagan 1994; Mieczkowski 1994; Granfield and Cloud; and Sterk-Elifson. Much of the information came from ethnographic studies revealing that even among the most marginalized and heaviest users, women were able to exercise a measure of control over many contexts of use (Mieczkowski 1994; Murphy et al. 1990; Dunlap 1992). Moreover, Maher and Curtis (1992) found that even women sex workers expanded their roles in the illicit drug economy after the crack epidemic forced them to adjust to changes in the sex market. They argue:

> Rather than one-dimensional characters defined by traditional gender roles, our research reveals that women act within a number of overlapping contextual environments that are socially, culturally, economically and psychologically constructed. (Maher and Curtis 1992:249)

Other evidence reveals the need to systematically examine the possibility that women from a range of backgrounds make conscious choices toward a lifestyle involving drug use along with a career in the illicit drug economy (Waldorf and Murphy 1995). Growing evidence of expanded roles for women in the illicit drug economy provides formidable testimony against traditional gender-role interpretations (Johnson et al. 1991; Mieczkowski 1994; Maher and Curtis 1992; Maher 1996; Waldorf and Murphy 1995).

This paper utilizes an exploratory community-based study of methamphetamine in California and Hawaii to disclose hidden worlds of drug use involving independent and active participation among women across diverse lifestyles and experiences. The hidden contexts of their social worlds link women's roles in the illicit drug economy to the rationales and norms guiding their experience. This hidden world also exposes a dynamic relationship between varied lifestyles and shifting patterns of involvement in the methamphetamine drug economy.

Study Design and Methodology

Research findings are drawn from a qualitative study of 450 moderate-to-heavy methamphetamine users in San Francisco, San Diego, and Honolulu, funded by the National Institute of Drug Abuse between 1991 and 1994 (Morgan et al. 1994). It compared gender, race, environmental, and cultural differences, to rationales, patterns, and problems of methamphetamine use. These three geographical areas were selected because the incidence of methamphetamine use and problems was substantially high in each site and because evidence from existing data revealed that the modes of methamphetamine use differed significantly in each locale

As a qualitative study, the objective was to reconstruct the reality of the social worlds of identified methamphetamine users in a naturalistic environment. The aim was to attain "members' knowledge" and thus identify and examine the meaning of naturally occurring social actions within given social worlds (Geertz 1973). This core strength of the ethnographic method is particularly important when involving women drug users. A key component in this process of discovery involved women interviewers in all three sites who had substantial past drug use experience but limited knowledge of the academic literature. Importantly, these interviewers did not fit the stereotype of the illicit drug-using woman. Although they had been heavy users previously, and had engaged in drug-dealing activity, these women eventually stopped their drug use, obtained college degrees, and were no longer involved with drugs.

Lifestyle Contexts

Women respondents fit a wide array of user lifestyles, both within and across study sites. Four primary lifestyle contexts were identified: outlaw, floater, welfare mom, and citizen (Morgan et al. 1994). In a previous article we examined each of these lifestyles in relation to criminal activity in the illicit methamphetamine economy (Morgan and Joe n.d.). We found that most of these women experienced more than one lifestyle change during different periods in their drug-use careers. An examination of these lifestyles enabled us to understand the contexts of experience among our female methamphetamine users. We also found that success in the illicit marketplace was not measured necessarily in terms of dollars, but frequently in terms of a person's ability to control his/her environment. It highlighted the need to examine closely the similarities and differences in the normative values guiding women in various lifestyle and economic contexts.

Outlaw Lifestyle

A definition of an outlaw lifestyle context involved respondents who were significantly immersed in deviant activities and marginal lives. The character of that lifestyle differed significantly depending on whether these women viewed themselves as "victims" or as "survivors." Those women who saw themselves trapped within an outlaw lifestyle were at the lowest level of economic and social status. With histories of chaotic and very unstable lives, many were merely existing, and often homeless. Most were barely surviving as hustlers and/or prostitutes. Some had lost touch with reality and with mainstream society. They generally engaged in the lowest level of dealing—as traders or sellers. Many of these women were heavy users for whom dealing to make a living and to maintain a steady supply has become a way of life. One respondent who has been in and out of jail for prostitution, drugs, and other crimes states, for example:

> I do speed to stay up so I can sell speed to make money, but I sold speed to make money so I can by more speed. I don't make a living at all. I'm existing, I don't call this living now! (056)

Other women respondents were able to maintain more control, because they saw an outlaw way of life as a chosen lifestyle. They perceived themselves as long-time survivors who have managed to earn respect on the streets as either dealers or hustlers. A 27-year-old white IV user from the Tenderloin has been using drugs and has been a sex worker since coming to San Francisco at age 17. She claims she has a good life, even though she's been busted a few times, is currently on parole, and only makes about $5,000 per year.

> Yes. I'm 27 years old now and I'm told that I look younger now than when I did when I 1st came here to the city! I don't work the streets, I don't have too many hassles, I have pretty good friends and I support myself pretty well. There's no hassles, I don't have to go out there and do whatever. I don't hurt for nothing cause it comes easy. Do that, because it's there. As long as you're not burning nobody, it pays for itself. (018)

In San Francisco, there were over 12 women over 40 who had been using drugs continuously for over 2 decades. The matriarch of this sub-sample was a 59-year-old survivor of the Beat Generation who has been injecting speed and heroin for over 40 years prior to the interview. She began selling liquid ampules in the 1950s, helped run the first bathtub crank operation, ran a bordello, spent many years in prison, and claimed to have given Lenny Bruce his first hit of speed. Most of this activity happened in and around the Tenderloin were she continued to live and sell methamphetamine. At the time of the interview, although living on about $10,000 per year, she described a life of excitement and adventure. In another example, a 48-year-old woman from Richmond with over 26 years experience, admitted to having held a legitimate job for only 6 months out of the previous 20 years.

Almost all women leading outlaw lifestyles had criminal arrest histories and bad rip-off experiences. Many in this outlaw lifestyle were often former high-level dealers, distributors, or cooks. Some of these women were born into this lifestyle, especially those from second or third generation outlaw biker families. The parents of a 29-year-old respondent had long criminal histories, including murder. She began IV meth use at the age of 12 with the assistance of her mother. Since then, she has been a high-level cook, distributor, and a dealer. At the time of the interview she reported a history of violent crime which resulted in several prison terms. She has been diagnosed with many medical problems, some inherited from her mother's drug use while pregnant, and others a product of her own drug abuse. Nonetheless, she sums up her life and her speed use with fatalistic enthusiasm:

> What do I hope for the future? To live every day like it's my last day, cramming everything in that I possibly can into one. Live happy . . . speed is a part of my life. It always will be. Speed, I don't use it because I need it or because my body craves for it it's a must or I'm just gonna go crazy if I don't get that hit. I couldn't care less if I do have it, I couldn't care less if I don't have it. If I've got it at the time, well, damn, that's sufficient. But if I don't have it, I'm not gonna stress over it. It's not something that I must have and that I won't live without it. But I sure do fucking like it. [Both laugh] It makes you feel good. It takes things away, makes you not think about things, the inedible things. It makes you kind of surpass all that. And then when you come down, you'll think about it. Sometimes you get depressed when you think about it. But basically, it takes away the edgy feeling that you have, your everyday "Enough, stress," your yesterday's life. It takes away a lot of it. I won't never quit. I'll do it till the day I die and I hope somebody puts about a 60 unit hit in my IV bag when I'm sitting in the hospital. (092)

Citizen Lifestyle

Those in our sample leading citizen lifestyles were women living totally within mainstream society regardless of the amount of drug used or level of drug dealing. Able to retain, or return to, a measure of stability and respectability, they were found living in good neighborhoods. They had money, often a husband and family, and sometimes a legitimate job. They were either high-level dealers with methamphetamine as major source of income or part-time sellers to limited selected friends and clientele while maintaining regular employment.

The most successful women dealers led citizen lifestyles, and actively participated in mainstream society. For example, in control of a successful high-level illicit business for a number of years, one women "citizen" from San Diego began dealing as a way to get off welfare after her marriage failed. She started 6 years prior to the interview with a quarter ounce and now earns over $50,000 per year working only 1 hour a day.

> I stayed within my goals, basically. . . . For about six years now. [you're careful?] Yes, I'm not a stupid person. I don't go around doing stupid things. I don't walk around telling people I have drugs for sale, I don't have people sitting out in front of my house, I don't have traffic in and out of my house . . . I control the people that I sell to. They get it for other people and I never see the people they sell it to. [In a typical week] . . . if it's been flowing real good, $1500 to $2000. With my kids I have to sneak around and do what I'm doing without me taking too much time. So I have to do a little each day . . . I try to put what's important first. Kids eat and do homework first. But this runs a real close second. I deal an hour or so each evening. (297)

Another woman from San Diego was able to maintain a professional job while dealing in partnership with her husband. Increasing level of use led to her husband's arrest, and the loss of her job. She began to supplement her unemployment benefits by continuing the dealing business that she now controls completely. She now believes this has enabled her to control her use as well.

It was common to find that a citizen lifestyle was not automatically a successful or middle-class one. Many working-class women in all sites began dealing in order to pay for use so that they could maintain at least the appearance of normalcy. For example, since losing her job as an accountant, one woman of mixed-Hawaiian ethnicity described the current level of dealing with her husband:

> . . . it's just to break even. Most of the time we don't. You break down your product and put aside what you're gonna smoke and what part to make your money back. (415)

Many of the women in our study fell in between these two extremes. One we defined as a floater lifestyle where their drug use led in and out of mainstream life. Although able to escape full long-term participation in deviant careers, women in this lifestyle seem unable to establish stable lives, relationships, or responsibilities. Another was labeled welfare-mom lifestyle category. These women were full-time participants in a subculture found in impoverished suburban communities where methamphetamine had been part of daily life, often for generations (Morgan et al. 1994). A welfare-mom lifestyle encompassed women's experiences as residents of public housing complexes often selling methamphetamine to pay for personal use and to provide a little extra income.

ECONOMIC CONTEXTS

The type of economic activity involving illicit drug dealing varied considerably, from small-time hustling to large-scale distribution. This activity, in turn, was linked to fluid contexts of user lifestyles. They combine to help shape a complex environment connecting patterns and consequences of use to shifts and variations for women's experience in the illicit drug economy. Analysis from our qualitative interview transcripts identified over 100 female methamphetamine users who reported a wide range of experiences as active participants in the illicit drug economy. These ranged from trading favors, to large-scale manufacture on a continuum encompassing diverse economic contexts. Importantly, these were rarely static roles, as many women moved from one role to another throughout their drug-use careers.

Although a number of women reported trading favors in exchange for the drug, few stated that these were sexual favors. This is particularly significant in light of the popular belief, especially from studies of female crack users, that sexual favors characterize a woman's role in the trade and barter system involving illicit drugs. More often, women acted as runners, or performed various other chores for dealers. Several women traded items they had stolen, or sold other drugs to obtain methamphetamine. A middle-aged black woman with 30 years experience using methamphetamine found that changing preferences forced her to sell crack; although she strongly dislikes the drug, it's the best way for her to earn money to buy methamphetamine.

Many women limited illicit drug activity to partnerships with their boyfriends or husbands. This was most common among female users in Honolulu and found least among women in San Francisco. In order to be successful, however, women either took over the partnership, or left the relationship to become independent entrepreneurs. For example, a San Diego woman was unable to control her methamphetamine use until she took over the business after her husband went to jail. She is now controlling the business, her own use as well as his:

> I probably have more trouble keeping D. to a daily ration, cause he's not handling it, he's not realizing how much of it he's doing. So I just dole him a certain amount and say "hey man, that's it, after that you start paying for it." Cause I'm not going to end up owing him money out of my pocket, that's not why I'm doing this. D. understands where I'm coming from. (345)

Most women dealers in the study sold methamphetamine on a part-time basis. The two major rationales were to supplement their legitimate income, and/or to pay for their personal use or at least keep the cost down. Among outlaw users in the inner-city, dealing was often episodic and more accurately described a service linking buyer and seller. The description from a sex worker in the Tenderloin is typical:

> I have a request and someone calls and says they want a certain quantity. I call someone and find out if they have it at the price and amount the buyer wanted. The person with the money comes to my house with the money. The person with the product could come over and they'd do their deal without me. They would try it out, if it was fine, the person would buy it and leave. Not more than 1 person at a time. (014)

Almost a fourth of female respondents in California, especially in San Francisco, reported receiving their main source of income from the illicit drug economy. They ranged from highly successful tightly organized enterprises that included, but were not always limited to, methamphetamine sales. There were a surprising number of women in our sample who had experience with large-scale manufacture and/or distribution. One woman from the Bay Area who dealt both heroin and speed with former boyfriends and her husband averaged about $2,000 per week for almost a decade. Women had a variety of roles at this level of economic activity, which ranged from directing the international importation of cocaine to assisting in the manufacture of methamphetamine. Many involved in manufacturing methamphetamine were involved in a partnership with their husband or boyfriend. For example, a female respondent from eastern San Diego County was the cook and her boyfriend the lookout. She described a high level of weekly production:

> About 5 lbs. per week. We'd distribute to certain people and they'd sell at their levels. We had people who bought one or 2 lbs. at a time! What was the factor was how fast did we want to run out! We moved as much as we made! There was certain people we'd give pure stuff to, but, the majority of the time, we cut at least 25% to half, 50% cut. (255)

Most manufacturers lasted only for limited periods, sometimes less than a year. Some women, however, managed to maintain this level of activity for a number of years. For most, the stressful lifestyle, the violence, gang rivalries, and arrests, marked the experience for these women.

HIDDEN CONTEXTS

Economic and Lifestyle Mobility

It is important to note that the majority of female respondents engaging in drug sales did not limit their economic activity to trading or bartering for drugs. Those who did were most often in Honolulu where women tended to be more attached to traditional relationships with men and into close family and kinship networks. Most commonly, this level tended to represent an early stage in what became a multifaceted drug-selling career. In each study site, women often began by dealing from a subordinate role in these relationships, only to eventually control or even take over the enterprise even in Honolulu, where dealing involves the entire family. One woman reported how she was introduced to Ice through her stepson, began buying from his dealer, and ended up being his supplier (424). These women stated they felt they needed to take on a more active role with both their and their partner's increasing use, which often led to an increasing inability of the males to handle the responsibility connected to their business enterprise. Many women began in partnership with their husbands, and took over the business after the husband's business or using habits threatened to shut it down. One Bay Area female dealer stated that after she took over the business from her husband,

> It boomed, it just exploded! F. knew everybody but just never had it together enough because he wasn't a business type person. He was Mr. Charisma, party-party, happy-happy. I'm more grounded and business like. That's all he needed and we really clicked. (059)

Similarly, we found that many of these social lifestyles were not static. Women reported life histories usually that included experiences in two or more user-type categories. Some females, for example, began using when they were prostitutes or on welfare, and eventually began leading citizen lifestyles with increasing success. Many others reported moving in the opposite direction, from citizens or successful outlaws to welfare moms or to very marginalized existences.

Examination of these active participants uncovered a complex web connecting drug use and drug dealing experiences over the course of their drug use career. Many women lived a number of user lifestyles while they moved among several different levels of economic activity throughout various stages in their involvement with methamphetamine. One respondent, for example, ran away from home and began shooting heroin at age 13. She eventually married a dentist (who was also a large-scale drug distributor), became a housewife and mother, and also helped with the drug business.

Six years later she ended up divorced, homeless, and steadily abused, living in the city's worst neighborhood. Now, however, she has become one of the toughest, most assertive and respected women in this subculture, running a successful small-scale drug dealing enterprise. So, at various times in her life, she has been an outlaw, a citizen, and a floater user type; she has also engaged in large-scale drug distribution and in low-level drug dealing.

Consequently, we found the contextual relationship integrating patterns and consequences of use with activity in the illicit drug enterprise to involve several interrelated issues, which often shift during different stages in their methamphetamine using careers, with various combinations. Importantly, women were more likely than male dealers in our study to experience upward mobility in the illicit drug economy. Two possible reasons are revealed in major findings from the study. One involved the issue of self-control, and the other the concept of professional pride and ethics. They are interwoven.

SELF-CONTROL

One of the most important findings to emerge from this study concerned beliefs and attitudes toward addiction and loss of control. More specifically, this research revealed divergent and specific definitions for the concepts of "control" and "addiction" among our respondents. Although 60% of our sample reported "ever" feeling addicted to methamphetamine, only 48% stated they had lost control over their use. Women were more likely to report feeling addicted than men (63% compared to 60%). And, they were less likely to report losing control over their use (45%) than men (50%). The major exception is in Honolulu, where a higher percentage of male respondents reported an addiction to methamphetamine (58%) than loss of control over their use of the drug (57%).

Analysis from the qualitative data suggests a significant number of our respondents considered themselves to be "controlled addicts." These definitions were particular to the specific context of their drug experience (cf. Morgan and Beck n.d.). Although all respondents had very strong feelings about being either in control, or out of control, over their use, women were much more likely to describe having control over their use and of their lives while using methamphetamine. A woman in Honolulu feels men tend to lose control much easier than women:

> Sometimes they get like that because of being high. They do things that they really normally wouldn't do. It can get your mind screwed up. It's a good high, but you gotta know how to handle it. A lot of my friends say it's strong enough for a man, but it's made for a woman! A lot of guys get really crazy. Even my boyfriend, he gets crazy off it. Real scary. I don't really want to smoke with him, cause he scares me. He thinks I'm hiding things! I get high all my life, nobody ever bummed my trip! That's bullshit. (459)

Our findings suggest that there were differences in the way male and female respondents talked about self-control. Women were more likely to stress the ways they managed to stay in control. Men, on the other hand, were more likely to discuss ways they had lost control, and were far less likely than women to talk about how to manage to stay in control. For women users the ability to maintain control was a very important issue:

> I have this phobia of something controlling me! Don't get me wrong, this controls me but, I have some kind of hold on it and I refuse to give in. If I have been up two or three days and I feel like I'm getting detached, I don't want to let myself get to the point where it could get any further. You hear these horror stories all the time and I refuse to let myself get to that point. So, if there's any warning signs of any kind, I listen to them and I'll go lay down and go to sleep. If I need to eat then I'll eat . . . I don't want to hallucinate, I don't want to be in la-la land where I have no control over what I'm seeing . . . I know when I get these little signs that it's time to go to bed, then I don't let it over power me. I never let it come before my rent or food! Ever since I started doing it, I made money on it so I can have my own personal stash, without it coming out of money that is already spent. (297)

PROFESSIONAL PRIDE AND ETHICS

Another major finding is the degree to which women dealers discussed the concept of pride in their accomplishments as dealers. The most frequently mentioned reasons centered on their achievements in operating and maintaining a competent business enterprise. Many referred to the bias and difficulties in dealing with men suppliers and customers in particular. One woman, for example, articulated the reasons behind a strong degree of pride in her profession and in her ability to run a successful business enterprise:

> I'm a good dealer. I don't cut my drugs. I have high quality drugs in so far as it's possible to get high quality drugs. I want to be known as somebody who sells good drugs, but doesn't always have them, opposed to someone who always has them and sometimes the drugs are good. (075)

Another female respondent who was formerly a high level dealer provides another example:

> Successful people will always succeed and I don't care if they do drugs or not! There are successful people and they will always succeed. I feel I'm one of those. I turned a real bad situation around, like getting fired from my government job of 15 and a half years invested! I turned that around from a very negative thing. . . . turned that around to be a positive. That was to take my money out and buy a business, re-vamp it and have a really fine, workable, collectable antique store that is respected and people desire to shop there. (322)

Importantly, all female dealers tended to have an ability to articulate an ethic that structures both their enterprise and which also speaks to their lives overall. One woman provided an eloquent example:

> It all depends on what your ethics and morality is. Mine is that I don't burn people, I don't fuck with the drugs. The drugs I sell are the drugs I do. If I see that somebody is not handling their drugs, I will not sell to them! I may not tell them that I think they're not handling their drugs because if they aren't, they aren't going to listen to me

anyway! But, it's easy for me to disappear from their venue for a while. More ethics and morality is that you don't talk about who you buy from, you don't talk about who you sell to. You do an honest and clean business. You don't fuck people up. You never sell people their first line! I expect the people that I buy from to be as honest as I am! (075)

INCORPORATING CONTROL AS A DEALING STRATEGY

The issues of self-control and control over others are dynamically interconnected in the lives of these women. We found a notable difference related to the issue of control over one's life framing the context of methamphetamine dealing. For many women dealers, the illicit drug business gave them the power to leave and/or control their husbands/boyfriends. It allowed them to choose their friends, to control their lifestyle, and to maximize their talents relative to their resources. Most of them considered themselves satisfied realists. One woman dealer who prides herself on her ability to have control over her life stated:

It's my business, my rules! And don't argue with me about it! I get to choose who knows I sell speed. I'm healthier now than I was when I started. I think it's because I don't have any moral dilemma over my drug use. I do drugs for a reason which I have rationalized to myself . . . they are now part and parcel with the whole thing. (075)

A basic rule mentioned by many serious dealers concerns protection of their home environment.

. . . never sold to people from within my home. I would have to go out of my home environment to give it to them or to sell to them. I can't bring people that are using a drug that causes such instability in their life into my home and expect my own lifestyle to stay stable. That would be a paradox within itself! (293)

A major reason for protecting the home, was to separate family life and business, especially women dealers with small children. Faced with living on welfare after her divorce, one woman

returned to dealing and now reports making over $60,000 per year. She states:

I make money from sales, I never do what I need to sell. I won't take it from my kids or my rent or my food or bills, I wanted off welfare, so I began to sell again. I'm not a stupid person. I don't go around doing stupid things. I don't walk around telling people I have drugs for sale, I don't have people sitting out in front of my house, I don't have traffic in and out of my house. (297)

A San Diego woman who dealt in pounds, did do business from home, but established a long list of rules to guide her customers:

They had to call first and tell me how much they wanted. They'd come by themselves. I had to know beforehand that they was bringing somebody new or I had to know the person they was bringing. Nobody ever sit out in cars in front of the house. They had to come in. Nobody could hang out in front of my house. Once the deal was made, there was no hanging around, partying. You had to leave. I wasn't a party person with business, this was a business! I gave you time to try the product out if you wanted to. If you liked it, you put your money down and left my house! I couldn't take the chance of being busted with all that stuff there. (281)

A woman of mixed Hawaiian ethnicity dealt in the heart of one of Honolulu's ice communities. Even though ice dealing was common, and use open, in her neighborhood, she set up rules to assure her participation was discreet.

I was so low profile, not even my own home had traffic. I had good rules and regulations that everyone had to abide by or they didn't get none! . . . no calls after 10:00 at night. No dealings after 10:00 at night. Call my house before you come over, don't just show up. If you see me out in the streets and I'm not expecting you, don't expect to score, you won't get shit! And I'd cut their line! If I say "No more nothing" they know why. They disobeyed my rules. (542)

For some dealers the best method of avoiding an out-of-control use pattern, increasing their use, and thus decreasing profits, was to establish a business ethics involving customers. Other people only did business with people they knew, or who had been introduced by people they knew.

> Even now, my business is largely a matter of acquiring, despite the fact that I can move a lot if I want to, both in quality and in quantity of small amounts or big amounts, my business is a matter of me getting drugs and supplying a selection of a few people. I change the list of people I'm willing to supply regularly. (075)

Other dealers limited their sales to a higher class customer. One woman learned the hard way after a police raid based on a customer-turned-informant. After that she changed her habits and her buyers.

> I wasn't selling to anybody on the street. I was only selling and made only deliveries. Or, the people came during business hours. They were all professional people, they had money to spend on drugs, they were all employed or their husbands or spouses were employed and they did it together. That is who my clients were. People that maintained control, only! People that are able to get up and go to work every single day and still do drugs every day, I feel, can have a sense of control and maintain control. I charged them top dollar for the drug because I only had quality drugs. I wasn't that greedy. If I'm gonna sell a drug, it better be the best. These were professionals who didn't mind paying for the best. (322)

Although we found this sample of women to be motivated by three normative guidelines: pride, ethics, and self-control, none were, in themselves, linked to gender relations.

SUMMARY—CITIZENS AND OUTLAWS

Among the varied and rich discoveries about women found in this study, several emerge as crucial building blocks toward establishing a realistic portrayal of women and illicit drugs. They strip away the primary rationales used as the basis for much of the misconceptions now found in the literature. These include the contention that women have less self-control over their drug use than men; that women suffer more from guilt and low self-esteem than men; and that women use and sell drugs in a subordinate relationship to men. Fundamentally, this study revealed that women's lives in the illicit drug economy are as varied as men, and that they are much more hidden. For example, we found women could simultaneously experience lives as "citizens"—that is, good wives and mothers—as well as "outlaws"—users and sellers of illicit drugs. The experience was almost always dynamic and fluid.

Taken together, these findings clearly underscore the need to move beyond the overwhelming urge of the past decade to focus exclusively on cocaine use (predominantly crack cocaine). It is equally imperative to move beyond the traditional hidden populations of disenfranchised minority women in large concentrated urban environments (cf. Maher and Curtis 1992; Dunlap and Johnson 1996). Our knowledge of the role of women in the illicit drug economy has been limited to existing structural relationships and conditions most common in these environments. Consequently, much available information on women in the illicit drug economy, by the nature of these limitations, is restricted to street-level low-status involvement often centered around sex workers (Ratner 1993; Maher 1996; Fagan 1994).

In reality, both the use and selling of drugs takes place within social worlds that are much broader and more complex. Most illicit drug users, male or female, do not live in large inner-city environments (Substance Abuse and Mental Health Services Administration 1995). They are not members of impoverished minority populations, and are not primarily crack cocaine users (Substance Abuse and Mental Health Services Administration 1995). Most illicit drug users are not outlaws living on the extreme marginal edge of society.

Furthermore, we found that women's involvement with the illicit drug economy was governed by logical and coherent motives. The research also provided evidence that women who use—and sell—illicit drugs represent a widely diverse population. And finally, the life histories of these women users revealed drug-dealing careers spanning several decades rather than being a recent phenomenon. This demonstrates our need to uncover a previously unknown population of women dealers, to locate and then chart the boundaries of this unknown hidden population.

REFERENCES

Curlee, J.
1967   Alcoholic women. *Bulletin of the Menninger Clinic* 31:154–163.

Dunlap, E., and B. D. Johnson
1996   Family and human resources in the development of a female crack seller career: Case study of a hidden population. *Journal of Drug Issues* 26(1): 175–198.

Ettorre, E.
1992   *Women and substance use.* New Brunswick, N.J.: Rutgers University Press.

Fagan, J.
1994   Women and drugs revisited: Female participation in the cocaine economy. *Journal of Drug Issues* 24:179–225.

Geertz, C.
1973   *The interpretation of cultures.* New York: Basic Books.

Hirsh, J.
1962   Women and alcoholism. In *Problems in addiction,* ed. W. C. Bier. New York: Fordham University Press.

Johnson, B. D., A. Hamid, and H. Sanabria
1991   Emerging models of crack distribution. In *Drugs and crime: A reader,* ed. T. Mieczkowski, 56–78. Boston: Allyn and Bacon.

Maher, L.
1992   Punishment and welfare: Crack cocaine and the regulation of mothering. *Women and Criminal Justice.* 3:35–70.

Maher, L.
1996   Hidden in the light: Occupational norms among crack-using street-level sex workers. *Journal of Drug Issues* 26(1):143–173.

Maher, L., and R. Curtis
1992   Women on the edge of crime: Crack cocaine and the changing contexts of street-level sex work in New York City. *Crime, Law and Social Change* 18:221–258.

Mieczkowski, T.
1992   *Drugs, crime and social policy.* Boston: Allyn and Bacon.

Mieczkowski, T.
1994   The experiences of women who sell crack: Some descriptive data from the Detroit Crack Ethnography Project. *Journal of Drug Issues* 24:227–248.

Morgan, P., J. Beck, K. Joe, D. McDonnell, and R. Guiterrez
1994   *Ice and other methamphetamine use.* Final report to the National Institute on Drug Abuse, National Institute of Health. Washington, D.C.: U.S. Government Printing Office.

Morgan, P., and K. Joe
n.d.   Uncharted terrain: Contexts of experience among women in the illicit drug economy. *Women and Criminal Justice.* Forthcoming.

Ratner, M. S.
1993   Sex, drugs and public policy: Studying and understanding the sex-for-crack phenomenon. In *Crack pipe as pimp: An ethnographic investigation of sex-for-crack exchanges,* ed. M. S. Ratner, 1–36. New York: Lexington Books.

Substance Abuse and Mental Health Services Administration (SAMSA)
1995   *National household survey on drug abuse: Main findings 1992.* United States Department of Health and Human Services. Washington D.C.: Government Printing Office.

Waldorf, D., and S. Murphy
1995   Perceived risks and criminal justice pressures on middle class cocaine sellers. *Journal of Drug Issues* 25(1): 11–32.

Woodhouse, L.
1992   Women with jagged edges: Voices from a culture of substance abuse. *Qualitative Health Research* 2:262–281.

Part VIII

DESISTANCE FROM CRIME

Participation in criminal activity does not necessarily mean that an offender will continue constantly to offend. That is, criminality oscillates throughout one's lifetime; changes vary depending on the shifting circumstances for that individual. Leaving a criminal lifestyle is very much dependent on an adherence to conventional values, with continuing association with conventional others as well as opportunities for reintegration within society. Desistance from criminal activity may be related to fear of incarceration, aging, risks involved to that person, and an introspective look at that person's life coupled with a strong desire to change. In short, offenders have many obstacles to face when they desire to exit from the criminal world. They have to join a conventional world, which is often reluctant to accept them, and they have to conform to a lifestyle and value system that is disdained by the criminal subculture. In addition, they experience difficulty gaining meaningful employment, especially if they are unskilled. The following article selections focus on the trials and tribulations that criminal offenders face in their attempts to join the conventional world. Their stories are real and often inspiring.

The majority of criminals do not want to spend their entire lives engaged in crime. Patricia Adler discusses the process by which offenders burn out of criminal behavior. Her research over several years with upper-level drug dealers and smugglers analyzes the shifts and oscillations in their criminal careers. She concludes that after a considerable amount of time as active marijuana smugglers, many participants in this criminal activity begin to perceive that the rewards gained do not exceed the drawbacks of the profession. Often the thrills, excitement, and monetary rewards that drug runners first experienced in the initial stages of smuggling begin to turn to paranoia due to people they know in the business getting arrested, along with their own perceived increased risk of apprehension. Adler goes on to explain that after years of using drugs, together with transporting them, this lifestyle begins to take its toll on them physically, and the conventional world that they once looked upon negatively begins to look a lot better. The author discusses those difficulties that dealers have in quitting and how often they return to the easy, high-spending world they have so much difficulty leaving. This leads to patterns of desistance characterized by attempts at quitting and returning because exiting is so difficult for them.

Sommers, Baskin, and Fagan examine exiting criminality in their study of how women offenders attempt to leave their criminal lifestyle. Researching a population of women who had a history of involvement in violent street crime, the authors analyze those distinct circumstances women experience once they have decided to desist from further illegal behavior. The authors point to varied factors that female offenders experience in their criminal activity that have caused for them a genuine desire to get out of a life of crime, the method they utilize to accomplish this change, and once they have left, how they sustain the difficult task of not returning to the criminal environment that they worked so hard to leave. This research project illustrates the complex process associated with desistance from crime and the struggles that such change causes.

In the final article dealing with desistance from crime, Meisenhelder explores the notion of certification, which he defines as the process by which people perceive ex-offenders as normal. According to the author, certification is an informal process that results in the acceptance of a former criminal as a person who has reentered conventional society and, above all, can be trusted. Certification, claims Meisenhelder, is a way in which former offenders attempt to deny a perception of themselves by conventional others as anything but a conventional person. In order to sustain a certified identity, ex-offenders must maintain normal relationships with noncriminals over an extended period of time and must convince others with whom they associate of their desistance from crime and their commitment to a reformed social identity as opposed to their former one as unworthy criminals.

29

The "Post" Phase of Deviant Careers: Reintegrating Drug Traffickers

Patricia A. Adler

Participation in a criminal enterprise does not necessarily mean that one continues this lifestyle forever. Pat Adler, in this article, shows how careers in crime oscillate over time. A successful transition or desistance out of a life of crime requires an adherence to a conventional life through associations with conventional others and opportunities for reintegration. In her study of upper-level drug dealers and smugglers over several years, Adler found that the negative drawbacks to living this type of existence exceed the rewards. Drug dealers experienced paranoia because associates were being arrested, and their risk of detection increased over time. However, those who are involved in the drug business at a high level find it difficult to leave this high-spending lifestyle. They are reluctant to give it up. Dealing and consuming drugs over time takes its toll physically, and they begin to see the conventional world as an alternative. The author explains that their attempts at desistance are characterized by many starts and returns as they experience quitting with much trepidation.

S cholars have noted that individuals' involvement in deviant or criminal activities displays many of the characteristics of legitimate careers, often compromising a beginning, an ascension, a peak, a decline, and an exit. The stages of these careers most often discussed in the literature are the entry and exit periods because they mark the boundaries of people's involvement in deviant worlds (Luckenbill and Best, 1981). Yet for many individuals, these careers encompass only a brief phase in their life span as they age and move out of deviance (Shover, 1985). What then happens to them? How do they make the transition into conventional society and its legitimate economy?

The issue of reintegrating deviants has received scant sociological attention. Braithwaite (1989) discussed the role of others in shaming individuals out of brief forays into deviance; Brown (1991) studied how people capitalize on their former deviant status and experiences to forge counseling careers; and Ebaugh (1988) analyzed the effects of previous roles on people's subsequent lives. Only a few studies (cf. Chambliss, 1984; Shover, 1985; Snodgrass, 1982) traced the lives of former deviants with the intent of discovering the paths they have taken and the effects of their former activities on their subsequent lives.

In this article, I offer a follow-up glimpse into the lives of 10 former upper level drug dealers and smugglers I studied in the 1970s. In my earlier writings, I described their active criminal careers and the attempts they made, often temporary and unsuccessful, to exit the drug world after years of trafficking and to find another way of making a living (Adler, 1985; Adler and Adler, 1983). Those members of my original sample whom I was able to locate in 1991 were all involved in other pursuits. Although they were, to varying degrees, reintegrated into more mainstream society, their lives had been indelibly affected by their years in trafficking. This, then, formed the postdealing phase of their deviant careers.

I begin by describing my return to the field and the activities of my former subjects whom I was able to trace. I then discuss a range of factors that affected their reintegration into mainstream society and the legitimate economy once they finally ended their careers in dealing. I conclude by considering those factors affecting the reintegrative stage of the deviant career.

METHODS

Between 1974 and 1980, I had the fortunate opportunity to be allowed entry into a community of upper level drug dealers and smugglers in a southwestern region of the United States. For 6 years my husband, Peter Adler, and I lived among these drug traffickers as friends, neighbors, and confidants, gathering ethnographic data on their hopes and fears, motivations, life style, business operations, and careers.

More than 10 years have passed since I moved away and left behind these drug traffickers and their scene. Over those years, I have maintained contact with a couple of my closest respondents and through them have managed to stay somewhat current with a few more. Change has come rapidly to the drug scene since that time; many transformations have occurred in the popularity of various drugs, the strategies of law enforcement, and the structure and organization of the drug business. Ten years out seemed like a good time to return to the field to see what had become of my former friends and subjects and to consider the effects of their experiences in drug trafficking.

SOUTHWEST COUNTY REVISITED

Dave, my key informant, was the first person I contacted when I went back to Southwest County in 1991. From 1972 to 1976, he had worked as a money launderer for a major smuggler and then branched out into dealing in large quantities on his own. He had brokered a ton of marijuana at a time or up to 60 pounds of cocaine per shipment. Over time, however, he began consuming too much, his marriage fell apart, and he started dealing with more disreputable associates until he was eventually abandoned by most of his former connections. I found him in a dilapidated motel, scrounging money on a near-daily basis, and sharing a room with a 26-year-old transient he had met while operating a surf shop in Florida. He was down on his luck and was keeping a low profile, avoiding people to whom he owed money. Through him I was able to find other former respondents and to hear about a couple more I could not find. Given the covert, unsteady nature of the business, in which people commonly leave no forwarding address, I was not surprised to encounter such a small group out of my original sample.

I knew what Dave had been up to since I left Southwest County. He had bottomed out of the drug market after losing so much money and reneging on so many fronts that no one would do business with him. He tried to get back into the real estate business in which he had worked as an agent before he met the Southwest County dealers and smugglers. Unfortunately, his license had been revoked when he was sent to prison, and his numerous attempts over the years to petition the real estate board for reinstatement had been repeatedly turned down. He then transferred his entrepreneurial buying and selling skills first to the flea market and county fair business, where he sold a variety of temporarily hot but inexpensive items, and then to the import business, buying legitimate goods (mostly clothing) from Mexicans he had met while dealing. During these years, he lived on the road for several months at a time, buying beat-up old vans and crisscrossing the country with his merchandise. He lived out of cheap motels, ate greasy foods, and bartered with other road hustlers and salespeople.

He eventually opened up a series of his own surf stores under a variety of assumed names (for the credit rating), but these went bankrupt one after another. He never became as successful financially as he had been during the early drug years because he never found a product that consistently enjoyed the same high level of consumer demand, and his careless business practices failed to improve.

However, Dave was getting older. After years of consuming drugs in large quantities and failing to take care of himself, he was feeling tired and less resilient. His two boys, whom he had dragged around the country from shop to shop, were in their early 20s, living on their own, and trying to go to college part-time in between partying and chasing girls. They were torn between following their parents' lifestyle of one scam after another and trying to find something more stable.

In his mid-40s, Dave was no longer a party animal. He could not tolerate the effects of cocaine any more. After years of overuse, he found that it depressed and confused him, causing him difficulty in forming words and physical gestures. He explained:

> I just don't get the euphoria anymore. If I ever get some and I do it, then I become sorry right away that I did it. So because of that I have to really keep away from the people that do it.

Dave was definitely out of the drug business now, even as a sideline. This also signified a major change in his lifestyle. Nearly all of the people he had associated with had been "coke-heads" of one sort or another.

Dave was not only tired, he was bored. He confided:

> I miss being on the road. I haven't been out of town for 9 months, and I'm stuck here staring at the walls of this motel room. I have nowhere to go. I have all this [surfing] merchandise in storage, but when I try to open up a store, I can't break even. Even last Christmas season. The economy is just terrible.

Four other people whom I tracked down through Dave were also out of the drug business. Marty quit after being busted and then stayed away from his former dealing friends. He was a big dealer and had gotten "hot" while I knew him. He had been one of those people whose name showed up on a drug agent's list, so he knew they were watching him. He vowed to "be careful," but he was not careful enough. He had a new wife when he got busted, and her feelings, along with the weight of all these factors, made him decide to retire for good. He described his attitude following his arrest:

> It was like my life was smashed. It was all I could do to just hang in there. I was smoking three packs of cigarettes a day. So I just said to myself, "Marty, you just have to be Mr. Joe 'Good Citizen.' You just have to get your life together." When you stand up there in front of that judge, you really feel the full weight of the law.

Marty had been a schoolteacher before he began dealing, and he eventually settled down with his wife into a similar steady job.

Two others, Ted, a former pilot-dealer, and Ben, a major smuggler, ended up "hustling for a living." They had escaped without ever having been arrested, which put them among the fortunate ones. Now they were in business for themselves, pursuing whatever deals they could while never finding a permanent stake in any line of legitimate work. Ted worked on screenplays for a while, which he attempted to peddle in Hollywood, and then made some money traveling around the country taking slides of railroad train cars, which he sold to collectors. Ben opened a restaurant with some partners that closed after 6 months and then tried to broker Zodiac rafts to specialty groups with little better fortune. It especially surprised me that Ben ended up with so little investment to transfer into a legitimate world. He had been the most successful trafficker I knew, the one who introduced Dave to the business. Ben, who had owned car dealerships, restaurants, and other ventures during his heyday, had nothing to show for it. However, he had had a long decline. His best years were during the era of commercial marijuana importation. When the cocaine trade rose to prominence, he began to indulge in overuse and in reclusive behavior, and he became less careful about his business dealings. Ben stayed in the business for longer than anyone I knew. During his later years, people did business with him as a favor based on who he was and what he meant to them. Over time, these people, too, quit the business, and there was no one left who knew Ben from the old days. Eventually, his money evaporated, and his wife left him.

The fourth person, Barney, was in the airplane business. A "trust-funder" from an upper-middle class background, Barney had never held a steady, legitimate job. After college, he and his wife, Betty, moved to Southwest County, and they used the freedom afforded them by his family income to avoid the shackles of employment. After a short while, they were drawn into the fast life by former college friends as well as new friends. Barney had always been interested in flying, and he used his free time to get his pilot's license, transporting family and friends around the country for vacations. Within 2 years of his entry into the drug crowd, he was recruited by a smuggler to fly runs for him. He worked transporting cocaine and spent his time partying for several years but got caught in a near-bust in South America. Reacting quickly, he was able to escape, one step ahead of the drug agents. Terrified by this close call, he returned home and fled Southwest County with Betty, and, by this time, their two children. They moved to a new state and established themselves, and Betty got a job. Barney could not bring himself to resort to employment, however, or to quit hanging around airports. He eventually managed to create an entrepreneurial

business for himself, buying and selling used airplanes. He supplemented this unsteady income with money from his continuing trust fund. He also changed his lifestyle, cutting down on his drug use significantly.

With the help of Dave's boys, I then tracked down Dave's ex-wife, Jean. Along with her sister, Marsha, Jean had become a major cocaine dealer in her own right after she divorced Dave. Her situation had subsequently gone through more ups and downs. She had divorced Jim, her second husband and successful dealing partner, blaming the erosion of their relationship on drugs:

> Too much cocaine. We were always doing it. Seemed like we needed to do it to have a good time. But then we were doing it separately from each other. He would go into his study, where he kept his secret stash and toot it up. I would go into the bathroom, where I kept mine, and toot. Eventually we got so wired that we weren't connecting with each other at all.

After the divorce, they each tried to stay in the dealing business separately, but neither was successful. Together they had complemented each other, she providing the hard-driving business acumen and security-conscious rules, he serving as the nice guy to round out her hard edges and frame their operation with class and generosity. When they tried to operate alone, Jean drove people away by being too demanding and Jim lost money by being too lackadaisical. He eventually moved to Hollywood. Starting from scratch, he managed to establish a successful venture catering to "the stars." Catering had been his and Jean's former legal front during their heaviest dealing years, and they had always groomed it to serve as their line of work after retirement.

In the catering business, they had been partners with another couple, Bobby and Sandy, who had owned the local fish store. Jean and Jim, regular customers of the fish store, became friendly with Bobby and Sandy and then discovered that they were heavy cocaine dealers. Jean was struggling to establish herself as a dealer after breaking up with Dave, and she welcomed this connection. Jean and Jim first bought cocaine from Bobby and Sandy, but their rise to trafficking in larger quantities was so rapid that they were soon dealing on the same level, fluidly selling back and forth as availability and connections dictated. They also did business with Jean's sister, Marsha, who sold to her younger circle of friends. With Jean and Sandy's shared interest in food, they launched the catering business together and were soon joined by Jim. The catering business was fun and successful for several years. At this time, money was rolling in from the drug dealing so well that they did not need the extra income, and so the work was sporadic and seasonal. The catering venture came undone through drugs. Around 1980, both couples began to freebase rather heavily. Bobby used up all of his and Sandy's money on a 2-month run and then checked into a rehabilitation clinic. Sandy moved out and returned to live with her parents. They divorced. Jim and Jean had also gotten to the point at which they considered their drug use out of control. They were sneaking behind each other's backs to snort cocaine at all hours. They went on freebasing runs that made them forget their obligations. They could not keep up with the catering business.

Upon my return to Southwest County, I learned through Jean that Bobby had since moved to Hawaii and opened a fish store. Sandy was living in San Francisco, married a chef, and was cooking in his restaurant. Jean and Marsha had had a terrible fight over Marsha's boyfriend, Vince, and had not spoken for several years. On a lead, I was able to locate Marsha in another state, now married to her former boyfriend. He was still working as an artisan, selling blown glass to stores, in fairs, and in mall shows. She worked in an office but helped out with his business during the months preceding Christmas, when they did most of their yearly business.

After Jean split up with Jim, she hit the bar scene for a while, picking up men and drinking a lot. She moved into a remote area and got a job in bar where she had once worked right after her divorce from Dave. Over the years, she drifted from bar to bar, waitressing, bartending, and serving as a bar, restaurant, or country club manager. She was fired from most of these jobs once for having her hand in the till, other times for irregular attendance, and was arrested

several times for drunk driving. When I spoke to her, she was employed at a country club as a bartender, working the club's catering events on the side. By her own admission, age had not dampened her ability to party; she could still get loose and have a good time far better than most of her contemporaries. After several years, she had settled into an on-again-off-again relationship with Cliff, a man who had a substance abuse problem. After drinking, he would get physically abusive in their relationship. She fled on several occasions, but she always returned because she had nowhere else to go. During their good times she spoke of getting married; during their bad times she hid from everyone.

Reflecting back on her dealing years in reference to her current life, she had mixed feelings:

> I wouldn't change a thing from the past, even if I could. Those years were great. Having as much money as we wanted, never having to worry about spending it. I really had a good time and I learned a lot from it. Not many people get to do all the things we did, and we did a lot of crazy things back then. But I'm not looking for the end of the rainbow anymore, all the scams and loose money. It was a wonderful thing to always have money, but we paid a price. I wouldn't want to live like that anymore. Now I like the comfort of my life. I like having a steady job. I like not having to worry about going to jail, having a driver's license that's legal.

My revisitation thus yielded results that were surprising in some ways and not in others. Direct or indirect follow-ups of 10 of my major subjects showed them all to be out of the drug business and involved in other ventures. As I originally described the scene, all of my subjects had been between the ages of 25 and 40 years. The people whom I returned to find were now well into their 40s and 50s and retired. Although their level of involvement in the drug business (carrying with it a greater potential for insulation and profit), coupled with the exit barriers they encountered (the treadmill of the dealing lifestyle and the removal of their credentials for legitimate professional work), might have hypothetically enabled or induced them to remain with the activity for even longer, they quit. Some quit with the help of Narcotics

Anonymous or various detoxification programs. Some quit because they had near-arrest escapes. Some were killed while involved in dangerous work. Most of the people I knew, however, just quit on their own. They quit, first, because like Waldorf, Reinarman, and Murphy's (1991) cocaine users, their troubles, from the physical burnout, to the diminished excitement, to the outright paranoia associated with their activities mounted. They quit, second, because the rewards of dealing, from the thrills, to the power, to the money, to the unending drugs, became less gratifying. They finally decided, either at some major turning point or more commonly over a gradual period, that they were tired of or unable to traffick in large quantities any longer.

Their attempts at getting out were not all successful. Many factors continued to hold them to the drug world and undermined their success in the legitimate world (see Adler and Adler, 1983). Their exits, then, tended to be fragile and temporary followed by periods of relapse into dealing. Reintegration formed the mirror image to the shifts and oscillations they made out of dealing: a series of forays into the legitimate world, many of them unsuccessful and temporary, but that were often followed by subsequent re-endeavors. Each attempt at reintegration, however, brought them further back into society and away from the insulated world of the fast life. Yet even once made, their reattachment to conventional society was problematic because of their many years out of the mainstream economy.

Factors Affecting Reintegration

What factors affected these people's lives and employment in society subsequent to their exit from the drug trade? Some were able to reintegrate more readily than others, finding a steady line of work. Others floundered, moving from activity to activity, unsuccessful and unsatisfied. Their final exit from the drug trade also saw them move into a variety of different venues characterized by distinct patterns. Factors affecting dealers' reintegration were rooted in the periods before, during, and after their dealing careers.

Predealing

One element influencing dealers' ability to ultimately reintegrate themselves into legitimate society was the age of onset in illicit activities. Individuals who became active in drug trafficking or other aspects of the underground economy at a young age remove themselves from pathways to options of legitimate success. Like Williams's (1989) cocaine kids, they drop out of school and fail to accumulate years of experience toward work in a lawful occupation. Their attitudes are also shaped by their early drug world experiences, so that they lose patience for legitimate work and seek the immediate reward of the scam and quick fix. When they become disenchanted with drug trafficking or are scared enough to think about quitting, they have no reasonable alternatives to consider. In contrast, individuals who enter the drug world after they have completed more schooling or have established themselves in lawful occupations have more options to pursue.

Very few of the dealers I studied entered the drug world at an early age. In fact, of my original sample, only five people went directly from high school into supporting themselves through drug money. None of the members of my follow-up sample experienced the early-onset pattern. Those whom I recontacted had all been in their late 20s and early 30s before becoming drawn into the drug world. Nearly all of them had been to college, and over half had graduated. Nine had been married, and four had borne children. This introduced an element of stability and responsibility into their lives and gave them some years of investment in the legitimate world that they could draw on in their future.

Related to this were the prior interests and skills individuals developed in the legitimate economy before they began to earn most of their income from dealing drugs. Very often, maturation and growth involves an identity-forming process in which individuals gradually narrow the range of career or occupational options to those in which they are interested. During this time, they may begin to pursue one or more of these avenues and gain knowledge or experience in these areas. Such occupational experience is later helpful in aiding dealers' reintegration because it can offer them an area to which they can return, an educational foundation, or a base of legitimate working experience that provides some transferable knowledge of and confidence about the legitimate world.

In my follow-up sample, Dave had worked as a real estate agent for the 4 years immediately before entering the drug business. He had previously held jobs as an automobile mechanic, an appliance salesperson, and the editor and publisher of a surfing magazine. Jean had been a homemaker and mother for many years, with sales experience in retail stores. Jim had pursued a career as a photographer, holding a job on the staff of a major national news magazine for many years. Marty had been a teacher in the secondary school system, and Bobby and Sandy had owned a fish store. Later, a couple of them returned to these early roots. For instance, Bobby and Marty resumed their original jobs as a retail fish store owner and a teacher, respectively, and Dave spent several years operating surf shops on both coasts.

A third factor affecting dealers' reintegration was the social class in which they were born and raised. When people grow up and become accustomed to a certain standard of living, they are reluctant to engage in downward social mobility. One of the noteworthy aspects of the members of my sample is their middle-class background. They are likely representative of a more widespread, hidden, middle-class population that is involved in the illicit drug trade on either a full- or part-time basis (cf. Morley, 1989; Rice, 1989). Such people move into drug trafficking to enhance their middle-class, materialist lifestyle, and when they leave the fast-money world their ties to the middle-class lifestyle force them to reintegrate into legitimate society more quickly. All of my subjects had grown up in upper-middle-class, middle-class, or lower-middle-class backgrounds, with parents engaged in a variety of occupations from educators to career military to manufacturers and sales.

CONCURRENT ACTIVITIES

The manner and style in which individuals comported their lives during the active phase of their trafficking careers also affected their later efforts at reintegrating. One of the most salient features toward this end was the dealers' degrees of outside involvement. A large number of drug dealers are engaged in other ventures in addition to trafficking. At both the upper and lower levels of the drug trade, individuals can participate in trafficking in either a part- or full-time manner. For example, Reuter et al. (1990) found that most of the arrested lower level crack dealers they surveyed in their Washington, DC sample were employed in full-time legitimate jobs but moonlighted as dealers to supplement their legitimate incomes. Similarly, at the upper levels, a whole coterie of accountants, bankers, lawyers, pilots, and other legitimate businesspeople are involved part-time in the drug economy, arranging smuggling runs or providing illicit services to full-time traffickers (Morley, 1989; Rice, 1989). These individuals who deal only part-time are likely to have more interpersonal, occupational, and economic factors tying them to society. When they renounce their dealing, they have fewer reintegration obstacles to face because they are already more fully integrated into the legitimate economy.

In contrast, those who deal full-time, like Williams's (1989) youthful Dominican-American dealers, Hamid's (1990) Caribbean-American dealers, and my white upper level dealers and smugglers, have more likely renounced their investment in socially sanctioned means of surviving financially. Of the 65 subjects in my original sample, nearly one third remained involved in their previous jobs for a significant period of time while they were dealing in large amounts. This was the case for all but one member of my follow-up sample as well. Eventually, they all renounced those jobs, however, and withdrew to dealing or smuggling as their primary activity and main source of economic support. Abandoning legitimate jobs or career tracks makes it significantly harder to re-enter these lines after several years away, and only Marty, the teacher, who remained in the classroom well into his dealing career, was able to subsequently find another job in his profession.

Yet, even while engaged in full-time trafficking, individuals can become involved with legitimate front businesses on the side. Because they were making so much money, nearly two thirds of my original sample had pretensions of being involved in a legitimate business (for the purpose of protecting themselves from the Internal Revenue Service). This figure roughly applies to the follow-up group as well. Jean, Jim, and Sandy worked in the catering business, Ben owned an automobile dealership, and Marsha ran an antique store. These occupations kept these people partially tied to the legitimate economy and made it subsequently easier for them to re-enter that economy on a serious basis. Maintaining some connection to the lawful world of work, as Meisenhelder (1977) noted, also implies a lifestyle commitment to keep some regular business hours. For these people, then, reintegration did not require as much of a lifestyle transformation as it did for those who had not worked at all. Thus, Jim eventually began his own successful catering business and Sandy worked in a restaurant, skills they had acquired in their legitimate front businesses.

Knowledge and experience about legitimate work that was potentially useful to individuals' later reintegration could come not only from outside involvements but from trafficking-related skills as well. For example, Barney took his piloting skills and turned them into an airplane-oriented business. At the very least, the dealers and smugglers I studied became educated and trained in handling money, working on credit, calculating profits and expenses, and living with the uncertainty of entrepreneurial business. Those with the discipline and business acumen to become successful in the drug world were often able to recreate some semblance of this outside arena. Others with less reputable and reliable approaches, who had survived in the drug business primarily on the selling strength of the product, did not usually fare as well.

Instrumental aspects of these dealers' lives were not the only significant factors affecting their later reintegration. The strength of these traffickers' outside associations were important as well. This included interpersonal relationships with their children, parents, siblings, close friends, and other family members. Such associations were important because they kept these dealers integrated, to some degree, into mainstream society. These upper level dealers and smugglers trod a delicate line, as they lived inside a conventional society and yet insulated themselves within it. That is, they ate at the same restaurants, sent their children to the same schools, and lived in the same housing developments as the other people, yet they kept their social contacts with those outside the drug world to a minimum. For protection, they removed themselves from the inquisitive prying of people who would not accept their occupation and lifestyle. Some Southwest County smugglers and dealers went for long periods, then, without seeing former friends and relatives. Others, though, kept in touch with their most important associates, whether they lived locally or at a distance. Through these ties, they remained connected to individuals and social worlds outside of dealing. These associations would be crucial for them to draw on in their re-entry to the legitimate world.

Such outside associates are not likely to "steer individuals away" from their deviance, as Braithwaite (1989) suggested, nor are they likely to provide a ratio of definitions favorable to the law and thereby reorient dealers' and smugglers' normative attitudes, as Sutherland's differential association theory holds. Rather, they hold traffickers, to greater or lesser degree, from totally removing themselves from society and provide a bridge back into society when these individuals feel an internal push to re-enter it.

Both outside involvements and outside associations serve as the type of "bonds to society" described by Hirschi's control theory. Although people diminish these bonds during their careers in trafficking, they do not cut them entirely. They then reach out to strengthen them during their attempts to move out of dealing and reintegrate. Other deinsulating factors serve as bonds to society as well, maintaining these dealers' connections to the mainstream and easing them back. This included ties like sports and hobbies (cf. Irwin, 1970) and could have included potential others such as religion. The dealers and smugglers in my sample all held onto some vestiges of their sport and hobby interests, rooting for their favorite teams, pursuing sports such as tennis or skiing, and indulging themselves in collecting things such as antiques or travel mementos. Religion, however, was not a significant part of their lives. Some individuals had come from religious backgrounds, attending parochial schools and church regularly, but this ended even before the onset of their dealing careers. Although most of my follow-up subjects continued their sport and hobby interests after they quit dealing, none returned to religion.

A final factor characterizing dealers' active career behavior that affected their success and type of reintegration was the degree of organizational sophistication associated with their trafficking activity. Drug traffickers' involvement in deviant associations may follow a continuum of organizational sophistication, beginning with lone operators ensconced in a collegial subculture at the lowest end and ascending to the loose associations of the crack house crews, with the more organized smuggling rings, the tighter and more serious delinquent gangs, and the deadly Colombian cartels or organized crime families at the highest end of the spectrum. We know from the broader study of deviance that individuals who are members of more organizationally sophisticated associations are more likely to be tied to those groups instead of integrated into society in a number of ways (see Best and Luckenbill, 1982).

The drug traffickers in my follow-up group, much like those from my original sample, represented a mixture of both dealers and smugglers. As such, their involvement in criminal organizations ranged from the lone operator to the member or leader of a smuggling ring. Although they were clearly drawn into this occupation and lifestyle, they made no lifetime commitment to the pursuit. Detaching and reintegrating into society required a major change of master status but had fewer unbreakable side bets.

POSTDEALING

Drug traffickers' success at reintegrating into society was also affected by several factors they could encounter subsequent to their involvement in dealing. As they oscillated back and forth between their phases of dealing and quitting, the availability of legitimate opportunities seriously affected their permanence of retiring. As Shover (1985) noted in his study of thieves, finding a satisfying job could tie an individual to a line of activity. A positive experience at a legitimate job can draw an individual back to more conventional peer associations, reinforce nondeviant identity, occupy significant amounts of time, and diminish the motivation to return to dealing. Although "straight" jobs were often looked upon with disdain by Southwest County dealers in their younger days, they were more likely to view them favorably at this later age. This was reflected in Jean's comments about her changed attitudes toward her current lifestyle and work.

The importance of opportunity structures illustrates the value of Cloward and Ohlin's differential opportunity theory for reintegrating drug traffickers. Those who had tried to oscillate out of dealing but could never find anything to support themselves in their hedonistic lifestyle returned to the drug world. Yet each time they attempted to quit reflected a greater dissatisfaction with the dealing life. After a while, a more somber job opportunity, even Jean's bartending and waitressing job, appeared attractive.

Some drug traffickers were aided in their societal reintegration by outside help. As Braithwaite (1989:100–01) noted, friends, associates, and acquaintances can aid former deviants' reintegration through "gestures of forgiveness" or "ceremonies to decertify offenders as deviant." Dealers may thus remove themselves from the scene, find a new life, make new friends, or meet a spouse (cf. Shover, 1985) who may help them start over. This begins the process of rebuilding the social bonds that tie individuals to legitimate society. For instance, Marty was strongly influenced in his move to reintegrate back into society by his new wife, who was opposed to his dealing activity, and Marsha was forcibly pulled from the dealing subculture by her boyfriend, Vince, who had never liked her hanging around with the dealing crowd.

The extent and type of dealers' reintegration into society were ultimately affected by their adaptability to the organizational world. In my revisitation to Southwest County, I observed a continuum, with those whose experiences in the days of the wheeling and dealing and the big money had left them permanently unsuited for work as an employee in the organizational or bureaucratic world at one end and those who sought and obtained jobs at the other. Some of my subjects could never stoop to getting a job. They had entered the dealing world to secure freedom for their "brute being," and they would not endure the shackles of becoming an employee. Others were willing to get back into the working world, even if it meant taking a job at the bottom. Interestingly, finding oneself in "dire straits," as Dave and several others had, proved an insufficient inducement to an entrepreneurial, freedom-seeking person to get a job. They all had their limits, below which they would not stoop.

None of the wheeler-dealers, then, entered the confines of the straight "work-a-day" employee world they had either fled or disdained in the first place. Like many legitimate entrepreneurs, they could not imagine themselves punching a clock or working for someone else. Having tasted the excitement of the drug world, the straight world seemed boring. For them, staying within the world of independent business was associated with the potential for freedom and adventure. They could still dream of making the big killing and retiring. This also enabled them to avoid the awkwardness of trying to explain on a resume how they had been earning a living during their dealing years. Like Dave and Barney, then, they became petty lawful entrepreneurs, leaving their glory days behind them.

In contrast, like Jean, some former dealers sought out a variety of jobs. Several of them had worked as employees before entering the drug world, whereas others tried to put together a legal business front that required them to work some regular hours during their dealing years. They did not feel uncomfortable, then, but rather enjoyed the assurance of a steady job and a predictable life. For them, quitting dealing and working became associated with security, domesticity, and freedom from paranoia. Their experiences with life as an employee, however, were never quite as predictable or as steady as the average worker.

DISCUSSION

Yet much crime is committed by individuals who begin criminal activities in their relative youth, much like drug traffickers, without intending to remain criminals all their lives. They enter these activities thinking they will make a lot of money and retire into some less dangerous line of work. Studies of deviant careers, in fact, show that many criminals and deviants (especially those who have never been incarcerated) naturally burn out, bottom out, grow out, and quit (Harris, 1973; Irwin, 1970; Waldorf, 1983; Waldorf, Reinarman and Murphy, 1991). Once they have made the decision to exit deviance, their success depends largely on their ability to reintegrate into society.

My research suggests that they return because they have evolved through the typical phases of their dealing careers and, like their peers, progress past the active stage into the inactive stage. With variations on the theme unique to each individual, dealers experience a progression through their early entry and involvement in the drug world, a middle period in which they rise and experience shifts in their level and style of operation, an exit phase in which they suddenly or gradually withdraw from the drug world, and the last phase, in which they readapt themselves to the nondeviant world. Their eventual return to conventional society requires a process of reintegration, which is affected by the structural factors described here. They reintegrate, then, more because of "push" than "pull" factors, because the career of involvement in drug trafficking moves them beyond the point at which they find it enjoyable to the point at which it is wearing and anxiety provoking. Only once they have made the decision to leave the drug world, either temporarily or finally, do they reactivate their abandoned ties to the network of conventional society's attachments. This occurs, as Shover (1985) noted, after they change their orientational (self-conceptions, goals, sense of time, tiredness) and interpersonal (ties to people or activities) foci, finding it preferable to detach from their deviant and criminal commitment and to reblend with the conventional society.

Individuals' postdealing lives are thus profoundly affected by their years in the drug world. The attitudes, values, and lifestyle they adopted during the active phase of their dealing careers remain nascent within them. Most are straight for pragmatic rather then ideological or moral reasons. The "quick buck" and the "sweet" deal thus remain embedded within their vocabulary of motives. Although these individuals may be too old to keep up with their former drug-using pace or to return to the fast life, many still enjoy a touch of hedonism. In an era when most middle-aged people are former marijuana smokers, party drinkers, and general revelers, these extraffickers still like to have adventures. It remains a part of their lifestyle and new identity, carried over from earlier times. Thus, although they have shed the dealing occupation, many retain some proclivity for deviant attitudes and lifestyles. They are post-dealers but not completely reformed deviants. They live near the fringes of conventional society, trying to draw from both within and outside of it.

REFERENCES

Adler, P.A. 1985. *Wheeling and Dealing.* New York: Columbia University Press.

Adler, P.A. and P. Adler. 1983. "Shifts and Oscillations in Deviant Careers: The Case of Upper-Level Drug Dealers and Smugglers." *Social Problems* 31:195–207.

Braithwaite, J. 1989. *Crime, Shame, and Reintegration.* Cambridge, England: Cambridge University Press.

Brown, J.D. 1991. "The Professional Ex: An Alternative for Exiting the Deviant Career." *The Sociological Quarterly* 32:219–30.

Chambliss, W.J. 1984. *Harry King: A Professional Thief's Journey.* New York: Wiley.

Ebaugh, H.R. 1988. *Becoming an Ex.* Chicago: University of Chicago Press.

Hamid, A. 1990. "The Political Economy of Crack-Related Violence." *Contemporary Drug Problems* 17:31–78.

Harris, M. 1973. *The Dilly Boys.* Rockville, Maryland: New Perspectives.

Irwin, J. 1970. *The Felon.* Englewood Cliffs, New Jersey: Prentice-Hall.

Luckenbill, D. and J. Best. 1981. "Careers in Deviance and Respectability: The Analogy's Limitations." *Social Problems* 29:197–206.

Meisenhelder, T. 1977. "An Exploratory Study of Exiting from Criminal Careers." *Criminology* 15:319–34.

Morley, J. 1989. "Contradictions of Cocaine Capitalism." *The Nation:* 341–47.

Reuter, P. 1983. *Disorganized Crime.* Cambridge: MIT Press.

Reuter, P., R. McCoun, and P. Murphy. 1990. *Money from Crime.* Santa Monica: Rand.

Rice, B. 1989. *Trafficking.* New York: St. Martin's.

Shover, N. 1983. "The Latter Stages of Ordinary Property Offenders' Careers." *Social Problems* 31:208–18.

———. 1985. *Aging Criminals.* Newbury Park, California: Sage.

Snodgrass, J. 1982. *The Jack-Roller at Seventy: A Fifty-Year Follow-up.* Lexington, Massachusetts: Lexington Books.

Waldorf, D. 1983. "Natural Recovery from Opiate Addiction: Some Social Psychological Processes of Untreated Recovery." *Journal of Drug Issues* 13:237–80.

Waldorf, D., C. Reinarman and S. Murphy. 1991. *Cocaine Changes.* Philadelphia: Temple University Press.

Williams, T. 1989. *The Cocaine Kids.* Reading, Massachusetts: Addison-Wesley.

30

GETTING OUT OF THE LIFE: CRIME DESISTANCE BY FEMALE STREET OFFENDERS

IRA SOMMERS, DEBORAH R. BASKIN, AND JEFFREY FAGAN

Those who have lived and functioned in a criminal subculture often find exiting from this form of existence difficult at best. Sommers, Baskin, and Fagan studied how women who were actively involved in a criminal life attempt to leave their past existence and move to a more conventional one. Female participants who had been involved in violent street crime and were trying to transition out faced many obstacles in their desire to go straight. The subjects studied were 30 women who had prior criminal histories, but had desisted from criminal activity for a minimum of 2 years. Interviews focusing on their life histories were conducted, with three being the number of prior incarcerations for the study group. The desire for a lifestyle transition out of crime was based on fear of incarceration, and these women attempted to establish relationships with conventional others while simultaneously leaving their criminal associates.

Studies over the past decade have provided a great deal of information about the criminal careers of male offenders. (See Blumstein et al. 1986 and Weiner and Wolfgang 1989 for reviews.) Unfortunately, much less is known about the initiation, escalation, and termination of criminal careers by female offenders. The general tendency to exclude female offenders from research on crime and delinquency may be due, at least in part, to the lower frequency and comparatively less serious nature of offending among women. Recent trends and studies, however, suggest that the omission of women may seriously bias both research and theory on crime.

Although a growing body of work on female crime has emerged within the last few years, much of this research continues to focus on what Daly and Chesney-Lind (1988) called generalizability and gender-ratio problems. The former concerns the degree to which traditional (i.e., male) theories of deviance and crime apply to women, and the latter focuses on what explain gender differences in rates and types of criminal activity. Although this article also examines women in crime, questions of inter- and intragender variability in crime are not specifically addressed. Instead, the aim of the paper is to describe the pathways out of deviance for a sample of women who have sig- nificantly invested themselves in criminal social worlds. To what extent are the social and psychological processes of stopping criminal behavior similar for men and women? Do the behavioral antecedents of such processes vary by gender? These questions remained unexplored.

Specifically, two main issues are addressed in this paper: (1) the role of life events in trig- gering the cessation process, and (2) the rela- tionship between cognitive and life situation changes in the desistance process. First, the crime desistance literature is reviewed briefly. Second, the broader deviance literature is drawn upon to construct a social-psychological model of cessation. Then the model is evaluated using life history data from a sample of female offenders convicted of serious street crimes.

THE DESISTANCE PROCESS

The common themes in the literature on exiting deviant careers offer useful perspectives for developing a theory of cessation. The decision to stop deviant behavior appears to be preceded by a variety of factors, most of which are nega- tive social sanctions or consequences. Health problems, difficulties with the law or with main- taining a current lifestyle, threats of other social sanctions from family or close relations, and a general rejection of the social world in which the behaviors thrive are often antecedents of the decision to quit. For some, religious conver- sions or immersion into alternative sociocultural settings with powerful norms (e.g., treatment ideology) provide paths for cessation (Mulvey and LaRosa 1986; Stall and Biernacki 1986).

... A model for understanding desistance from crime is presented below. Three stages characterize the cessation process: building resolve or discovering motivation to stop (i.e., socially disjunctive experiences), making and

publicly disclosing the decision to stop, and maintaining the new behaviors and integrating into new social networks (Stall and Biernacki 1986; Mulvey and Aber 1988). These phases ... describe three ideal-typical phases of desis- tance: "turning points" where offenders begin consciously to experience negative effects (socially disjunctive experiences); "active quit- ting" where they take steps to exit crime (public pronouncement); and "maintaining cessation" (identity transformation):

Stage 1 **Catalysts for change**
Socially disjunctive experiences
- Hitting rock bottom
- Fear of death
- Tiredness
- Illness

Delayed deterrence
- Increased probability of punishment
- Increased difficulty in doing time
- Increased severity of sanctions
- Increasing fear

Assessment
- Reappraisal of life and goals
- Psychic change

Decision
- Decision to quit and/or initial attempts at desistance
- Continuing possibility of criminal participation

Stage 2 **Discontinuance**
- Public pronouncement of decision to end criminal participation
- Claim to a new social identity

Stage 3 **Maintenance of the decision to stop**
- Ability to successfully renegotiate identity
- Support of significant others
- Integration into new social networks
- Ties to conventional roles
- Stabilization of new social identity

Stage 1: Catalysts for Change

When external conditions change and reduce the rewards of deviant behavior, motivation may build to end criminal involvement. That process, and the resulting decision, seem to be associated with two related conditions: a series of negative, aversive, unpleasant experiences with criminal behavior, or corollary situations where the positive rewards, status, or gratification from crime are reduced. Shover and Thompson's (1992) research suggests that the probability of desistance from criminal participation increases as expectations for achieving rewards (e.g., friends, money, autonomy) via crime decrease and that changes in expectations are age-related. Shover (1983) contended that the daily routines of managing criminal involvement become tiring and burdensome to aging offenders. Consequently, the allure of crime diminishes as offenders get older. Aging may also increase the perceived formal risk of criminal participation. Cusson and Pinsonneault (1986, p. 76) posited that "with age, criminals raise their estimates of the certainty of punishment." Fear of reimprisonment, fear of longer sentences, and the increasing difficulty of "doing time" have often been reported by investigators who have explored desistance.

Stage 2: Discontinuance

The second stage of the model begins with the public announcement that the offender has decided to end her criminal participation. Such an announcement forces the start of a process of renegotiation of the offender's social identity (Stall and Biernacki 1986). After this announcement, the offender must not only cope with the instrumental aspects (e.g., financial) of her life but must also begin to redefine important emotional and social relationships that are influenced by or predicated upon criminal behavior.

Leaving a deviant subculture is difficult. Biernacki (1986) noted the exclusiveness of the social involvements maintained by former addicts during initial stages of abstinence. With social embedment comes the gratification of social acceptance and identity. The decision to end a behavior that is socially determined and supported implies withdrawal of the social gratification it brings. Thus, the more deeply embedded in a criminal social context, the more dependent the offender is on that social world for her primary sources of approval and social definition.

The responses by social control agents, family members, and peer supporters to further

criminal participation are critical to shaping the outcome of discontinuance. New social and emotional worlds to replace the old ones may strengthen the decision to stop. Adler (1992) found that outside associations and involvements provide a critical bridge back into society for dealers who have decided to leave the drug subculture. With discontinuance comes the difficult work of identify transformations (Biernacki 1986) and establishing new social definitions of behavior and relationships to reinforce them.

Stage 3: Maintenance

Following the initial stages of discontinuance, strategies to avoid a return to crime build on the strategies first used to break from a lengthy pattern of criminal participation: further integration into a noncriminal identity and social world and maintenance of this new identity. Maintenance depends in part on replacing deviant networks of peers and associates with supports that both censure criminal participation and approve of new nondeviant beliefs. Treatment interventions (e.g., drug treatment, social service programs) are important sources of alternative social supports to maintain a noncriminal lifestyle. In other words, maintenance depends on immersion into a social world where criminal behavior meets immediately with strong formal and informal sanctions.

Despite efforts to maintain noncriminal involvement, desistance is likely to be episodic, with occasional relapses interspersed with lengthening of lulls in criminal activity. Le Blanc and Frechette (1989) proposed the possibility that criminal activity slows down before coming to an end and that this slowing down process becomes apparent in three ways: deceleration, specialization, and reaching a ceiling. Thus, before stopping criminal activity, the offender gradually acts out less frequently, limits the variety of crimes more and more, and ceases increasing the seriousness of criminal involvement.

Age is a critical variable in desistance research, regardless of whether it is associated with maturation or similar developmental concepts. Cessation is part of a social-psychological transformation for the offender. A strategy to stabilize the transition to a noncriminal lifestyle requires active use of supports to maintain the norms that have been substituted for the forces that supported criminal behavior in the past.

METHODS

Constructing a sample of women who have desisted from crime poses several challenges. . . . Through the use of snowball sampling, 30 women were recruited and interviewed from January to May 1992. An initial pool of 16 women was recruited through various offender and drug treatment programs in New York City. Fourteen additional women were recruited through chain referrals. To be included in the study, the women had to have at least one official arrest for a violent street crime (robbery, assault, burglary, weapons possession, arson, kidnapping) and to have desisted from all criminal involvement for at least 2 years prior to the interview. Eligibility was verified through official arrest records and through contact with program staff for those women participating in treatment programs (87%).

The life history interviews . . . were open-ended, in-depth, and audio-taped. The open-ended technique created a context in which respondents could speak freely and in their own words. Furthermore, it facilitated the pursuit of issues raised by respondents during the interview but not recognized previously by the researcher. Use of this approach enabled us to probe for information about specific events and provided an opportunity for respondents to reflect on those events. As a result, we were able to gain insight into their attitudes, feelings, and other subjective orientations to their experiences. Finally, tape-recording the interviews allowed the interviewer to adopt a more conversational style, devote complete attention to the respondent, and concentrate on the discussion. By not being preoccupied with notetaking, interviewers were able to create a more relaxed atmosphere for the participants.

All interviews were conducted by the first two authors in a private university office. They were typically held in the morning or early afternoon and lasted approximately 2 hours. A stipend of $50 was paid to each respondent. In total, 30 interviews were conducted, comprising approximately 60 hours of talk.

With regard to self-reported drug and criminal histories, the data indicate that the study respondents had engaged in a wide range of criminal and deviant activities. All of the respondents were experienced drug users. Eighty-seven percent were addicted to crack, 70% used cocaine regularly, and 10% were addicted to heroin. Of the 30 women interviewed, 19 (63%) reported involvement in robbery, 60% reported involvement in burglary, 94% were involved in selling drugs, and 47% were at some time involved in prostitution. A great deal of variation was found with regard to incarceration history. The median number of incarcerations was 3 (the mean was 4.29); however, 17% of the women ($n = 5$) had never been incarcerated.

Patterns of offending for 17 of the 30 women (57%) resemble those of a sample of addicts studied by Anglin and Speckhart (1988, p. 223) in that "while some criminality precedes the addiction career, the great majority is found during the addiction career." It seems that for these women, the cost of their increasing substance use was a major influence to engage in offending, especially robbery and selling drugs. For the remaining 13 women (43%), drug abuse appears not to be as etiologically important to violent behavior, in spite of the high correlations between these two behaviors. Instead, addiction seems to be part of a more generalized lifestyle (Peterson and Braiker 1980; Collins et al. 1985) in which involvement in violent criminal careers precedes, yet may be amplified by, addiction.

RESULTS

Resolving to Stop

Despite its initial excitement and allure, the life of a street criminal is a hard one. A host of severe personal problems plague most street offenders and normally become progressively worse as their careers continue. In the present study, the women's lives were dominated by a powerful, often incapacitating, need for drugs. Consequently, economic problems were the most frequent complaint voiced by the respondents. Savings were quickly exhausted, and the culture of addiction justified virtually any means to get money to support their habits. For the majority of the women, the problem of maintaining an addiction took precedence over other interests and participation in other social worlds.

People the respondents associated with, their primary reference group, were involved in illicit behaviors. Over time, the women in the study became further enmeshed in deviance and further alienated, both socially and psychologically, from conventional life. The women's lives became bereft of conventional involvements, obligations, and responsibilities. The excitement at the lifestyle that may have characterized their early criminal career phase gave way to a much more grave daily existence.

Thus, the women in our study could not and did not simply cease their deviant acts by "drifting" (Matza 1964) back toward conventional norms, values, and lifestyles. Unlike many of Waldorf's (1983) heroin addicts who drifted away from heroin without conscious effort, all of the women in our study made a conscious decision to stop. In short, Matza's concept of drift did not provide a useful framework for understanding our respondents' exit from crime.

The following accounts illustrate the uncertainty and vulnerability of street life for the women in our sample. Denise, a 33-year-old black woman, has participated in a wide range of street crimes including burglary, robbery, assault, and drug dealing. She began dealing drugs when she was 14 and was herself using cocaine on a regular basis by age 19.

> I was in a lot of fights: So I had fights over, uh, drugs, or, you know, just manipulation. There's a lot of manipulation in that life. Everybody's tryin' to get over. Everybody will stab you in your back, you know. Nobody gives a fuck about the next person, you know. It's just when you want it, you want it. You know, when you want that drug, you know, you want that drug. There's a lot of lyin', a lot of manipulation. It's, it's, it's crazy!

Gazella, a 38-year-old Hispanic woman, had been involved in crime for 22 years when we interviewed her.

> I'm 38 years old. I ain't no young woman no more, man. Drugs have changed, lifestyles have changed. Kids are killing you now for turf. Yeah, turf, and I was destroyin' myself. I was miserable. I was . . . I was gettin' high all the time to stay up to keep the business going, and it was really nobody I could trust.

Additional illustrations of the exigencies of street life are provided by April and Stephanie. April is a 25-year-old black woman who had been involved in crime since she was 11.

> I wasn't eating. Sometimes I wouldn't eat for two or three days. And I would . . . a lot of times I wouldn't have the time, or I wouldn't want to spend the money to eat—I've got to use it to get high.

Stephanie, a 27-year-old black woman, had used and sold crack for 5 years when we interviewed her.

> I knew that, uh, I was gonna get killed out here. I wasn't havin' no respect for myself. No one else was respecting me. Every relationship I got into, as long as I did drugs, it was gonna be constant disrespect involved, and it come . . . to the point of me gettin' killed.

When the spiral down finally reached its lowest point, the women were overwhelmed by a sense of personal despair. In reporting the early stages of this period of despair, the respondents consistently voiced two themes: the futility of their lives and their isolation.

Barbara, a 31-year-old black woman, began using crack when she was 23. By age 25, Barbara had lost her job at the Board of Education and was involved in burglary and robbery. Her account is typical of the despair the women in our sample eventually experienced.

> . . . the fact that my family didn't trust me anymore, and the way that my daughter was looking at me, and, uh, my mother wouldn't let me in her house any more, and I was sleepin' on the trains. And I was sleepin' on the beaches in the summertime. And I was really frightened. I was real scared of the fact that I had to sleep on the train. And, uh, I had to wash up in the Port Authority.

The spiral down for Gazella also resulted in her living on the streets.

> I didn't have a place to live. My kids had been taken away from me. You know, constantly being harassed like 3 days out of the week by the Tactical Narcotics Team (police). I didn't want to be bothered with people. I was gettin' tired of the lyin', schemin', you know, stayin' in abandoned buildings.

Alicia, a 29-year-old Hispanic woman, became involved in street violence at age 12. She commented on the personal isolation that was a consequence of her involvement in crime:

> When I started getting involved in crime, you know, and drugs, the friends that I had, even my family, I stayed away from them, you know. You know how you look bad and you feel bad, and you just don't want those people to see you like you are. So I avoided seeing them.

For some, the emotional depth of the rock bottom crisis was felt as a sense of mortification. The women felt as if they had nowhere to turn to salvage a sense of well-being or self-worth. Suicide was considered a better alternative than remaining in such an undesirable social and psychological state. Denise is one example:

> I ran into a girl who I went to school with that works on Wall Street. And I compared her life to mine and it was like miserable. And I just wanted out. I wanted a new life. I was tired, I was run down, looking bad. I got out by smashing myself through a sixth-floor window. Then I went to the psychiatric ward and I met this real nice doctor, and we talked every day. She fought to keep me in the hospital because she felt I wouldn't survive. She believed in me. And she talked me into going into a drug program.

Marginalization from family, friends, children, and work—in short, the loss of traditional life structures—left the women vulnerable to chaotic street conditions. After initially being overwhelmed by despair, the women began to question and reevaluate basic assumptions about their identities and their social construction of the world. Like Shover's (1983) male property offenders, the women also began to view the criminal justice system as "an imposing accumulation of aggravations and deprivations" (p. 212). They grew tired of the street experiences and the problems and consequences of criminal involvement.

Many of the women acknowledged that, with age, it is more difficult to do time and that the fear of incurring a long prison sentence the next time influenced their decision to stop. Cusson and Pinsonneault (1986, p. 76) made the same observation with male robbers. Gazella, April, and Denise, quoted earlier, recall:

Gazella: First of all, when I was in prison I was like, I was so humiliated. At my age [38] I was really kind of embarrassed, but I knew that was the lifestyle that I was leadin'. And people I used to talk to would tell me, well, you could do this, and you don't have to get busted. But then I started thinking why are all these people here. So it doesn't, you know, really work. So I came home, and I did go back to selling again, but you know I knew I was on probation. And I didn't want to do no more time.

April: Jail, being in jail. The environment, having my freedom taken away. I saw myself keep repeating the same pattern, and I didn't want to do that. Uh, I had missed my daughter. See, being in jail that long period of time, I was able to detox. And when I detoxed, I kind of like had a clear sense of thinking, and that's when I came to the realization that, uh, this is not working for me.

Denise: I saw the person that I was dealing with— my partner—I saw her go upstate to Bedford for 2 to 4 years. I didn't want to deal with it. I didn't want to go. Bedford is a prison, women's prison. And I couldn't see myself givin' up 2 years of my life for something that I knew I could change in another way.

As can be seen from the above, the influence of punishment on these women was due to their belief that if they continued to be involved in crime, they would be apprehended, convicted, and incarcerated.

For many of the women, it was the stresses of street life and the fear of dying on the streets that motivated their decision to quit the criminal life. Darlene, a 25-year-old black woman, recalled the stress associated with the latter stage of her career selling drugs:

> The simple fact is that I really, I thought that I would die out there. I thought that someone would kill me out there and I would be killed; I had a fear of being on the front page one day and being in the newspaper dying. I wanted to live, and I didn't just want to exist.

Sonya, a 27-year-old Hispanic woman, provided an account of what daily life was like on the streets:

> You get tired of bein' tired, you know. I got tired of hustlin', you know. I got tired of livin' the way I was livin', you know. Due to your body, your body, mentally, emotionally, you know. Everybody's tryin' to get over. Everybody will stab you in your back. Nobody gives a fuck about the next person. And I used to have people talkin' to me, "You know, you're not a bad lookin' girl. You know, why you don't get yourself together."

Perhaps even more important, the women felt that they had wasted time. They became acutely aware of time as a diminishing resource (Shover 1983). They reported that they saw themselves going nowhere. They had arrived at a point where crime seemed senseless, and their lives had reached a dead end. Implicit in this assessment was the belief that gaining a longer-range perspective on one's life was a first step in changing. Such deliberations develop as a result of "socially disjunctive experiences" that cause the offender to experience social stress, feelings of alienation, and dissatisfaction with her present identity (Ray 1961).

Breaking Away From the Life

Forming a commitment to change is only the first step toward the termination of a criminal career. The offender enters a period that has been characterized as a "running struggle" with problems of social identity (Ray 1961, p. 136). Successful desisters must work to clarify and strengthen their nondeviant identity and redefine their street experience in terms more compatible with a conventional lifestyle. The second stage of the desistance process begins with the public announcement or "certification" (Meisenhelder 1977, p. 329) that the offender has decided to end her deviant behavior. After this announcement, the offender must begin to redefine economic, social, and emotional relationships that were based on a deviant street subculture.

The time following the announcement was generally a period of ambivalence and crisis for the study participants, because so much of their lives revolved around street life and because they had, at best, weak associations with the conventional world. Many of the women remembered the uncertainty they felt and the social dilemmas they faced after they decided to stop their involvement in crime.

Denise: I went and looked up my friends and to see what was doing, and my girlfriend Mia was like, she was gettin' paid. And I was livin' on a $60 stipend. And I wasn't with it. Mia was good to me, she always kept money in my pocket when I came home. I would walk into her closet and change into clothes that I'm more accustomed to. She started calling me Pen again. She stopped calling me Denise. And I would ride with her knowing that she had a gun or a package in the car. But I wouldn't touch nothin'. But that was my rationale. As long as I don't fuck with nothin. Yeah, she was like I can give you a grand and get you started. I said I know you can, but I can't. She said I can give you a grand, and she kept telling me that over and over; and I wasn't that far from taking the grand and getting started again.

Barbara: After I decided to change, I went to a party with my friend. And people was around me and they was drinkin' and stuff, and I didn't want to drink. I don't have the urge of drinking. If anything, it would be smokin' crack. And when I left the party, I felt like I was missing something—like something was missing. And it was the fact that I wasn't gettin' high. But I know the consequences of it. If I take a drink, I'm gonna smoke crack. If I, uh, sniff some blow, I'm gonna smoke crack. I might do some things like rob a store or something stupid and go to jail. So I don't want to put myself in that position.

At this stage of their transition, the women had to decide how to establish and maintain conventional relationships and what to do with themselves and their lives. Few of the women had maintained good relationships with people who were not involved in crime and drugs. Given this situation, the women had to seek alternatives to their present situation.

The large majority of study participants were aided in their social reintegration by outside help. These respondents sought formal treatment of some kind, typically residential drug treatment, to provide structure, social support, and a pathway to behavioral change. The women perceived clearly the need to remove themselves from the "scene," to meet new friends, and to begin the process of identity reformation. The following account by Alicia typifies the importance of a "geographic" cure:

I love to get high, you know, and I love the way crack makes me feel. I knew that I needed long-term, I knew that I needed to go somewhere. All away from everything, and I just needed to away from everything. And I couldn't deal with responsibility at all. And, uh, I was just so ashamed of the way that I had, you know, became and the person that I became that I just wanted to start over again.

Social avoidance strategies were common to all attempts at stopping. When the women removed themselves from their old world and old locations, involvement in crime and drugs was more difficult.

April: Yeah, I go home, but I don't, I don't socialize with the people. I don't even speak to anybody really. I go and I come. I don't go to the areas that I used to be in. I don't go

there anymore. I don't walk down the same blocks I used to walk down. I always take different locations.

Denise: I miss the fast money; otherwise, I don't miss my old life. I get support from my positive friends, and in the program. I talk about how I felt being around my old associates, seeing them, you know, going back to my old neighborhood. It's hard to deal with, I have to push away.

Maintaining a Conventional Life

Desisters have little chance of staying out of the life for an extended period of time if they stay in the social world of crime and addiction. They must rebuild and maintain a network of primary relations who accept and support their nondeviant identity if they are to be successful (the third stage of this model). This is no easy task, since in most cases the desisters have alienated their old nondeviant primary relations.

To a great extent, the women in this study most resemble religious converts in their attempts to establish and maintain support networks that validate their new sense of self. Treatment programs not only provide a ready-made primary group for desisters, but also a well-established pervasive identity (Travisano 1970), that of ex-con and/or ex-addict, that informs the women's view of themselves in a variety of interactions. Reminders of "spoiled identities" (Goffman 1963) such as criminal, "con," and "junkie" serve as a constant reference point for new experiences and keep salient the ideology of conventional living (Faupel 1991). Perhaps most important, these programs provide the women with an alternative basis for life structure—one that is devoid of crime, drugs, and other subcultural elements.

The successful treatment program, however, is one that ultimately facilitates dissociation from the program and promotes independent living. Dissociation from programs to participate in conventional living requires association, or reintegration, with conventional society. Friends and educational and occupational roles helped study participants reaffirm their non-criminal identities and bond themselves to conventional lifestyles. Barbara described the assistance she receives from friends and treatment groups:

> . . . a bunch of friends that always confronts me on what I'm doing' and where I'm goin', and they just want the best for me. And none of them use drugs. I go to a lot like outside support groups, you know. They help me have more confidence in myself. I have new friends now. Some of them are in treatment. Some have always been straight. They know. You know, they glad, you know, when I see them.

In the course of experiencing relationships with conventional others and participating in conventional roles, the women developed a strong social-psychological commitment not to return to crime and drug use. These commitments most often revolved around renewed affiliations with their children, relationships with new friends, and the acquisition of educational and vocational skills. The social relationships, interests, and investments that develop in the course of desistance reflect the gradual emergence of new identities. Such stakes in conventional identity form the social-psychological context within which control and desistance are possible (Waldorf et al. 1991).

In short, the women in the study developed a stake in their new lives that was incompatible with street life. This new stake served as a wedge to help maintain the separation of the women from the world of the streets (Biernacki 1986). The desire to maintain one's sense of self was an important incentive for avoiding return to crime.

Alicia: I like the fact that I have my respect back. I like the fact that, uh, my daughter trusts me again. And my mother don't mind leavin' me in the house, and she don't have to worry that when she come in her TV might be gone.

Barbara: I have new friends. I have my children back in my life. I have my education. It keeps me straight. I can't forget where I came from because I get scared to go back. I don't want to hurt nobody. I just want to live a normal life.

Janelle, a 22-year-old black woman, started dealing drugs and carrying a .38-caliber gun when she was 15. She described the ongoing tension between staying straight and returning to her old social world:

> It's hard, it's hard stayin' on the right track. But letting myself know that I'm worth more. I don't have to go in a store today and steal anything. I don't deserve that. I don't deserve to make myself feel really bad. Then once again I would be steppin' back and feel that this is all I can do.

Overall, the success of identity transformations hinges on the women's abilities to establish and maintain commitments and involvements in conventional aspects of life. As the women began to feel accepted and trusted within some conventional social circles, their determination to exit from crime was strengthened, as were their social and personal identities as noncriminals.

DISCUSSION

The primary purpose of this study was to describe—from the offenders' perspective—how women embedded in criminal street subcultures could end their deviance. Desistance appears to be a process as complex and lengthy as that of initial involvement. It was interesting to find that some of the key concepts in initiation of deviance—social bond, differential association, deterrence, age—were important in our analysis. We saw the aging offender take the threat of punishment seriously, reestablish links with conventional society, and sever association with subcultural street elements.

Our research supports Adler's (1992) finding that shame plays a limited role in the decision to return to conventional life for individuals who are entrenched in deviant subcultures. Rather, they exit deviance because they have evolved through the typical phases of their deviant careers.

In the present study, we found that the decision to give up crime was triggered by a shock of some sort (i.e., a socially disjunctive experience), by a delayed deterrence process, or both. The women then entered a period of crisis. Anxious and dissatisfied, they took stock of their lives and criminal activity. They arrived at a point where their way of life seemed senseless. Having made this assessment, the women then worked to clarify and strengthen their nondeviant identities. This phase began with the reevaluation of life goals and the public announcement of their decision to end involvement in crime. Once the decision to quit was made, the women turned to relationships that had not been ruined by their deviance, or they created new relationships. The final stage, maintaining cessation, involved integration into a nondeviant lifestyle. This meant restructuring the entire pattern of their lives (i.e., primary relationships, daily routines, social situations). For most women, treatment groups provided the continuing support needed to maintain a nondeviant status.

The change processes and turning points described by the women in the present research were quite similar to those reported by men in previous studies (Shover 1983, 1985; Cusson and Pinsonneault 1986). Collectively, these findings suggest that desistance is a pragmatically constructed project of action created by the individual within a given social context. Turning points occur as a "part of a process over time and not as a dramatic lasting change that takes place at any one time" (Pickles and Rutter 1991, p. 134). Thus, the return to conventional life occurs more because of "push" than "pull" factors (Adler 1992), because the career of involvement in crime moves offenders beyond the point at which they find it enjoyable to the point at which it is debilitating and anxiety-provoking.

Considering the narrow confines of our empirical data, it is hardly necessary to point out the limits of generalizability. Our analysis refers to the woman deeply involved in crime and immersed in a street subculture who finds the strength and resources to change her way of life. The fact that all the women in this study experienced a long period of personal deterioration and a "rock bottom" experience before they were able to exit crime does not justify a conclusion that this process occurs with all offenders. Undoubtedly, there are other scenarios (e.g., the occasional offender who drifts in and out of crime, the offender who stops when criminal involvement conflicts with commitments to conventional life,

the battered woman who kills) in which the question of desistance does not arise. Hence, there is a need to conceptualize and measure the objective and subjective elements of change among various male and female offender subgroups.

Furthermore, the evidence presented here does not warrant the conclusion that none of the women ever renewed their involvement in crime. Because the study materials consist of retrospective information, with all its attendant problems, we cannot state with certainty whether desistance from crime is permanent. Still, it is also clear that these women broke their pattern of involvement in crime for substantial lengths of time and have substantially changed their lives.

REFERENCES

Adler, Patricia. 1992. "The 'Post' Phase of Deviant Careers: Reintegrating Drug Traffickers." *Deviant Behavior* 13: 103–126.

Anglin, Douglas, and George Speckhart. 1988. "Narcotics Use and Crime: A Multisample, Multimethod Analysis." *Criminology* 26: 197–234.

Biernacki, Patrick A. 1986. *Pathways from Heroin Addiction: Recovery Without Treatment.* Philadelphia: Temple University Press.

Biernacki, Patrick A., and Dan Waldorf. 1981. "Snowball Sampling: Problems and Techniques of Chain Referral Sampling." *Sociological Methods and Research* 10: 141–163.

Blumstein, Alfred, Jacqueline Cohen, Jeffrey A. Roth, and Christy A. Visher. 1986. *Criminal Careers and Career Criminals.* Washington, DC: National Academy Press.

Collins, J., R. Hubbard, and J. V. Rachal. 1985. "Expensive Drug Use and Illegal Income: A Test of Explanatory Hypotheses." *Criminology* 23: 743–764.

Cusson, Maurice, and Pierre Pinsonneault. 1986. "The Decision to Give Up Crime." In *The Reasoning Criminal: Rational Choice Perspectives on Offending*, edited by Derek Cornish and Ronald Clarke. New York: Springer-Verlag.

Daly, Kathy, and Meda Chesney-Lind. 1988. "Feminism and Criminology." *Justice Quarterly* 5: 101–143.

Faupel, Charles. 1991. *Shooting Dope: Career Patterns of Hard-Core Heroin Users*. Gainesville: University of Florida Press.

Goffman, Erving. 1963. *Stigma: Notes on the Management of Spoiled Identity*. Englewood Cliffs, NJ: Prentice-Hall.

Hirschi, Travis, and H. C. Selvin. 1967. *Delinquency Research: An Appraisal of Analytic Methods*. New York: Free Press.

Le Blanc, Marc, and M. Frechette. 1989. *Male Criminal Activity from Childhood Through Youth: Multilevel and Developmental Perspective*. New York: Springer-Verlag.

Matza, David. 1964. *Delinquency and Drift*. New York: Wiley.

Meisenhelder, Thomas. 1977. "An Exploratory Study of Exiting from Criminal Careers." *Criminology* 15: 319–334.

Mulvey, Edward P., and John F. LaRosa. 1986. "Delinquency Cessation and Adolescent Development: Preliminary Data." *American Journal of Orthopsychiatry* 56: 212–224.

Petersilia, Joan, Peter Greenwood, and Marvin Lavin. 1978. *Criminal Careers of Habitual Felons*. Washington, DC: Law Enforcement Assistance Administration, U.S. Department of Justice.

Peterson, M., and H. Braiker. 1980. *Doing Crime: A Survey of California Prison Inmates*. Santa Monica, CA: Rand.

Pickles, Andrew, and Michael Rutter. 1991. "Statistical and Conceptual Models of 'Turning Points' in Developmental Processes." In *Problems and Methods in Longitudinal Research: Stability and Change*, edited by D. Magnusson, L. Bergman, G. Rudinger, and B. Torestad (pp. 110–136). New York: Cambridge University Press.

Ray, Marsh. 1961. "The Cycle of Abstinence and Relapse Among Heroin Addicts." *Social Problems* 9: 132–140.

Shover, Neil. 1983. "The Latter Stages of Ordinary Property Offenders' Careers." *Social Problems* 31: 208–218.

———. 1985. *Aging Criminals*. Newbury Park, CA: Sage.

Shover, Neil, and Carol Thompson. 1992. "Age, Differential Expectations, and Crime Desistance." *Criminology* 30: 89–104.

Stall, Ron, and Patrick Biernacki. 1986. "Spontaneous Remission from the Problematic Use of Substances: An Inductive Model Derived from a Comparative Analysis of the Alcohol, Opiate, Tobacco, and Food/Obesity Literatures." *International Journal of the Addictions* 2: 1–23.

Travisano, R. 1970. "Alteration and Conversion as Qualitative Different Transformations." In *Social Psychology Through Symbolic Interaction*, edited by G. Stone and H. Farberman (pp. 594–605). Boston: Ginn-Blaisdell.

Waldorf, Dan. 1983. "Natural Recovery from Opiate Addiction: Some Social-Psychological Process of Untreated Recovery." *Journal of Drug Issues* 13: 237–280.

Waldorf, Dan, Craig Reinerman, and Sheila Murphy. 1991. *Cocaine Changes*. Philadelphia: Temple University Press.

Weiner, Neil, and Marvin E. Wolfgang. 1989. *Violent Crime, Violent Criminals*. Newbury Park, CA: Sage.

Weis, Joseph G. 1989. "Family Violence Research Methodology and Design." In *Family Violence*, edited by Lloyd Ohlin and Michael Tonry (pp. 117–162). Chicago: University of Chicago Press.

31

Becoming Normal: Certification as a Stage in Exiting From Crime

Thomas Meisenhelder

Meisenhelder terms the final process of exiting from crime "certification." The author explains that the formal completion or last phase in leaving a life of crime requires the ex-offender to fully attain a social identity: that is, transitioning from a criminal identity to that of a social identity as a conventional person. Certification, he claims, is the verification that the offender has truly changed. What is essential for certification to take place is the recognition of this reform by members of the conventional community who have to proclaim publicly and certify that the ex-offender is now considered to be a law-abiding member of society. This, claims the author, represents the last stage in the process of desistance from crime.

The final aspect of the process of exiting from a career in crime can be termed "certification," which is briefly described as follows:

The formal completion of a successful exiting project requires a symbolic component, certification. This final phase in exiting was required in order for the individual fully to achieve a social identity as a noncriminal. Certification is simply the social verification of the individual's "reform." Some recognized member(s) of the conventional community must publicly announce and certify that the offender has changed and that he is now to be considered essentially noncriminal (Meisenhelder, 1977:329).

The instigation of an exiting project by a criminal offender depends on the individual's decision to attempt to go "straight." This decision seems to be at least in part based on the individual's personal identity as a noncriminal. However, change within criminal careers also includes the social recognition (as perceived by ex-offenders themselves) of their abandonment of criminal activity. Thus change is not complete when individuals fully commit themselves to conventionality, rather, commitment must be supplemented by the social recognition of their reform. The process by which exiting from crime is conceived of as socially recognized is called certification. By certification, individuals convince themselves that they have convinced others to view them as conventional members of the community. They perceive by the reactions of others that they are defined as being largely conventional. They begin to feel trusted; that is, they feel that their contemporaries are likely to see them as normal and noncriminal. They no longer feel suspect (Matza, 1969:195). Certification, then, completes the exiting, or change, process by solidifying the self-concept of the ex-offender.

It is important to note that along with the impression of conventionality certification often produces an increased probability of noncriminal patterns of behavior. A social identity as noncriminal may regenerate and support the changed felon's commitment to conventionality. This may be so for several reasons. First, certification produces an integrated conception of the self as normal, and this congruence of self may facilitate the abandonment of crime (see Lofland, 1969:281). In addition, certification through the testimony of others often parallels the formation of strong interpersonal ties between the ex-offender and conventional others. These social bonds provide the changed criminal with meaningful reasons for staying "straight" (Meisenhelder, 1977).

In sum, certification is a process of social interaction through which the ex-offender's social identity is changed to one of noncriminality. Certification signals the end of exiting as a stage in the individual's biography. In the following pages this outline is filled out by a detailed description of how certification is experienced by the offenders themselves.

METHODS

The research reported here has been conducted within a frame of reference that Matza (1969) has referred to as the "appreciative stance." Simply put, this point of view requires that the sociologist strive to grasp the experiential life-world of his or her subjects. In this case the stance of appreciation directs the researcher to attempt to understand the role of the process of certification within the larger exiting project and criminal career of the offender.

In line with this approach the data presented in this paper were collected during tape-recorded interview conversations with 25 male prison inmates (see Meisenhelder, 1977). Inmates were chosen if they manifested offense histories composed entirely or largely of property crimes. Then each of the 48 selected inmates was briefly introduced to the general project, assured of its confidentiality, and asked to volunteer to be interviewed. It should be noted that participation in the study offered no advantage to the respondents, and they were clearly aware of the purely academic interests and position of the researcher.

THE SOCIAL AND PERSONAL CONTEXT OF CERTIFICATION

The social self is a central notion in modern sociology. In order to understand the process of certification, the self must be thought of as a dual identity. That is, each person is conscious of himself or herself in terms of both a social and a personal identity. The self is both socially imputed and personally constructed.

Personal identity refers to the manner in which the individual defines his or her self. The concept concerns the judgmental, emotional, and cognitive categories into which the individual places himself or herself. Prior to certification, the typical respondent's personal identity was shakily noncriminal. He had constructed and committed himself to a personal identity as basically a conventional member of society. The men realized that they had been "in trouble"; yet they did not conclude from this concept that they were criminal types of persons. These feelings are revealed in the following statement taken from the interviews:

> Contrary to what you, they might think, I ain't no hard-down criminal. I just consider myself as an average person. But there is one difference, I got into some troubles, I've been to the joint.

Indeed the plan to exit from crime is in large part founded on this sense of the self as noncriminal.

> I try to be straight. If the people give me half a chance, I'll take it. I'm just like any other person, looking for stuff like settling down, having me an old lady, couple of kids.

Our sense of self entails more than a subjective definition of ourselves. Identity is also constructed on the basis of received evidence; that is, our self is socially imputed to us through our perception of the reactions of others. An awareness of social reactions results in a social identity, which may be defined as the self imputed to the individual by others (See Hawkins and Tiedeman, 1975:243). Laing and his co-workers (1966:6) refer to this as "my view of your view of me." Social identity is my perception of what type of person the other thinks I am.

Logically, then, one's sense of personal identity may conflict with one's perception of his or her social identity. As discussed earlier, the ex-felon felt that he was basically conventional, but he realized that many other people saw him as a criminal type. The respondents related that they sensed that many people "on the street" believed that they were simply, once and for all, criminals. Others are perceived by the ex-offender as seeing him as less than normal and not to be trusted. They believed that others expect him to behave and live in a fundamentally deviant fashion. These sentiments are evident in the following interview excerpt:

> You are no longer considered as being a man. Now that I've committed a crime, I got incarcerated, I don't get no respect. You just a plain criminal in their eyes. People in the streets are down on you all the time.

At best, the exiting ex-felon felt that others saw him as a deviant feigning conventionality (see also Matza, 1969:174).

Thus, the exiting ex-criminal experiences disagreement between his personal identity and other people's definition of his self. His social identity remains that of a criminal while his personal identity is noncriminal. Certification is essentially a process of constructing a conventional, or noncriminal, social identity for and by the ex-offender. Just as negative reactions may result in a criminal social identity, certification as an adjudication of conventionality labels the changing ex-offender as noncriminal.

CERTIFICATION

Certification is the final social verification of the ex-offender's substantial reform. It can be described as a set of communicative actions through which the ex-offender impresses others with the conventionality of his motives, values, and personality (see Brim and Wheeler, 1966:42). In certification, some recognized members of the conventional community publicly announce and verify that the ex-felon need no longer be considered a criminal but rather should now be treated as a normal member of the social group.

In order to impress others with their successful reform and to achieve certification, the respondents employed some adept forms of self-presentation and impression management. For instance, although they believed that change is an activity conceived and achieved by the individual, they were well aware that it must be finally verified through the testimony of others. In their own words, although all the men believed that "you rehabilitate yourself," they also knew that "you got to appear normal . . . It keeps them off your back," and "you have to, I had to, find someone to change me. To make me look good."

Further, the men related that the certification process involved manipulation of the image of oneself held by others rather than by any "real" personal or personality change (see Becker, 1964:42). Certification, then, can be seen as a specific instance of the general process that Goffman (1959) has described as the presentation of self. It too involves procedures, strategies, and tools that are used to present oneself as a particular sort of person. Here, the ex-offender must present his social audience with verbal and behavioral evidence of his normality. This evidence most typically includes others or a presentational team (Goffman, 1959:104) and social settings (Goffman, 1959:22). That is, by appearing with conventional others and in conventional places the exiting felon is able to demonstrate his change from crime. Likewise, by appearing with him others can testify to the individual's noncriminality and socially certify that he may be trusted to behave in a normal fashion. The phenomenological logic of certification seems to be that if one's associates seem reasonably normal and if they further seem to define one as trustworthy, others should and will do likewise. In short, as Schultz (1962) has noted, others assume that they would see things as one's close associates do if they were in their place. If he achieves certification and the construction of a noncriminal social identity, the individual no longer feels stigmatized. He has a chance to become anonymously conventional. He is now, for all practical purposes, typically conventional, and he has successfully (if temporarily) exited from crime.

In the following pages, a closer analysis of the certification process is presented. This analysis is ordered according to the most frequently mentioned tools of certification: other persons, physical settings or places, and formal agencies of correctional reform.

CERTIFICATION AGENCIES

Certification may be achieved through participation in the rehabilitation programs of various social service and criminal justice agencies. Lofland (1969:288) has noted that these agencies provide the public with a reasonable accounting of their clients' criminality and subsequent reform. That is, they justify seeing the ex-offender as noncriminal.

However, it is easy to exaggerate the actual importance of these agencies to the ex-offender. The interviews reported here indicate that only a very few of the respondents used these agencies to demonstrate their reform. Further, most of these men used the programs in a manner far removed from their stated purposes.

> I tried the church thing, I'm not really religious, but I was gonna try it out. I'm gonna make people think that, like, I trying to change, which I was in my own way.
>
> Sure there are programs to help you, but it's, most people get into them account of it looks

good. They see that you're in the program, they don't know that you ain't getting nothing out of it. They think you are changed.

As these comments suggest, formal agencies are often used purely as techniques of self-presentation.

From the point of view of the ex-offender, an important shortcoming of these agencies is that they are most often designed to effect some change in the personal identity of their clients. To the contrary, the exiting ex-offender already possesses a conventional personal identity and merely wants to use the program as ceremonial evidence for others in order to construct a similar social identity. Thus the ex-felon is forced to deceitfully manipulate his involvement in the program in order to demonstrate his commitment to change and in order to publicly announce and account for his presumed reform. The exiting experiences of "Ron," described below, illustrate the role of rehabilitation agencies in the certification process.

Ron is a middle-aged black male who has lived most of his years in the southwest and in California. As a youth he became involved in petty theft and some semiorganized shoplifting. As a result he has a fairly long juvenile arrest record that includes one substantial incarceration. As an adult Ron graduated into a career of property offenses, particularly in theft and burglary. He experienced one significant period of exiting that lasted about 14 months.

A prison-administered self-improvement program was a significant factor in Ron's exiting experiences. He clearly felt that this program both convinced him that he alone was responsible for his actions *and* changed his image for others: "You're trying to determine a new image so to speak, whether knowingly or unknowingly." Ron changed his image in the minds of others through his everyday behavior and through his participation and acceptance of the self-improvement course. While attending the program, which he did twice, Ron began to construct a new social identity. Besides altering his style of adjustment to prison toward that of a "square john" (Irwin, 1970) and beginning to associate with other model prisoners, Ron methodically catered to the image he presented to his environment. As he himself put it:

I was very methodic about it. I didn't break no rules. And a lot of people I had known objected to how I began to react to situations. I knew what I was doing.

Ron's case provides us with an example of an individual consciously using an agency program to achieve certification. By his participation in the self-improvement course Ron loudly announced to his social environment at the time that he was a changed man. Admittedly a weak and partial case, Ron's is the exception rather than the rule among the respondents. More often certification was achieved through the informal testimony of conventional others and conventional places.

CERTIFYING SETTINGS

Certification of change may also occur via the physical places where the ex-offender can be observed to spend his time. As he continually appears in conventional surroundings, others may begin to define the person as non-criminal. As Lofland (1969:238) has suggested, "an Actor's places serve to communicate a part of what he 'is.'"

It seems that in the everyday world one's appearance in deviant places leads to the imputation of personal deviance, and one's appearance in conventional places indicates that one is normal or conventional. Thus, as revealed in the comments below, many of the respondents consciously attempted to avoid deviant surroundings and to put themselves in more normal social settings.

> Some places get you in trouble just by your being there . . . The wrong place at the wrong time.
>
> I always made it my business to try and get away from bad places. Places where I shouldn't be. I try to be more skeptical about where I would be.

A setting of primary import for certification is the person's place of employment. Appearance in a legitimate occupational setting for 8 hours a day, 5 days a week, is one of the most effective techniques of self-presentation available to the exiting ex-offender. The men interviewed were clearly cognizant of this fact. One respondent related that a conventional job made people accept him because

> They know what your job consists of, know what you is supposed to do. They think you is all cut and dry.

As Weber (1958) realized long ago, our culture seems to equate work with moral worth, and the changing offender's appearance in a conventional work setting strongly implies that he is worthy of trust. In short, he becomes as familiar as his job. Of course, work also gives the ex-offender practical economic aid and resources as he attempts to disengage from crime. But, beyond these practicalities, the work place also may lead others to impute conventionality to the ex-felon.

Places of leisure are also available as ways to achieve a certified non-criminal social identity. Being visible in churches, civic associations, and other socially recognized sites for the constructive use of leisure time can help the individual to become verified as having reformed. On the other hand, the exiting offender must not be seen frequenting pool halls, certain taverns, or other settings characterized by conventional others as the site of loitering, wasting time, and looking for trouble. Places such as these lead to the imputation of deviant and criminal social identity. As one respondent phrased it these kinds of places make "people look hard on you" or, even more to the point, "Pool halls and stuff, they's frustrating and full of trouble. And you looks bad there." Finally, as implied by the above, conventional places of leisure aid the exiting offender in a more practical way as a result of the fact that they are less likely to present the individual with the temptation to return to crime.

In sum, physical settings assist in the achievement of certification by providing evidence that leads others to label the ex-offender as noncriminal. That is, they can announce and verify the conventionality of those that work and play within them. The case of "Bob" highlights the role of places in the process of certification.

Bob is a middle-aged white male whose most frequent offense is passing and writing bad checks. He began his career in crime late in life (at the age of 30), and since then he has been incarcerated twice for passing bad checks. From 1965 to 1969, Bob "turned over a new leaf and went straight." In discussing these years of his life Bob continually pointed to the significance of his social and physical surroundings as a tool for achieving certification.

Bob stressed the negative impact of parole, for instance, as a symbolic place that is by definition reserved for criminals. Parole then involves the social imputation of criminality to the parolee. As Bob himself put it:

> Parole . . . is not beneficial to that man whatever. They (parole agents) come around, they'll sneak around, they peep on him. In some cases harass a man and cause him to do things he wouldn't normally do. If people, if they find out you're on parole they don't want to be around you. Parole doesn't help a man.

For these reasons Bob (and many other parolees) must actually skip parole in order to achieve certification as a noncriminal.

After "running from parole," Bob achieved certification through his appearance in more conventional environments. He intentionally avoided known "bad" places where he might find the opportunity to "get into trouble" and where he surely would find the social imputation of deviance. Instead he began to associate with a more or less conventional group of people and to frequent conventional places. He "got involved with work." Beyond its mere economic effects, Bob's job was a place wherein he would be likely to be seen by others as normal and noncriminal.

> I was a cook in this restaurant in a small town. I enjoyed my work and the people I met were friendly. They saw me as a good cook, that's all.

Thus, through his appearance in the restaurant, Bob was able to achieve some degree of certification as a conventional person. His new associates and acquaintances did not know of his criminal past or present (he had skipped parole) and seeing him everyday "at work," they simply assumed that he was just like any other person in the community. That is, they imputed to Bob a normal social identity.

To further the process of certification Bob also acquired a car and a stable residence. Again these aspects of his daily life provided Bob with conventional surroundings. Driving his car to and from work, spending his nonwork time at his home, Bob became an accepted member of the community. It is extremely important to note the correctional paradox explicit in Bob's case. All these sources of certification might have been denied Bob if he had remained under parole supervision. However, by fleeing his parole (and thus violating the law) he was able to leave a deviant symbolic environment and adopt a new set of social places in the conventional world. As he acquired a new community, a conventional job, a car, and a home, Bob was defined as normal. He had achieved certification. But, ironically, in order to accomplish all this, Bob was forced to violate the law once more. Finally, the parole violation caught up with Bob, and he was returned to prison.

Bob's case is an example of the flow of the process of certification when it is primarily assisted by the individual's appearance in conventional settings. Equally important to certification are the people whom one finds in such environments.

CERTIFYING OTHERS

Places are, of course, peopled by others as well as by the ex-offender. Most social settings are normatively defined as the proper locations for particular types of persons and activities. For instance, the post office is a place thought to be inhabited primarily by post office employees and their customers. The respondents indicated in the interviews that they felt that the people with whom they associated were extremely important for their eventual certification.

The single most important group of others for the changing ex-offender is his family. A conventional family, newly formed or renewed, provided the individual with a significant group of close associates that could overtly testify to his noncriminality and trustworthiness. The importance of the family for exiting from crime was continually mentioned during the interviews. For example,

They made me feel that I was doing right. I looked good while I was with them, living at home. And I stayed out of it for two years, with my family.

I got tired of the life, so I went back home. There I had a place to stay and people accepted me for myself. I was doing my best.

A family gives the changing ex-offender both a place that is recognized as conventional and others who announce to the public world that the family member has changed and is now to be considered noncriminal.

It's very important, it keeps society from looking down on you. If you married and got kids, they think: "Well, he's rehabilitated. He's married and settled down, not wild anymore." Some people use it as a front, for others like me it ain't no cover up.

Of course, kin also present the ex-felon with much practical assistance, including job opportunities, financial loans, and simply "a place to start off from." Emotionally, families provide individuals with a sense of belonging, of being at home in the conventional world. Within his family the ex-felon is able to at least partially overcome his feelings of strangeness as he confronts the conventional world. By caring for him his kin create relational ties that are most important during the trials of exiting. These ties in turn represent an investment in conformity that works to keep the ex-offender "straight." The family, then, is defined as one of the primary rewards of settling down (Meisenhelder, 1977).

The last nine months of my life, being with my wife and my son doing the things that normal people do is actually, seriously, have been the most important months of my life.

Besides these very important contributions to exiting, the family also provides the changing ex-offender with an effective means of self-presentation in order to achieve certification. Through his presence within a conventional family group, the individual announces that he is normal. At home and at ease in this most fundamental of social institutions, the ex-offender is seen by nonfamilial others as one to whom normality and respectability may be reasonably imputed (see Ball, 1970). The ex-felon's family

members testify that he can be trusted and that he is not a dangerous person.

Nonfamilial associates also help to certify the individual's reform. Many of the respondents indicated the importance of carefully selecting one's friends and companions. Bad associates lead to the social imputation of personal deviance. On the other hand, conventional associates create an impression of conventionality. Recognizing this common practice, many of the respondents attempted to steer clear of "bad crowds."

I quit hanging around with bad crowds, that helped me. Police and folks didn't look at me the way they used to. If you hang around with a bad crowd, they would follow you more.

I always made it my business to try and get away from those kind of people. I wouldn't come in contact with those that look bad whatsoever.

By avoiding bad persons and by cultivating relationships with conventional people at work and at home, the changing offender presents himself as a noncriminal. By their acceptance of him, conventional acquaintances certify that the ex-felon can be considered a conforming member of the community.

Scott is a middle-aged white male whose varied career in crime includes theft, automobile theft, and check passing. Prior to his most successful change he was deeply involved in "checks." Scott experienced a 3-year disengagement from crime that ended with incarceration for auto theft and parole violation. His change experience provides a nice example of certification through others.

Scott decided to "get out of the life" because he was tired of "being on the run," a frequently noted hazard of check writing as an occupation. Passing checks forces the criminal into a very mobile existence, and Scott grew tired of this type of existence:

This was a time when I just got exhausted and said, "This ain't worth it." I just up and quit, tore up my check book.

He then went to a small town where he was "not known at all" and attempted to establish a normal life. In the process of doing so, Scott was able to get a job at a service station, and importantly, he established a meaningful relationship with a woman and her child. He related this experience:

> I met a girl there that I became very interested in, her and her child. I decided to stay there, settle down with her. I was interested in her. She was something that I wanted to hold on to.

After he settled down in this family and became known in the area, Scott says that he began to feel accepted by the local community. People who knew him only as the new mate of "June" began to show that they trusted him with their friendship and with more material signs of acceptance. One such person loaned Scott a considerable sum of cash, and Scott's reaction to this event reveals that, to him, it was a meaningful sign of admission to the conventional world:

> He entrusted me with that money. I could have just took that money and gone, I didn't. And by doing that it let me know *I was accepted for what I was.*

In short, by being seen around the community with his new girlfriend, a lifelong resident, Scott too received trust and acceptance from his new neighbors. Further, his own perception of these reactions from others led him to believe that he was no longer judged to be a criminal. The reactions of others, then, certified Scott's change. This is reflected in the comments below,

> I met some pretty nice people in that town. My wife's friends and all, and they went out of their way to make me feel at home. That was a place, a nice place to live, nice people. If I hadn't been violated, I'd still be working and living there.

Obviously, Scott felt that he succeeded in becoming an accepted member of a conventional community. His own recognition of the trusting reactions of others led him to believe in his own change and further pushed him along the path of exiting from crime. Here certification was clearly the result of relationships with conventional others.

SUMMARY

In the preceding sections of the paper, the process of certification was reviewed by holding the tools or means of achieving certification analytically separate. In actuality, of course, some combination of places, others, and to a lesser extent agencies is evident in the total process. The ex-offender's continuing presence in conventional places and with conventional others forms an imputational process similar to that which Lofland (1969:227) has termed an "informal elevation ceremony." Such ceremonies made up of the behavioral evidence that testify to the conventionality of the ex-felon lead more and more people to categorize him as a conventional societal member. Having been seen with a "good job," a "fine family," and "respectable friends," the individual is presumed to be someone who can be trusted, someone,

> who is normal, who upholds social order and is deserving of order-maintaining and sustaining gestures of worth-acknowledgement from like-minded persons in society (Ball, 1970:339).

At this point, then, exiting is at least temporarily complete. The actor's self is consistently conventional, and he has developed a pattern of behavior and a style of life that, for the time being, are noncriminal.

Thus, certification is the final contingency in the abandonment of criminal behavior. Now the individual's social and personal identities join into a consistent sense of the self as noncriminal. The actor sees himself as normal and feels that others see him in the same fashion. However, it must be noted that certification is a process and as such is changeable and impermanent. Remission remains a possibility:

> It's a habit, like smoking. I thought I had it beat at one time. I was doing pretty good; got into a job that I loved doing, had a family, people began to accept me for myself, staying out of trouble. I thought I had it beat. I stayed out for one year and three months. Had a divorce, lost my job, easiest thing for me to do was go back to my old habits. What I knew best, what I knew how to do.

CONCLUSIONS

This paper has explored certification as the process by which people label ex-offenders as normal. It is an informal process of reality construction that results in the acceptance of the ex-felon as one who can be trusted. Certification revolves around the dual self of the actor and the use of others, places, agencies, and activities as means of self-presentation. These results in a sense confirm Davis' (1961) earlier conclusions concerning "deviance disavowal" among the visibly handicapped by revealing a similar process in the career of criminals. Certification is a way in which the stigmatized ex-offender attempts to resist a societal imputation of essential deviance and to deny that he is anything but a conventional person. The ex-felon, too, may be granted "fictional acceptance" at the onset of social encounters (Davis, 1961:126). He also uses settings, activities, and particularly others to encourage the other to identify with him through role-taking and thereby legitimate the ex-offender's non-criminal identity. Any social reality is bounded by horizons of more or less open possibility. Through the process of certification, the individual has relegated criminality to the fringes of his life-world, but it does exist there as well as in his personal biography. Thus, criminal behavior remains, as it does for us all, a possibility.

REFERENCES

Ball, D.W. 1970. "The Problems of Respectability." Pp. 326–71 in Jack Douglas (ed.), *Deviance and Respectability.* New York: Basic Books.

Becker, H. 1963. *Outsiders.* New York: Free Press.

———. 1964. "Personal Change in Adult Life." *Sociometry* 27:40–53.

Berger, P. and T. Luckmann. 1968. *The Social Construction of Reality.* Garden City, New York: Doubleday.

Brim, O. and S. Wheeler. 1966. *Socialization After Childhood.* New York: Wiley.

Davis, F. 1961. "Deviance Disavowal." *Social Problems* 9:120–32.

Goffman, E. 1959. *The Presentation of Self in Everyday Life.* Garden City, New York: Doubleday.

Hawkins, R. and G. Tiedeman. 1975. *The Creation of Deviance.* Columbus, Ohio: Charles Merrill.

Irwin, J. 1970. *The Felon.* Englewood Cliffs, New Jersey: Prentice-Hall.

Laing, R.D., H. Phillipson, and A.R. Lee. 1966. *Interpersonal Perception.* New York: Harper & Row.

Lofland, J. 1969. *Deviance and Identity.* Englewood Cliffs, New Jersey: Prentice-Hall.

Matza, D. 1969. *Becoming Deviant.* Englewood Cliffs, New Jersey: Prentice-Hall.

Meisenhelder, T. 1977. "An Exploratory Study of Exiting from Criminal Careers." *Criminology* 15:319–34.

Schutz, A. 1962. *Collected Papers I.* The Hague: Nijhoff.

Weber, M. 1958. *The Protestant Ethic and the Spirit of Capitalism.* New York: Scribner's.

Author Index

Abel, G. G., 55
Adler, P. A., 224, 230, 312, 316, 324, 330, 331
Agar, M., 224, 227
Akamatsu, J. T., 101
Alexander, D., 38
Allen, J., 26, 31
Anderson, A., 158
Anderson, E., 63, 64, 83, 196, 230
Anglin, D., 325
Armstrong, L., 72
Athens, L., 45, 46
Avary, D. W., 21, 23

Bailey, E., 36
Baker, A., 36
Ball, D. W., 339, 340
Ball, R. A., 200
Baril, M., 62
Baskin, D., 252, 268
Baumer, R., 72
Beck, J., 300, 302
Becker, H., 335
Becker, H. S., 238, 288
Benetin, J., 101
Bennet, J., 36
Bennett, T., 76, 83
Best, J., 168, 169, 311, 319
Biernacki, P., 183, 238, 258, 323, 324, 330
Birnbaum, J., 54
Bissell, L., 145
Bjerregaard, B., 211
Black, D., 54, 186
Blakey, R., 15, 20, 21, 23
Blanchard, E. B., 55
Blark, R., 158
Blau, R., 200
Block, C. R., 182, 187
Block, R., 182, 187
Blumer, H., 124
Blumstein, A., 322

Blumstein, P., 128
Borges, S. S., 130
Bouhoutsos, J., 101
Bourdieu, P., 66, 197, 198, 229, 235, 252, 256
Bowker, L. H., 214, 215
Bragg, W., 34
Braiker, H., 325
Braithwaite, J., 312, 318, 319
Brett, P., 34
Brewer, D., 34
Brim, O., 335
Brown, J. D., 312
Brown, L. S., 101, 111
Brownstein, H. H., 268
Buell, M., 122, 123
Bureau of Justice Statistics, 71
Burgess, A. W., 54, 55, 57, 58
Burr, A., 223, 224
Bushnell, J., 34

Caldarella, P., 200
Cameron, M., 25, 31
Campbell, A., 210, 211, 214, 256
Carlson, K., 223
Carr, C., 36
Casey, J. H., 196
Casey, J. H., Jr., 238
Caspar, J. D., 173
Castleman, C., 34
Caulfield, S., 2
Chalfant, H., 34
Chambliss, W. J., 26, 32, 158, 312
Chappell, D., 20
Chesney-Lind, M., 2, 211, 212, 322
Chin, K., 229
Ching, S., 35, 36
Clear, T. R., 122
Clinard, M., 1
Cloward, R., 192, 193, 196, 197

Subject Index

About the Contributors

Patricia Adler is a professor of sociology at the University of Colorado—Boulder.

D'Aunn Wester Avary (deceased) was director of Quality Management at Texas Tech University Regional Health Science Center.

Deborah Baskin is a professor and chair in the Department of Criminal Justice at California State University—Los Angeles.

Michael Benson is an associate professor of sociology at the University of Tennessee—Knoxville.

Henry Brownstein is Director of Narcotics and Alcohol Section, National Institute of Justice.

Phyllis Coontz is an associate professor in the Graduate School of Public and International Affairs at the University of Pittsburgh.

Susan Crimmins is with the National Development and Research Institutes, Inc.

Paul F. Cromwell is a professor of criminal justice and director of the School of Community Affairs at Wichita State University.

Dean Dabney is an associate professor in the Department of Criminal Justice at Georgia State University.

Kathleen Daly is a professor in the School of Justice Administration at Griffith University in Queensland, Australia.

Scott H. Decker is a professor of criminology and criminal justice at the University of Missouri—St. Louis.

Eloise Dunlap is a researcher with the National Development and Research Institutes, New York.

Jeffrey Fagan is a professor in the Center for Violence Research and Prevention in the School of Public Health at Columbia University.

Mary Ann Farkas is an associate professor in criminology and law studies at Marquette University.

Jeffrey Ferrell is a professor of sociology at Southern Methodist University.

Gilbert Geis is professor emeritus of criminology at the University of California at Irving.

John Hagedorn is an associate professor of criminal justice at the University of Illinois at Chicago.

Robert Hale (deceased) was an assistant professor of sociology and criminal justice at Southwestern Louisiana University.

Richard Hollinger is a professor of sociology at the University of Florida.

Bruce Jacobs is an associate professor of criminal justice at the University of Texas at Dallas.

Paul Jesilow is an associate professor of criminology, law, and society in the School of Social Ecology at the University of California at Irvine.

Karen Joe is an assistant professor of sociology at the University of Hawaii at Manda.

Bruce Johnson is a researcher with the National Development and Research Institutes, Inc., New York.

Jack Katz is a professor of sociology at the University of California, Los Angeles.

Sandra Langley is a researcher with the National Development and Research Institutes, New York.

Barbara Lex is a medical anthropologist. She is affiliated with the Harvard Medical School—McLean Hospital Alcohol and Drug Research Center.

Lisa Maher is with the National Drug and Alcohol Research Center at the University of New South Wales in Sydney, Australia.

Ali Manwar is a researcher with the National Development and Research Institutes, New York.

Amos Martinez is program administrator for the Mental Health Licensing Section, Colorado Department of Regulatory Agencies.

Thomas Meisenhelder is a professor of sociology at California State University at San Bernardino.

Tom Mieczkowski is a professor of criminal justice at the University of South Florida.

Eleanor Miller is an associate professor of sociology at the University of Wisconsin at Milwaukee.

Jody Miller is an assistant professor of criminology and criminal justice at the University of Missouri—St. Louis.

Patricia Morgan is an associate professor of Public Health at the University of California at Berkeley.

Sheigla Murphy is a medical sociologist with the Institute for Scientific Analysis in San Francisco.

James Olson is a professor of psychology at the University of Texas—Permian Basin.

Mark Pogrebin is professor and director of the Criminal Justice Program in the Graduate School of Public Affairs at the University of Colorado at Denver.

Henry Pontell is a professor of criminology, law, and society at the University of California at Irvine.

Eric Poole is a professor of criminal justice in the Graduate School of Public Affairs at the University of Colorado at Denver.

Melvin Ray is a professor of sociology at Cornell College in Iowa.

Kim Romenesko is an academic adviser at the University of Wisconsin—Milwaukee.

Ronald Simons is with the Department of Sociology at Iowa State University.

Ira Sommers is an associate professor of criminal justice at California State University—Los Angeles.

Barry Spunt is an associate professor of sociology at John Jay College of Criminal Justice, City University of New York.

Volkan Topalli is an assistant professor of criminal justice at Georgia State University.

Dan Waldorf (deceased) was a senior research associate at the Institute for Scientific Analysis, San Francisco.

Priscilla Kiehnle Warner is an associate professor of sociology at Northwestern University.

Ralph Weisheit is a professor of criminal justice at Illinois State University.

L. Edward Wells is a professor of criminal justice at Illinois State University.

Richard Wright is a professor of criminology and criminal justice at the University of Missouri—St. Louis.

Richard G. Zevitz is an associate professor of criminology and law studies at Marquette University.